AF598037

METHODS IN MOLECULAR BIOLOGY™

Series Editor
John M. Walker
School of Life Sciences
University of Hertfordshire
Hatfield, Hertfordshire, AL10 9AB, UK

For further volumes:
http://www.springer.com/series/7651

Stem Cell Migration

Methods and Protocols

Edited by

Marie-Dominique Filippi

*Division of Experimental Hematology and Cancer Biology,
Cincinnati Children's Hospital Medical Center, Cincinnati, OH, USA*

Hartmut Geiger

*Division of Experimental Hematology and Cancer Biology,
Cincinnati Children's Hospital Medical Center, Cincinnati, OH, USA*

*Department of Dermatology and Allergic Diseases, Aging Research,
University of Ulm, Ulm, Germany*

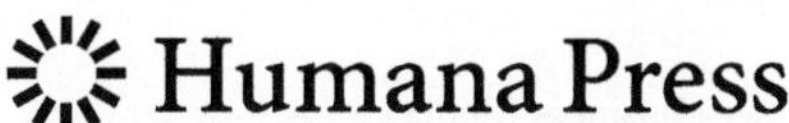

Editors
Marie-Dominique Filippi, Ph.D.
Division of Experimental Hematology
and Cancer Biology
Cincinnati Children's Hospital Medical Center
3333 Burnet Avenue
Cincinnati, OH 45229, USA
marie-dominique.filippi@cchmc.org

Hartmut Geiger, Ph.D.
Division of Experimental Hematology
and Cancer Biology
Cincinnati Children's Hospital Medical Center
3333 Burnet Avenue
Cincinnati, OH 45229, USA
and
Department of Dermatology and Allergic Diseases
Aging Research
University of Ulm, Ulm
Germany
hartmut.geiger@cchmc.org

ISSN 1064-3745 e-ISSN 1940-6029
ISBN 978-1-61779-144-4 e-ISBN 978-1-61779-145-1
DOI 10.1007/978-1-61779-145-1
Springer New York Dordrecht Heidelberg London

Library of Congress Control Number: 2011928372

Printed on acid-free paper

Humana Press is part of Springer Science+Business Media (www.springer.com)

Preface

Migration of stem cells is critical during early development and adult life for the organization of the embryonic body as well as tissue homeostasis and regeneration of organ function. During early development, the formation of functional organs depends on the migration of stem cells from the site of their specification toward the region where the corresponding organ develops. Similarly, the ontogeny of hematopoiesis is characterized by a temporal migration and thus spatial distribution of hematopoietic stem cells throughout embryogenesis until adulthood. During adult life, stem cells mostly migrate in response to tissue regeneration and thus are regarded as being central to regenerative medicine. Hematopoietic stem cells are the best characterized adult type of migrating stem cells. However, although still controversial, it is believed that in general tissue regeneration via stem cell migration and differentiation to a site of tissue injury is not limited to the hematopoietic system. Based on the concept of cancer stem cells, metastatic/migrating cancer stem cell might assume a critical role in the dissemination of the disease. Therefore, stem cell migration represents a great area of research in the fields of development, regenerative medicine, and cancer.

In spite of its importance in development, regeneration, and disease, research on migration of especially adult stem cells remained, until recently, difficult. This owes to the low frequency of these cells in vivo, problems in identifying and prospectively purifying tissue-specific stem cells near homogeneity, and mostly because of a lack of adequate technologies and protocols to study stem cell migration in vivo.

It is therefore the focus of this book to compile and highlight the standard and novel techniques that allow the studying of the migration of stem cells in a succinct manual. It includes protocols with respect to germ, neuronal, and hematopoietic stem cells, during development and adulthood with a clear emphasis on in vivo technologies. In addition, several developmentally conserved signaling pathways that have emerged as important control devices of stem cell migration are discussed, and the book reviews the in vitro approaches that are available to study these pathways. In summary, this book provides state of the art information on experimental techniques for studying stem cell migration both at a cellular and molecular level in development, regeneration, and disease.

We would like to thank Jessica Williams for her great administrative assistance.

Marie-Dominique Filippi
Hartmut Geiger

Contents

Contributors

ALISON L. ALLAN • *London Regional Cancer Program, London Health Sciences Centre, London, ON, Canada*

RONEN ALON • *Immunology Department, The Weizmann Institute of Science, Rehovot, Israel*

HASSAN AZARI • *McKnight Brain Institute, University of Florida, Department of Anatomical Sciences, Shiraz University of Medical Sciences, Gainesville, FL, USA*

SERENA BOVETTI • *Department of Human and Animal Biology, University of Torino, Torino, Italy*

JOSE A. CANCELAS • *Division of Experimental Hematology and Cancer Biology, Cincinnati Children's Hospital Medical Center, University of Cincinnati, Cincinnati, OH, USA*

GRANT A. CHALLEN • *Center for Cell and Gene Therapy, Baylor College of Medicine, Houston, TX, USA*

LOIC P. DELEYROLLE • *McKnight Brain Institute, University of Florida, Gainesville, FL, USA*

KYLE M. DRAHEIM • *Department of Cancer Biology, UMass Medical School, Worcester, MA, USA*

BRIAN DUDLEY • *Department of Genetics, School of Medicine, Case Western Reserve University, Cleveland, OH, USA*

YUXIN FENG • *Division of Experimental Hematology and Cancer Biology, Cincinnati Children's Hospital Medical Center, University of Cincinnati, Cincinnati, OH, USA*

MICHAEL J. FERKOWICZ • *Wells Center for Pediatric Research, Indiana University School of Medicine, Indianapolis, IN, USA*

PAUL S. FRENETTE • *Departments of Medicine, Gene and Cell Medicine, Tisch Cancer Institute, Immunology Institute and Black Family Stem Cell Institute, Mount Sinai School of Medicine, New York, NY, USA; Albert Einstein College of Medicine, Bronx, NY, USA*

HARTMUT GEIGER • *Division of Experimental Hematology and Cancer Biology, Cincinnati Children's Hospital Medical Center, Cincinnati, OH, USA; Department of Dermatology and Allergic Diseases, Aging Research, University of Ulm, Ulm, Germany*

POLINA GOICHBERG • *Immunology Department, The Weizmann Institute of Science, Rehovot, Israel*

MARGARET A. GOODELL • *Center for Cell and Gene Therapy, Baylor College of Medicine, Houston, TX, USA*

JOCHEN GRASSINGER • *Australian Stem Cell Centre, Clayton, VIC, Australia*

MATTHIAS GUNZER • *Institute for Molecular and Clinical Immunology, Otto von Guericke University, Magdeburg, Germany*

VÍT HERYNEK • *Department of Radiodiagnostic and Interventional Radiology, MR-Unit, Institute for Clinical and Experimental Medicine, Prague, Czech Republic*

DAVID A. HESS • *Robarts Research Institute, University of Western Ontario, London, ON, Canada*

PAVLA JENDELOVÁ • *Institute of Experimental Medicine ASCR, Prague, Czech Republic; Department of Neuroscience and Center for Cell Therapy and Tissue Repair, Charles University, Second Medical Faculty, Prague, Czech Republic*

RAHUL KASUKURTHI • *Division of Plastic and Reconstructive Surgery, School of Medicine, Washington University in St. Louis, St. Louis, MO, USA*

ANJA KÖHLER • *Institute for Molecular and Clinical Immunology, Otto von Guericke University, Magdeburg, Germany*

ORIT KOLLET • *Immunology Department, The Weizmann Institute of Science, Rehovot, Israel*

KFIR LAPID • *Immunology Department, The Weizmann Institute of Science, Rehovot, Israel*

TSVEE LAPIDOT • *Immunology Department, The Weizmann Institute of Science, Rehovot, Israel*

ANDREW S. LEE • *Molecular Imaging Program at Stanford (MIPS), Stanford University School of Medicine, Stanford, CA, USA*

PULIN LI • *Howard Hughes Medical Institute, Harvard Stem Cell Institute, Children's Hospital Boston, Harvard Medical School, Boston, MA, USA*

WILLIS X. LI • *Department of Medicine, University of California, San Diego, CA, USA*

KAREN KUAN-YIN LIN • *Center for Cell and Gene Therapy, Baylor College of Medicine, Houston, TX, USA*

WEI LIU • *Division of Experimental Hematology and Cancer Biology, Cincinnati Children's Hospital Medical Center, University of Cincinnati, Cincinnati, OH, USA*

DANIEL LUCAS • *Departments of Medicine, Gene and Cell Medicine, Tisch Cancer Institute, Immunology Institute and Black Family Stem Cell Institute, Mount Sinai School of Medicine, New York, NY, USA*

STEPHEN LYLE • *Department of Cancer Biology, UMass Medical School, Worcester, MA, USA*

CLAIRE MAGNON • *Departments of Medicine, Gene and Cell Medicine, Tisch Cancer Institute, Immunology Institute and Black Family Stem Cell Institute, Mount Sinai School of Medicine, New York, NY, USA*

FRANK C. MARINI • *Department of Stem Cell Transplantation and Cellular Therapy, Section of Molecular Hematology and Therapy, The University of Texas M. D. Anderson Cancer Center, Houston, TX, USA*

KATHLEEN MOLYNEAUX • *Department of Genetics, School of Medicine, Case Western Reserve University, Cleveland, OH, USA*

MICHAEL J. MURRAY • *Molecular Genetics and Evolution, Research School of Biological Sciences, Australian National University, Acton, ACT, Australia*

TERENCE M. MYCKATYN • *Division of Plastic and Reconstructive Surgery, School of Medicine, Washington University in St. Louis, St. Louis, MO, USA*
SUSIE K. NILSSON • *CSIRO Molecular and Health Technologies, C/O Australian Stem Cell Centre, Monash University, Clayton, VIC, Australia*
ADAM C. PUCHE • *Department of Anatomy and Neurobiology, School of Medicine, University of Maryland, Baltimore, MD, USA*
BRENT A. REYNOLDS • *McKnight Brain Institute, University of Florida, Gainesville, FL, USA*
LARA ROSSI • *Center for Cell and Gene Therapy, Baylor College of Medicine, Houston, TX, USA; Institute of Hematology and Medical Oncology "L. & A. Seràgnoli", University of Bologna, Bologna, Italy*
ROBERT SAINT • *Molecular Genetics and Evolution, Research School of Biological Sciences, Australian National University, Acton, ACT, Australia*
XUN SHANG • *Division of Experimental Hematology and Cancer Biology, Cincinnati Children's Hospital Medical Center, University of Cincinnati, Cincinnati, OH, USA*
FLORIAN A. SIEBZEHNRUBL • *McKnight Brain Institute, University of Florida, Gainesville, FL, USA*
LOUISE SILVER-MORSE • *Department of Biomedical Genetics, University of Rochester Medical Center, Rochester, NY, USA*
OLGA SIRIN • *Center for Cell and Gene Therapy, Baylor College of Medicine, Houston, TX, USA*
ERIKA L. SPAETH • *Department of Stem Cell Transplantation and Cellular Therapy, Section of Molecular Hematology and Therapy, The University of Texas M. D. Anderson Cancer Center, Houston, TX, USA*
EVA SYKOVÁ • *Institute of Experimental Medicine ASCR, Prague, Czech Republic; Department of Neuroscience and Center for Cell Therapy and Tissue Repair, Second Medical Faculty, Charles University, Prague, Czech Republic*
JIN-WU TSAI • *Integrated Program in Cellular, Molecular and Biophysical Studies, Department of Pathology and Cell Biology, Center for Neurobiology and Behavior, College of Physicians & Surgeons, Columbia University, New York, NY, USA*
YARON VAGIMA • *Immunology Department, The Weizmann Institute of Science, Rehovot, Israel*
RICHARD B. VALLEE • *Integrated Program in Cellular, Molecular and Biophysical Studies, Department of Pathology and Cell Biology, Center for Neurobiology and Behavior, College of Physicians & Surgeons, Columbia University, New York, NY, USA*
VINATA VEDAM-MAI • *McKnight Brain Institute, University of Florida, Gainesville, FL, USA*
JOSEPH C. WU • *Division of Cardiology, Department of Medicine, Molecular Imaging Program at Stanford (MIPS), Stanford University School of Medicine, Stanford, CA, USA*
MERVIN C. YODER • *Department of Pediatrics, Wells Center for Pediatric Research, Indiana University School of Medicine, Indianapolis, IN, USA*

Yi Zheng • *Division of Experimental Hematology and Cancer Biology, Cincinnati Children's Hospital Medical Center, University of Cincinnati, Cincinnati, OH, USA*
Leonard I. Zon • *Howard Hughes Medical Institute, Harvard Stem Cell Institute, Children's Hospital Boston, Harvard Medical School, Boston, MA, USA*

Part I

Stem Cell and Migration: An Overview

Chapter 1

Trafficking of Stem Cells

Claire Magnon*, Daniel Lucas*, and Paul S. Frenette

Abstract

Stem cells undergo regulated trafficking from the developmental stages to the adulthood. Stem cell migration is critical to organize developing organs and likely contributes postnatally to tissue regeneration. Here, we review the molecular mechanisms underlying migration of hematopoietic stem cells, neural stem cells, and primordial germ cells, revealing common operative pathways.

Key words: Stem cell, Trafficking, Development, Tissue regeneration

1. Introduction

Stem cells, fundamental component of tissue biology, are thought to reside in most adult tissues where they participate in organogenesis, homeostasis, and tissue repair throughout life. Stem cells have the unique ability to self-renew and differentiate into mature tissue cells. They reside in specific tissue microenvironments, also known as niches, which provide critical signals that maintain their status throughout life. Stem cell migration during ontogeny from embryonic tissues to definitive organs is critical for organogenesis and stem cell maintenance. In the adult, the migratory capacity is retained in at least some stem cell types, contributing to regeneration and replenishment of stem and differentiated cell pools. This phenomenon is replicated in clinical transplantation procedure where, for instance, after a simple intravascular injection, hematopoietic stem cells (HSC) can home to bone marrow niches and reconstitute all blood cell lineages. Understanding stem cell trafficking will be critical for the development of future targeted

*Both authors contributed equally to this work.

Marie-Dominique Filippi and Hartmut Geiger (eds.), *Stem Cell Migration: Methods and Protocols*,
Methods in Molecular Biology, vol. 750, DOI 10.1007/978-1-61779-145-1_1, © Springer Science+Business Media, LLC 2011

stem cell therapies (1–3). Here, we provide a brief overview about the migration of three stem cell types that will be discussed further in subsequent chapters, namely HSC, neural stem cells (NSC), and primordial germ cells (PGC).

2. Stem Cell Migration During Ontogeny

During development, both HSC and PGC originate at a distant location from their definitive niches. They must migrate to appropriate locations to maintain themselves and sustain differentiated progeny throughout the life of the organism. By contrast, NSC do not migrate by themselves but rather direct the migration of immature neurons to their proper location in the cortex.

2.1. Emergence and Migration of Hematopoietic Stem Cells During Ontogeny

Generation and expansion of HSC and their precursors occur during embryogenesis in a developmental process involving different sites at distinct times (4–6). Surprisingly, the site where the first HSC are formed remains controversial (7, 8). At embryonic day 7 (E7.5) in the mouse, the extra-embryonic yolk sac (YS) is a primitive hematopoietic site capable to produce hematopoietic precursors with short-term reconstitution capacity and to generate differentiated hematopoietic cells *in situ* (Fig. 1a) (4, 9, 10). A second wave of multilineage hematopoietic precursors emergence takes place in the intra-embryonic compartment, starting at E7-8 in the splanchnopleura (Sp) (Fig. 1a) and subsequently in the para-aortic splanchnopleura which becomes the aorta-gonad-mesonephros (P-Sp/AGM (Fig. 1a), E8.5-13) (11–13). It has been shown that multilineage precursors with long-term reconstitution capacity are generated in the Sp, rather than in the YS, which confers to the intra-embryonic hemogenic site the feature to carry out definitive hematopoiesis (12, 13). Later, at E9.5, the placenta harbors a larger pool of multipotential progenitors and HSCs than does the P-Sp/AGM and the YS (14–17). Finally, when the circulatory system becomes functional at E10.5, myeloerythroid progenitors and HSC may egress from the embryonic sites to seed the fetal liver (FL) (Fig. 1a) where they are nurtured temporally before homing definitely in the bone marrow (BM) at birth (18, 19).

The molecular mechanisms underlying the journey of HSC during development are still unclear. The presence of mesenchymal stem cells (MSC) in embryonic and fetal hematopoietic organs suggests their putative role in generation, maintenance, and differentiation of HSC in certain niches (19). The migration of HSC is thought to be controlled by chemoattractant and adhesive molecules. During development, deficiency in the chemokine stromal-cell-derived factor 1 (SDF-1, also named CXCL12) or its cognate receptor CXCR4 induces defects in hematopoiesis in the fetal liver and the bone marrow (20–22). Further,

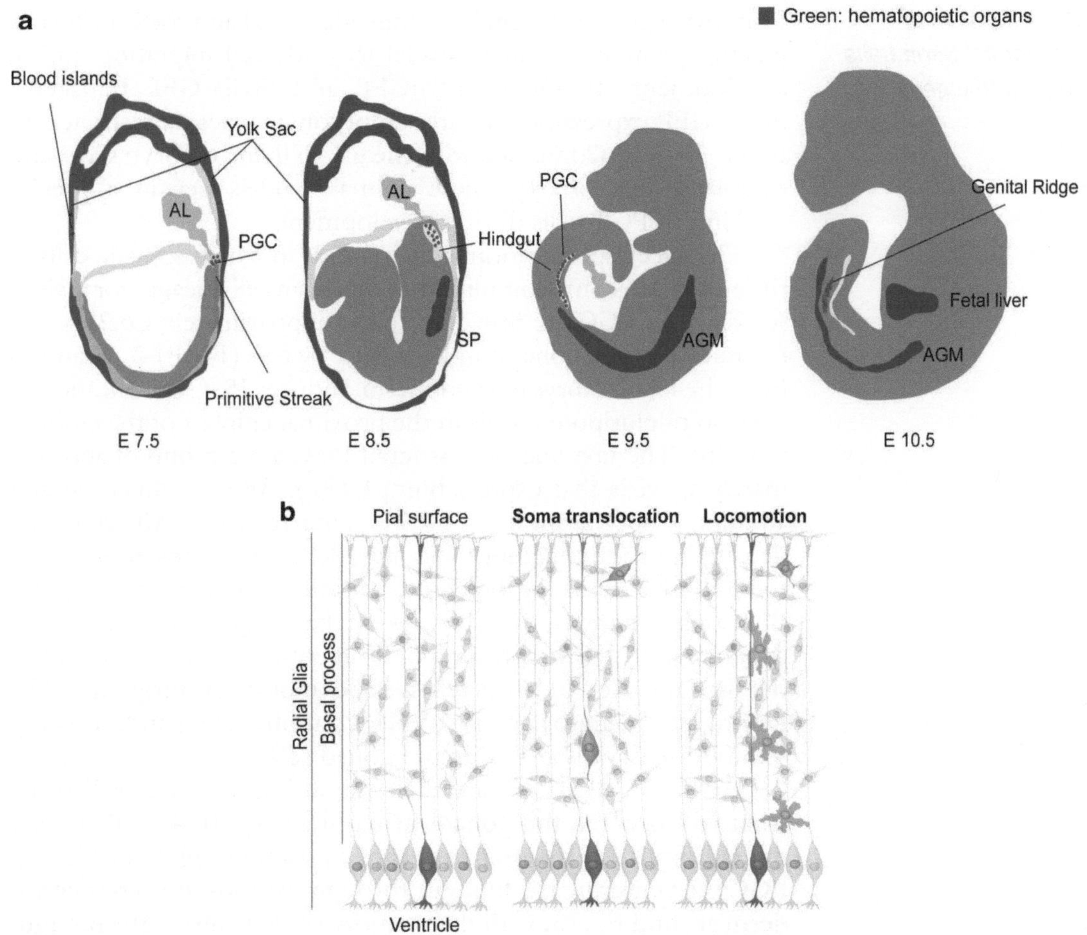

Fig. 1. Migration of HSC and PGC during development and role of radial glia. (**a**) Representation of the major organs containing HSC in the embryo (with the exception of the placenta), as well as the pathway of PGC migration at different developmental stages. *PGC* primordial germ cell, *SP* splancnopleura, *AGM* aorta-gonad-mesonephros, *AL* allantois. (**b**) Scheme showing the role of radial glia in directing neuroblast migration during development. Radial glia cells touch the surface of the ventricle and with a long basal process the pial surface. Newborn neuroblast can inherit this radial process and "pull up" through soma translocation to the pial region. In locomotion they do not inherit the radial process but will use it as a guideline to migrate toward the pial surface.

CXCL12-mediated signaling may regulate cell trafficking since its pattern of gene expressions correlates with the migration of hematopoietic progenitors between different hematopoietic organs (23, 24). The migratory response to CXCL12 toward the fetal liver is also enhanced in the presence of kit ligand (Ktl, also known as Steel Factor) (25). In parallel, HSC lose partially the expression of VE-cadherin, an endothelial adhesive protein, along their migration from the embryonic sites toward the fetal liver and the bone marrow (26). In addition, β1 integrins, cell-surface adhesion receptors which regulate cell migration, have been shown to be crucial for HSC traffic and homing upon the fetal liver colonization (27–29). Beta-6 integrins may be essential as a homing receptor only for fetal hematopoietic progenitors in the BM (29).

2.2. Migration of Primordial Germ Cells During Ontogeny

Primordial germ cells can be readily identified and tracked *in vivo*, making them an excellent model to study cell migration during development. In Oct4(DPE):GFP and Stella-GFP transgenic mice, GFP expression, at early embryonic stages, is restricted to PGC (30–32). Combined with the use of living embryo slices and time-lapse microscopy, these animal models permit a precise tracking of PGC cells during development.

Germ cell specification is the process in which somatic cells in the embryo become committed to the germ cell lineage (for review, see ref. 33). PGC are first detected at approximately E6.25 when the release of the bone morphogenetic factors (BMP)-2, -4 and -8 from the extraembryonic endoderm, induces PGC specification in a group of pluripotent cells in the proximal epiblast of the embryo (34–36). The first lineage-restricted PGC are a group of approximately six cells that express Blimp1 (37). After proliferation and additional specification at E7.25, approximately 40 PGC are detected in the extraembryonic mesoderm in the posterior end of the primitive streak, at the base of the allantois (Fig. 1a) (37, 38). Blimp1 is a transcriptional repressor that targets the somatic cell program in PGC by repression of Homeobox genes (37), allowing the initiation of a germ-cell-specific molecular program (33). During the migratory period, PGC continue to expand, reaching ~25,000 stem cells at E13 in the gonad (39).

At E7.25, PGC initiate their migration and they will reach what will become the gonads at day E11.5 (40, 41). PGC first migrate from the allantois toward the posterior epiblast (38, 42). PGC start entering the hindgut through the most posterior endoderm around E7.75, with the majority of PGC inside the hindgut at E8.5 (Fig. 1a) (42). The mechanism that regulates this part of the migration process is poorly understood. Expression of kit ligand by somatic cells in the allantois is necessary for both prevention of apoptosis and migration of PGC toward the hindgut. Although kit ligand is a chemotactic factor for PGC *in vitro* (43), PGC in *Kitl*-null embryos, albeit reduced in number and mobility, still migrate and enter the hindgut. This suggests that Kitl participates in, but does not direct, migration (38).

Concomitant with PGC migration the hindgut expands. Using Sox17 null mice, in which hindgut expansion does not occur, Hara et al. demonstrated that PGC cannot migrate into the hindgut of Sox17 null mice (42). These authors propose that morphogenetic changes in the hindgut results in the inclusion of the PGC within it (42). However, it is also possible that lack of hindgut expansion prevents the expression of one (or more) PGC chemoattractant(s).

Interferon-induced transmembrane proteins (IFTIM) have also been suggested to regulate PGC migration. Expression of the protein IFTIM1 in endodermal cells resulted in PGC chemorepulsion. On the contrary, expression of IFTIM3 on the same cells caused the migration of these cells toward PGC-rich regions (44). These data suggested that the activity of IFTIM proteins regulated the migration of the PGCs. However, the genetic deletion of the entire *Iftim* loci did not affect PGC migration or survival (45) suggesting the IFTIMs were not required for PGC migration.

Between E8.5–9.0, PGC move randomly inside the hindgut although they do not leave it (40). Between E9.0–9.5 PGC divide into two groups and start migrating toward the developing genital ridges (Fig. 1a), which contain the somatic precursors of the gonad, forming a network of migrating cells (41, 46). The PGC exit the hindgut from its dorsal part, and between E10.5 and E11.5 they reach the genital ridges (Fig. 1a) where they will coalesce with somatic cells to form the gonads (40). PGC remaining in the hindgut do not receive survival signals from steel-c-kit signaling and die through Bax-mediated apoptosis (47).

Several molecules participate in the regulation of these steps of migration. Again, Kitl-c-kit signaling is critical for migration and survival of PGC; at day 10.5 Kitl is downregulated in the hindgut but is expressed in the genital ridges (47). Kitl is also required for PGC mobility, but not directionality, in the hindgut (38). This suggests that upregulation and downregulation of Kitl by somatic cells may create a moving PGC niche throughout development (38). The chemokine CXCL12 and its receptor CXCR4 also regulate the colonization of the germinal ridge. CXCR4 is expressed by PGC (48) and deficiency in either CXCL12 or CXCR4 results in a delay in the speed of migration toward the genital ridges, causing a reduction on the number of PGC that will form the gonad (48, 49). However, PGC still migrate toward the genital ridges in *Cxcl12*$^{-/-}$ and *Cxcr4*$^{-/-}$ mice (48, 49), suggesting that although they promote migration they are not the sole molecular mechanism recruiting the PGC. Similarly, fibroblast growth factor (FGF), signaling through the FGFR2IIIb in the PGC is necessary for survival (50). In contrast, transforming growth factor α (TGF-α) signaling seems to negatively affect migration; PGC colonize more efficiently the genital ridges in TGF-α receptor-deficient mice (51).

Some of the adhesion mechanisms mediating PGC migration have been characterized. PGC deficient in integrin β1 do not colonize the hindgut (30). Mice deficient in connexin43 also show reduced migration speed and survival throughout the migration process, probably through β1 integrin dysfunction (52). E-cadherin also appears to be necessary for PGC specification and migration. Blockade of E-cadherin function in embryo cultures inhibits PGC appearance at E6.75 (53) and PGC condensation in the gonads at E11.5 after migration toward the genital ridges (54).

Despite the fact that several factors regulating migration have been identified, the identity of the molecule(s) that directs PGC migration in the mouse remain unknown suggesting that more than one factor may be acting in concert.

2.3. Radial Glia Cells Direct Neuronal Migration During Central Nervous System Development

Neural stem cells (NSC) are a population of cells residing in the central nervous system (CNS), that contribute to neuronal, astroglial, and olygodendrocytic lineages. NSC activity during development is found within a population of cells called "radial glia" (55) that has a dual function: cell production and direction of migration for newborn neurons (reviewed in refs. 56, 57). Radial glia cells have a bipolar cell body with processes interacting apically with the ventricle and basally with pial surface (Fig. 1a). During development, radial glia cells gives rise by asymmetric division to a new radial glia cell and a differentiated cell or committed progenitor. These cells are called basal progenitors or intermediate progenitor cells when they generate neurons (58–60), and olygodendrocyte progenitor cells (OPC) when they are committed to the oligondendrocytic lineage (61).

The radial process of the glia is used by the newborn neuron as a guide to climb from the ventricles toward the pial surface in a process called radial migration (reviewed in ref. 62). Disruption of the radial processes results in impaired migration without affecting neurogenesis (63, 64). Neuroblasts also use glia-independent pathways to move long distances during development in what is called "tangential migration" where cells move parallel to the surface of the ventricles (reviewed in ref. 62). Radial migration can be subdivided into two different modes of migration: locomotion (55, 65) and somal translocation (65, 66). In somal translocation, following the asymmetric division of the radial glia cells, the daughter neuroblast inherits the radial process that contacts the pial surface (Fig. 1b). The neuroblast then translocates its cell body toward the pial surface by "pulling" its soma up the radial process (65, 66). In locomotion (Fig. 1b), the neuroblast becomes multipolar and uses the radial process of the radial glia cell as a guideline toward the pial surface (65). Several molecules regulate this process: adhesion to the glial cells is mediated by astrotactin (67) and connexins 26 and 43 (68, 69). Signals mediated by reelin (70) and neuregulin (71, 72) direct migration, and cytoskeleton proteins like LIS1, dynein (73), and doublecortin (74) are also required for normal migration (for review, see refs. 62, 75). In contrast to other species, mammal radial glia cells are lost at the last stages of embryogenesis (76) and thus cannot direct migration in the adult.

3. Stem Cell Trafficking During Adulthood

3.1. HSC Trafficking During the Adult Life

During adulthood, blood cell homeostasis depends on HSC capacities of self-renewal, proliferation, and differentiation. The bone marrow is commonly known to be the physiologic reservoir of HSC, providing specialized niches where HSC lodge and engraft. However, HSC constitutively migrate out the BM toward the bloodstream in a circadian manner under steady state (77, 78). Conversely, circulating HSC have the unique skill to recirculate from the periphery and extravasate into the BM according to a process called "homing" (79–81). Specific trafficking and tropism of HSC toward the BM has been successfully used in clinical practice for hematopoietic stem cell transplantation (1–3).

3.1.1. Molecular Basis of HSC Homing

Stem cell homing to the BM involves a complex sequence of molecular events mediating first the recognition of stem cells by bone marrow endothelial cells of sinusoids (tethering, rolling, and arrest). Later, stem cells migrate through the endothelium and the bone marrow parenchyma toward their niches where they can survive for the entire life of the organism (Fig. 2). We will briefly overview the main molecular mechanisms related to this process.

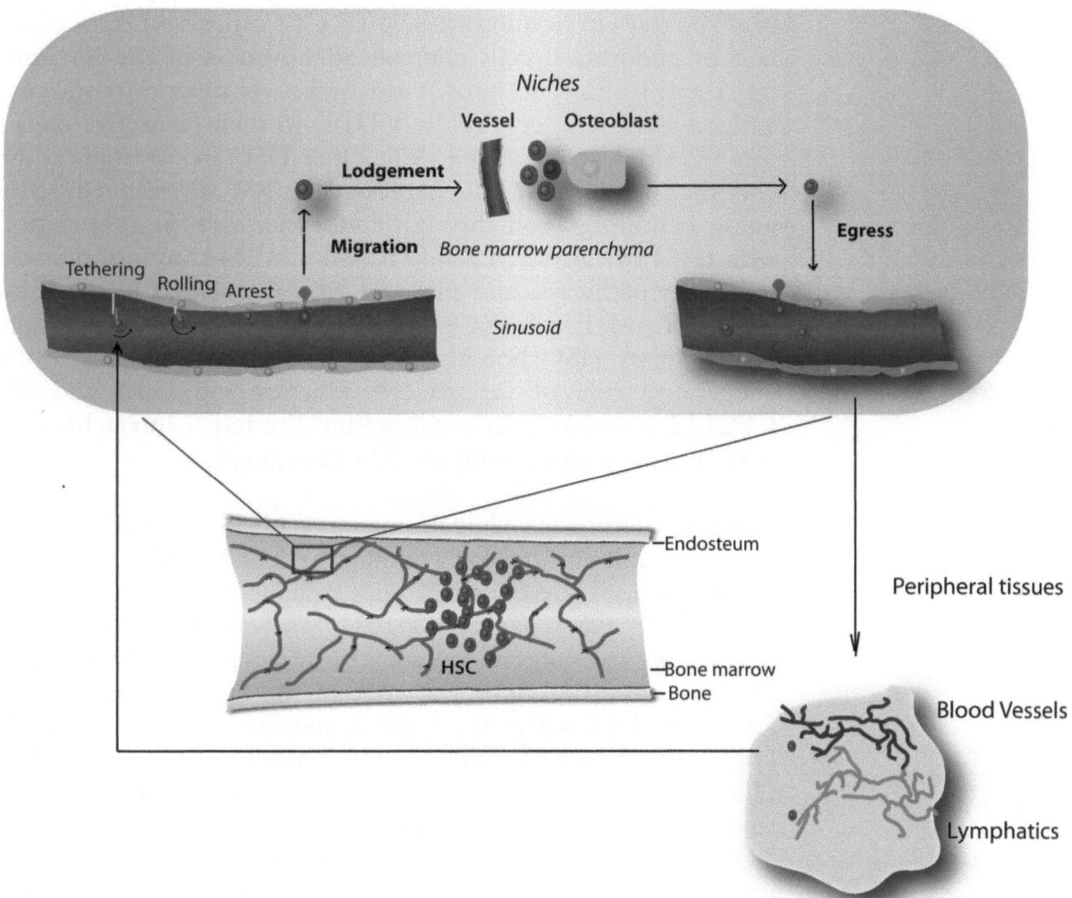

Fig. 2. Trafficking of HSC in the adult mice. HSC homing to the bone marrow is initiated by tethering and rolling interactions with bone marrow sinusoidal vessels. This allows HSC to arrest on the endothelium before migrating through bone marrow parenchyma under the guidance of chemoattractant signals. Then HSC are retained within endosteal and vascular niches where they lodge and engraft for survival and proliferation. During adult life, HSC continuously leave the bone marrow and re-enter the bloodstream to reach peripheral organs through the hematogenous or lymphatic circulation.

The initial steps involve a set of endothelial-progenitor cell receptor–counterreceptor interactions similar to those that govern the recruitment of mature leukocytes to sites of inflammation (82–85). Indeed, parallel contributions of P- and E-selectins, expressed on endothelial cells of the BM sinusoids, is known to be critical determinants of the HSC tethering and rolling along the endothelium (86, 87). This step requires an active fucosylation of the selectin ligand (PSGL-1) expressed on the surface of progenitors to augment their binding to E- and P-selectins on BM vasculature (88, 89). Importantly, it has been shown that the inhibition of the α4β1/VCAM-1 interaction highly compromises the proper rolling of progenitor cells, their arrest, and definitive homing within the BM (86, 87, 90). Further integrin-dependent interactions (α4β7 and β6) with endothelial cells regulate HSC homing significantly (91, 92).

The chemokine CXCL12 and its cognate receptor CXCR4 play key roles in homing to BM. CXC12 expression on the surface of endothelial cells controls adhesiveness of the integrins α4β1 and αLβ2 (also known as lymphocyte function-associated antigen-1, LFA-1, or CD11a/CD18) to their respective receptors VCAM-1 and ICAM-1 (93). Once HSC are arrested on the endothelial wall, CXCL12 expressed by BM stromal cells, may guide HSC migration through endothelial cells in cooperation with α4β1/VCAM-1, LFA-1/ICAM-1, CD44/hyaluronic acid molecular pathways, and Flt3 (94–96). Combined blockade of α4β1/VCAM-1 and CXCR4/CXCL12 interactions dramatically compromises HSC homing (97). These studies highlight the cooperative role of various cell adhesion molecules and the CXCL12 chemokine in orchestrating the initial interactions of HSC and progenitors with the BM vasculature.

3.1.2. HSC Lodgment and Retention

The mechanisms controlling the migration of HSC in the BM parenchyma are still unclear. After homing to the BM, HSC migrate to specific regions, described as endosteal or vascular niches that insure their maintenance (98–101). The endosteal location of progenitors was suggested by seminal studies published more than 30 years ago (102, 103). This concept was further supported by genetic analyses where the number and function of osteoblastic cells, were found to influence HSC numbers (99, 100, 104). HSC, which express the calcium-sensing receptor (CaR), respond to the high calcium ion concentration in the endosteal niche, allowing HSC lodgment close to osteoblasts (105). Osteopontin interacts with CD44 and β1integrins on HSC restricting the presence and proliferation of HSC on the endosteum (106, 107). CXCR4/CXCL12 axis and the angiopoietin/Tie2 signaling pathway may also contribute to the quiescence of HSC in the endosteal niches (108, 109). In addition, HSC engraftment in niches may require the guanine-nucleotide-binding stimulatory α (Gαs) subunit (110), suggesting that decision making about mobilization or retention/homing may result from the integration of a tightly controlled balance of Gαs (undefined receptor) and Gαi (CXCR4-mediated) activities (111). Concomitantly, the transmembrane form of stem cell factor (tm-SCF) may drive the lodgment of HSC in the endosteal niche (112). It has been argued that HSC, identified using the signaling lymphocyte attractant molecule (SLAM) family of markers (CD150+, CD244–, and CD48–), are instead located near sinusoids (113, 114). Recent imaging studies suggest a close relationship between blood vessels and endosteal surface (115, 116), blurring the lines between physical niches in the bone marrow.

3.1.3. Bone Marrow HSC Egress

In adult homeostasis, HSC continuously traffic from the BM niches throughout the peripheral circulation (Fig. 2). This phenomenon was highlighted using the parabiotic mouse model which lead to functional cross-engraftment of the BM of a lethally irradiated partner surgically conjoined with a nonirradiated mouse (117, 118). The role of HSC trafficking under steady state is not understood. It might supply and restore the local production of innate immune cells in peripheral organs under steady-state conditions, as shown in a study where HSC recirculated through the lymph to peripheral extramedullary tissues (119). This might also explain why HSC constitutively egress from the bone marrow to the bloodstream in a circadian manner reaching a peak during the resting period to potentially refill local hematopoietic niches (78). This phenomenon is tightly controlled by a local and cyclical release of noradrenaline from nerve terminals of the sympathetic innervation in the bone marrow. The adrenergic cues act on the β3 adrenergic receptor expressed on stromal cells, leading to a local downregulation of CXCL12 inducing HSC release (Fig. 3).

Successful autologous and allogeneic hematopoietic transplantations require large infusion of HSC capable to home, engraft, proliferate, and differentiate in the BM (1–3). For this reason, any further explorations of mechanisms underlying HSC egress under steady state or during mobilization might be valuable to improve the efficiency of clinical transplantations. Rather than harvesting HSC directly in the bone marrow, the vast majority of stem cell transplantation procedures use HSC and progenitors mobilized in the circulation using the hematopoietic cytokine granulocyte colony stimulating factor (G-CSF). Studies in mice have revealed that the G-CSF receptor (encoded by Csf3r) expression on the surface of HSC was not required to induce their release from the BM during a G-CSF regimen, suggesting that CSF3R-dependent signals act in trans (120). Subsequently, several other studies suggested that different enzymes (metallo- and/or serine proteases, or elastases) might represent the soluble

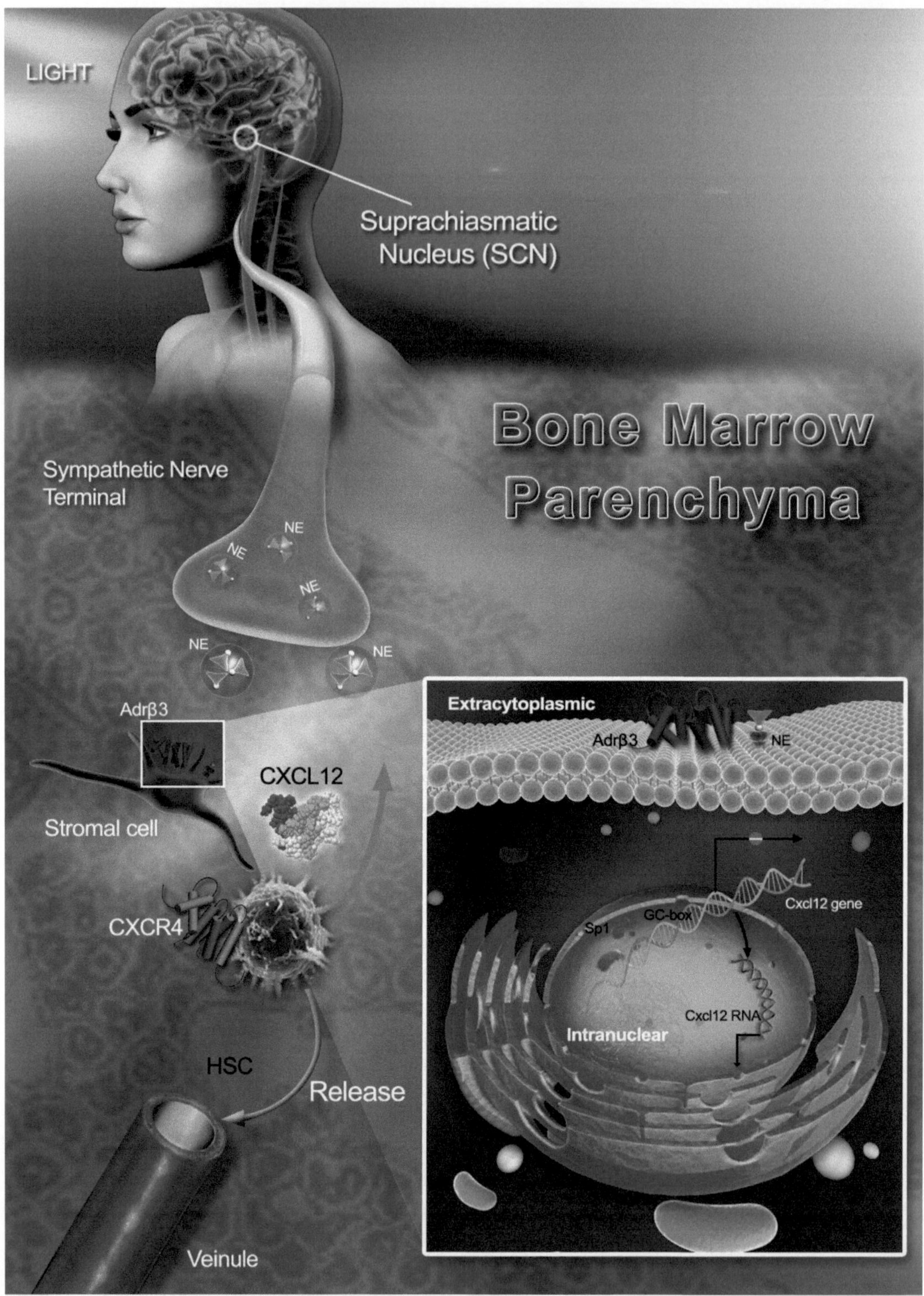

Fig. 3. Circadian regulation of HSC egress from bone marrow to the circulation. Physiological trafficking of HSC is regulated by the sympathetic nervous system in a circadian manner. Rhythmic secretion of noradrenaline activates the β_3-adrenergic receptor (Adrβ3) expressed on stromal cells in the BM, inducing degradation of Sp1 transcription factor, and downregulation of Cxcl12 transcription. Reproduced with permission from Magnon, C. and Frenette, P.S., Hematopoietic stem cell trafficking (July 14, 2008), StemBook, ed. The Stem Cell Research Community, StemBook, doi/10.3824/stembook.1.8.1, http://www.stembook.org.

"signal" released in G-CSF-induced mobilization (121–123). However, mobilization by G-CSF was normal in mice lacking virtually all neutrophil serine protease activity suggesting that other mechanisms must be involved (124). Recently, further studies have revealed that G-CSF-dependent mobilization of HSC is mediated by an adrenergic signal (125). Whether G-CSF modulates neural activity directly is currently unclear but possible since CSF3R is expressed on neurons. Additionally, G-CSF promotes neuronal survival (126) and protects dopaminergic neurons in a model of Parkinson's disease (127).

3.2. Migration of Neural Progenitors in the Adult Brain

In contrast to HSC where prospectively isolated cell subsets have clearly been shown at the clonal level to self-renew, it is not clear if the cells migrating in the adult central nervous system are *bona fide* NSC or more committed neuroblast progenitors. In the subventricular zone (SVZ), NSC give rise to neuroblasts that migrate toward the olfactory bulb following the rostral migratory stream (RMS) (Fig. 4). Neuroblasts in the RMS first migrate tangentially toward the olfactory bulb and once in it, they migrate radially. In this long journey, two different types of migration mechanisms have been identified: homophilic and vasophilic migration. In homophilic migration, chains of migrating neuroblasts move from the SVZ toward the olfactory bulb. Instead of using glia or axons as the substrate for migration, they use each others somas (128, 129). The migrating RMS neuroblasts are enclosed in specialized glial tubes composed from GFAP+ astrocytes (128). Several molecules have been implicated in facilitating cell–cell adhesion in homophilic migration. Deficiency in PSA-NCAM (polysialylated neural cell adhesion molecule) (130, 131) or β1 integrins (132) results in deficient migration. RMS migration is controlled by both chemorepulsion and chemoattraction. A gradient of the chemorepulsive factors Slit1 and 2 secreted by the

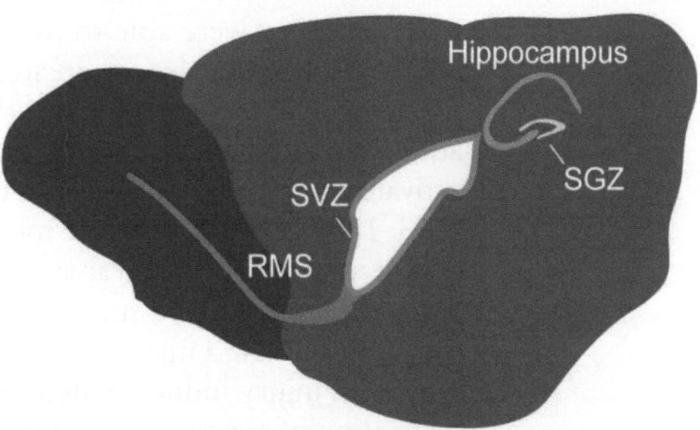

Fig. 4. NSC migration in adult brain. Sagittal representation of an adult mouse brain showing the subgranular zone (SGZ) of the hippocampus and the subventricular zone (SVZ) that contain NSC; the rostral migratory stream (RMS) is also exhibited.

lateral septum and the SVZ impulses the neuroblasts out of the SVZ toward the olfactory bulb (133–135) and is maintained by flow of the cerebrospinal fluid that directs the orientation of the migratory chains (136). Several factors act as chemoattractants for the migrating neuroblasts in the olfactory bulb, including netrin (137), neuregulin (138), ephrin (139), and brain-derived neurotrophic factor (BDNF) (140), although this last molecule is also involved in the regulation of vasophilic migration.

In vasophilic migration, as the name implies, the neural progenitors use blood vessels as guides for locomotion. Vasophilic migration was first reported in the olfactory bulb (141), and then detected in the RMS (142). The neuroblasts follow the vasculature although they never touch the endothelial cell; they are separated from it by very thin astrocytic end-feet that compartmentalize the migrating cells (141). BDNF, released by the endothelial cells, was shown to be one of the molecules that attracts neurons to the RMS (142).

In the subgranular zone (SGZ) of the hippocampus, newly formed neuroblasts are very close to their final destination and proliferate in clusters (143, 144). For migration, the neuroblast emits lateral processes allowing them to migrate tangentially. Then the neuroblast retracts the tangential process and emits radial processes that will become dendritic processes (143). This migration appears to depend on PSA (polysialic acid) expression by the neuroblast (144).

3.2.1. Migration of Neural Progenitors in the Adult Brain After Injury

The migration of cells with NSC properties is better characterized after brain injury. The discovery of multipotent murine (145, 146) and human (147) NSC opened the door to the possibility of using them for CNS regeneration. When transplanted into adult recipients these cells generated abundant astrocytes and olygodendrocytes, but very rarely neurons (146, 148, 149). Only when the recipient has received a CNS injury were these cells able to differentiate into neurons (149). In addition, NSC transplanted at distant places from an injury were able to migrate to the lesion and differentiate (150). Hence the lesion created an environment that recruited exogenous NSC and promoted their differentiation to neurons. In addition, lesions also recruit endogenous cells. Neurogenesis is activated in humans and rats in the SGZ (151, 152) and the SVZ (153, 154) after injury. In the mouse, endogenous newborn neurons appeared in the damaged areas of the cortex (155). These cells originated in the SVZ and migrated toward the area of damage where they differentiated into neurons (156).

Brain injury induces a neuroinflammatory response that causes generation of reactive astrocytes, activation of microglia (157, 158), and angiogenesis in the affected area (159). These events result in the release of chemokines and growth factors like CXCL12, angiopoietin-1, erythropoietin, MCP-1, BDNF, GDNF (160, 161), and

VEGF (162) that will increase neurogenesis in the SVZ and increase migration of neuroprogenitors toward the inflamed area.

The most common model for brain injury is stroke-induced ischemia (for review, see ref. 161). After ischemia, BrdU-labeling experiments indicate that proliferation is increased in the SVZ zone for approximately 2 weeks (155, 156). Neuroblast migration is detected approximately 1-week after the stroke. Reactive astrocytes extend processes that facilitate neuroblast migration (162–164). As in the RMS during normal neurogenesis, neuroblasts can form chains (156) and follow blood vessels toward the lesion (165, 166). Many of these blood vessels are generated *de novo* by angiogenesis after the stroke (166) and will release neurotrophic factors and chemokines to facilitate migration (160, 161).

The best characterized molecule that directs migration of NSC toward the site of injury is the chemokine CXCL12. Its receptor, CXCR4, is expressed in NSC in the SVZ and SGZ (167). After stroke injury, NSC migrate toward the ischemic area following a gradient of CXCL12 secreted by reactive astrocytes and endothelial cells in the ischemic areas (168). The role of CXCL12 in the migration of endogenous NSC was confirmed by administration of AMD3100, a CXCR4 antagonist, which partially blocked neuroblast migration toward the site of injury. This suggests that additional molecules are involved in directing this migration (164, 169). Another molecule that acts as a chemoattractant is angiopoietin-1 and its receptor Tie-2. Ang-1 is upregulated in the remodeling blood vessels after the lesion and its blockade reduces neuroblast migration from the SVZ (169), erythropoietin (73), MCP-1 (160), MIP-1α, GRO-α (170), and stem cell factor (162, 171) also act as chemoattractants for migrating neuroblasts.

One of the main limitations for the use of NSC in regenerative medicine is that only a minor fraction of the cells that reach the lesion are able to survive and differentiate into neurons (149, 156). A better understanding of the mechanisms that promote neuroblast recruitment and differentiation will be necessary before NSC therapy can be used to restore damaged neuronal circuits.

4. Stem Cell Trafficking: Different Cell Origins, Common Molecular Pathways

During development and adult life, some populations of stem cells have to migrate to distant locations to accomplish their functions in the establishment of embryonic tissues or in the regeneration of adult ones. This phenomenon involves chemoattractants, adhesion molecules, and specific pathways that direct migration to defined sites or niches. Among these molecules, the CXCL12/CXCR4 signaling pathway is arguably the most conserved mechanism in directing stem cell migration in the embryo and adult mammals (20–22). CXCL12 and its receptor CXCR4 are necessary for HSC migration and retention in the bone marrow throughout the life (23). CXCL12/CXCR4 axis also contributes to PGC migration in the mouse (48, 49) as well as in zebrafish (172–174). Similarly, during brain injury, NSC migrate toward CXCL12 released by reactive astrocytes (168). In addition to CXCL12 and CXCR4, other molecules have similar roles in different stem cells. Similarly, kit ligand and its receptor c-kit that contributes both to HSC (25) and PGC (38, 47) migration, and Angiopoietin-1 with its receptor Tie-2 regulate both HSC function (108) and NSC recruitment to vascular injuries (169). Understanding the molecular pathways that govern the complex trafficking of stem cells throughout the organism opens novel perspectives for clinical application of targeted stem cell therapies.

References

1. Bensinger, W., DiPersio, J.F., and McCarty, J.M. (2009) Improving stem cell mobilization strategies: future directions. *Bone Marrow Transplant* **43**, 181–95.
2. Cartier, N., Hacein-Bey-Abina, S., Bartholomae, C.C., Veres, G., Schmidt, M., Kutschera, I., Vidaud, M., Abel, U., Dal-Cortivo, L., Caccavelli, L., Mahlaoui, N., Kiermer, V., Mittelstaedt, D., Bellesme, C., Lahlou, N., Lefrere, F., Blanche, S., Audit, M., Payen, E., Leboulch, P., l'Homme, B., Bougneres, P., Von Kalle, C. Fischer, A., Cavazzana-Calvo, M., and Aubourg, P. (2009) Hematopoietic stem cell gene therapy with a lentiviral vector in X-linked adrenoleukodystrophy *Science* **326**, 818–23.
3. Cavazzana-Calvo, M., Hacein-Bey, S., de Saint Basile, G., Gross, F., Yvon, E., Nusbaum, P., Selz, F., Hue, C., Certain, S., Casanova, J.L., Bousso, P., Deist, F.L., and Fischer, A. (2000) Gene therapy of human severe combined immunodeficiency (SCID)-X1 disease *Science* **288**, 669–72.
4. Godin, I., and Cumano, A. (2002) The hare and the tortoise: an embryonic haematopoietic race *Nat Rev Immunol* **2**, 593–604.
5. Mikkola, H.K., and Orkin, S.H. (2006) The journey of developing hematopoietic stem cells *Development* **133**, 3733–44.
6. Orkin, S.H., and Zon, L.I. (2008) Hematopoiesis: an evolving paradigm for stem cell biology *Cell* **132**, 631–44.
7. Samokhvalov, I.M., Samokhvalova, N.I., and Nishikawa, S. (2007) Cell tracing shows the contribution of the yolk sac to adult haematopoiesis *Nature* **446**, 1056–61.
8. Medvinsky, A., and Dzierzak, E. (1996) Definitive hematopoiesis is autonomously initiated by the AGM region *Cell* **86**, 897–906.
9. Moore, M.A., and Metcalf, D. (1970) Ontogeny of the haemopoietic system: yolk sac origin of in vivo and in vitro colony forming cells in the developing mouse embryo *Br J Haematol* **18**, 279–96.
10. Toles, J.F., Chui, D.H., Belbeck, L.W., Starr, E., and Barker, J.E. (1989) Hemopoietic stem cells in murine embryonic yolk sac and peripheral blood *Proc Natl Acad Sci USA* **86**, 7456–59.
11. Muller, A.M., Medvinsky, A., Strouboulis, J., Grosveld, F., and Dzierzak, E. (1994) Development of hematopoietic stem cell activity in the mouse embryo *Immunity* **1**, 291–301.
12. Cumano ,A., Ferraz, J.C., Klaine, M., Di Santo, J.P., and Godin, I. (2001) Intraembryonic, but not yolk sac hematopoietic precursors, isolated before circulation, provide long-term multilineage reconstitution *Immunity* **15**, 477–85.
13. Cumano, A., Dieterlen-Lievre, F., and Godin, I. (1996) Lymphoid potential, probed before circulation in mouse, is restricted to caudal intraembryonic splanchnopleura *Cell* **86**, 907–16.
14. Gekas, C., Dieterlen-Lievre, F., Orkin, S.H., and Mikkola, H.K. (2005) The placenta is a niche for hematopoietic stem cells *Dev Cell* **8**, 365–75.
15. Ottersbach, K., and Dzierzak, E. (2005) The murine placenta contains hematopoietic stem cells within the vascular labyrinth region *Dev Cell* **8**, 377–87.

16. Rhodes, K.E., Gekas, C., Wang, Y., Lux, C.T., Francis, C.S., Chan, D.N., Conway, S., Orkin, S.H., Yoder, M.C., and Mikkola, H.K. (2008) The emergence of hematopoietic stem cells is initiated in the placental vasculature in the absence of circulation *Cell Stem Cell* **2**, 252–63.
17. Robin, C., Bollerot, K., Mendes, S., Haak, E., Crisan, M., Cerisoli, F., Lauw, I., Kaimakis, P., Jorna, R., Vermeulen, M., Kayser, M., van der Linden, R., Imanirad, P., Verstegen, M., Nawaz-Yousaf, H., Papazian, N., Steegers, E., Cupedo, T., and Dzierzak, E. (2009) Human placenta is a potent hematopoietic niche containing hematopoietic stem and progenitor cells throughout development *Cell Stem Cell* **5**, 385–95.
18. Johnson, G.R., and Moore, M.A. (1975) Role of stem cell migration in initiation of mouse foetal liver haemopoiesis *Nature* **258**, 726–8.
19. Mendes, S.C., Robin, C., and Dzierzak, E. (2005) Mesenchymal progenitor cells localize within hematopoietic sites throughout ontogeny *Development* **132**, 1127–36.
20. Ma, Q., Jones, D., Borghesani, P.R., Segal, R.A., Nagasawa, T., Kishimoto, T., Bronson, R.T., and Springer, T.A. (1998) Impaired B-lymphopoiesis, myelopoiesis, and derailed cerebellar neuron migration in CXCR4- and SDF-1-deficient mice *Proc Natl Acad Sci USA* **95**, 9448–53.
21. Nagasawa, T., Hirota, S., Tachibana, K., Takakura, N., Nishikawa, S., Kitamura, Y., Yoshida, N., Kikutani, H., and Kishimoto, T. (1996) Defects of B-cell lymphopoiesis and bone-marrow myelopoiesis in mice lacking the CXC chemokine PBSF/SDF-1 *Nature* **382**, 635–8.
22. Zou, Y.R., Kottmann, A.H., Kuroda, M., Taniuchi, I., and Littman, D.R. (1998) Function of the chemokine receptor CXCR4 in haematopoiesis and in cerebellar development *Nature* **393**, 595–9.
23. McGrath, K.E., Koniski, A.D., Maltby, K.M., McGann, J.K., and Palis, J. (1999) Embryonic expression and function of the chemokine SDF-1 and its receptor, CXCR4 *Dev Biol* **213**, 442–56.
24. Aiuti, A., Tavian, M., Cipponi, A., Ficara, F., Zappone, E., Hoxie, J., Peault, B., and Bordignon, C. (1999) Expression of CXCR4, the receptor for stromal cell-derived factor-1 on fetal and adult human lympho-hematopoietic progenitors *Eur J Immunol* **29**, 1823–31.
25. Christensen, J.L., Wright, D.E., Wagers, A.J., and Weissman, I.L. (2004) Circulation and chemotaxis of fetal hematopoietic stem cells *PLoS Biol* **2**, E75.
26. Taoudi, S., Morrison, A.M., Inoue, H., Gribi, R., Ure, J., and Medvinsky, A. (2005) Progressive divergence of definitive haematopoietic stem cells from the endothelial compartment does not depend on contact with the foetal liver *Development* **132**, 4179–91.
27. Hirsch, E., Iglesias, A., Potocnik, A.J., Hartmann, U., and Fassler, R. (1996) Impaired migration but not differentiation of haematopoietic stem cells in the absence of beta1 integrins *Nature* **380**, 171–5.
28. Potocnik, A.J., Brakebusch, C., and Fassler, R. (2000) Fetal and adult hematopoietic stem cells require beta1 integrin function for colonizing fetal liver, spleen, and bone marrow *Immunity* **12**, 653–63.
29. Qian, H., Georges-Labouesse, E. Nystrom, A., Domogatskaya, A., Tryggvason, K., Jacobsen, S.E., and Ekblom, M. (2007) Distinct roles of integrins alpha6 and alpha4 in homing of fetal liver hematopoietic stem and progenitor cells *Blood* **110**, 2399–407.
30. Anderson, R., Fassler, R., Georges-Labouesse, E., Hynes, R.O., Bader, B.L., Kreidberg, J.A., Schaible, K., Heasman, J., and Wylie, C. (1999) Mouse primordial germ cells lacking beta1 integrins enter the germline but fail to migrate normally to the gonads *Development* **126**, 1655–64.
31. Yoshimizu, T., Sugiyama, N., De Felice, M., Yeom, Y.I., Ohbo, K., Masuko, K., Obinata, M., Abe, K., Scholer, H.R., and Matsui ,Y. (1999) Germline-specific expression of the Oct-4/green fluorescent protein (GFP) transgene in mice *Dev Growth Differ* **41**, 675–84.
32. Payer, B., Chuva de Sousa Lopes, S.M., Barton, S.C., Lee, C., Saitou, M., and Surani, M.A. (2006) Generation of stella-GFP transgenic mice: a novel tool to study germ cell development *Genesis* **44**, 75–83.
33. Hayashi, K., de Sousa Lopes, S.M., and Surani, M.A. (2007) Germ cell specification in mice *Science* **316**, 394–6.
34. Lawson, K.A., Dunn, N.R., Roelen, B.A., Zeinstra, L.M., Davis, A.M., Wright, C.V., Korving, J.P., and Hogan, B.L. (1999) Bmp4 is required for the generation of primordial germ cells in the mouse embryo Genes Dev **13**, 424–36.
35. Ying, Y., Liu, X.M., Marble, A., Lawson, K.A., and Zhao, G.Q. (2000) Requirement of Bmp8b for the generation of primordial germ cells in the mouse *Mol Endocrinol* **14**, 1053–63.

36. Ying, Y., and Zhao, G.Q. (2001) Cooperation of endoderm-derived BMP2 and extraembryonic ectoderm-derived BMP4 in primordial germ cell generation in the mouse *Dev Biol* **232**, 484–92.
37. Ohinata, Y., Payer, B., O'Carroll, D., Ancelin, K., Ono, Y., Sano, M., Barton, S.C., Obukhanych, T., Nussenzweig, M., Tarakhovsky, A., Saitou, M., and Surani, M.A. (2005) Blimp1 is a critical determinant of the germ cell lineage in mice *Nature* **436**, 207–13.
38. Gu, Y., Runyan, C., Shoemaker, A., Surani, A., and and Wylie, C. (2009) Steel factor controls primordial germ cell survival and motility from the time of their specification in the allantois, and provides a continuous niche throughout their migration *Development* **136**, 1295–303.
39. Tam, P..P, and Snow, M.H. (1981) Proliferation and migration of primordial germ cells during compensatory growth in mouse embryos *J Embryol Exp Morphol* **64**, 133–47.
40. Molyneaux, K.A., Stallock, J., Schaible, K., and Wylie, C. (2001) Time-lapse analysis of living mouse germ cell migration *Dev Biol* **240**, 488–98.
41. Molyneaux K, and Wylie C. (2004) Primordial germ cell migration *Int J Dev Biol* **48**, 537–44.
42. Hara, K., Kanai-Azuma, M., Uemura, M., Shitara, H., Taya, C., Yonekawa, H., Kawakami, H., Tsunekawa, N., Kurohmaru, M., and Kanai, Y. (2009) Evidence for crucial role of hindgut expansion in directing proper migration of primordial germ cells in mouse early embryogenesis *Dev Biol* **330**, 427–39.
43. Farini, D., La Sala, G., Tedesco, M., and De Felici, M. (2007) Chemoattractant action and molecular signaling pathways of Kit ligand on mouse primordial germ cells *Dev Biol* **306**, 572–83.
44. Tanaka, S.S., Yamaguchi, Y.L., Tsoi, B., Lickert, H., and Tam, P.P. (2005) IFITM/Mil/fragilis family proteins IFITM1 and IFITM3 play distinct roles in mouse primordial germ cell homing and repulsion *Dev Cell* **9**, 745–56.
45. Lange, U.C., Adams, D.J., Lee, C., Barton, S., Schneider, R., Bradley, A., and Surani, M.A. (2008) Normal germ line establishment in mice carrying a deletion of the Ifitm/Fragilis gene family cluster *Mol Cell Biol* **28**, 4688–96.
46. Gomperts, M., Garcia-Castro, M., Wylie, C., and Heasman, J. (1994) Interactions between primordial germ cells play a role in their migration in mouse embryos *Development* **120**, 135–41.
47. Runyan, C., Schaible, K. Molyneaux, K., Wang, Z., Levin, L., and Wylie, C. (2006) Steel factor controls midline cell death of primordial germ cells and is essential for their normal proliferation and migration *Development* **133**, 4861–69.
48. Molyneaux, K.A., Zinszner, H., Kunwar, P.S., Schaible, K., Stebler, J., Sunshine, M.J., O'Brien, W., Raz, E., Littman, D., Wylie, C., and Lehmann, R. (2003) The chemokine SDF1/CXCL12 and its receptor CXCR4 regulate mouse germ cell migration and survival *Development* **130**, 4279–86.
49. Ara, T., Nakamura, Y., Egawa, T., Sugiyama, T., Abe, K., Kishimoto, T., Matsui, Y., and Nagasawa, T. (2003) Impaired colonization of the gonads by primordial germ cells in mice lacking a chemokine, stromal cell-derived factor-1 (SDF-1) *Proc Natl Acad Sci USA* **100**, 5319–23.
50. Takeuchi, Y., Molyneaux, K., Runyan, C., Schaible, K., and Wylie, C. (2005) The roles of FGF signaling in germ cell migration in the mouse *Development* **132**, 5399–409.
51. Chuva de Sousa Lopes, S.M., van den Driesche, S., Carvalho, R.L., Larsson, J., Eggen, B., Surani, M.A., and Mummery, C.L. (2005) Altered primordial germ cell migration in the absence of transforming growth factor beta signaling via ALK5 *Dev Biol* **284**, 194–203.
52. Francis, R.J., and Lo, C.W. (2006) Primordial germ cell deficiency in the connexin 43 knockout mouse arises from apoptosis associated with abnormal p53 activation *Development* **133**, 3451–60.
53. Okamura, D., Kimura, T., Nakano, T., and Matsui Y. (2003) Cadherin-mediated cell interaction regulates germ cell determination in mice *Development* **130**, 6423–6430.
54. Bendel-Stenzel, M.R., Gomperts, M., Anderson, R., Heasman, J., and Wylie, C. (2000) The role of cadherins during primordial germ cell migration and early gonad formation in the mouse *Mech Dev* **91**, 143–152.
55. Rakic, P. (1972) Mode of cell migration to the superficial layers of fetal monkey neocortex *J Comp Neurol* **145**, 61–83.
56. Kriegstein, A., and Alvarez-Buylla, A. (2009) The glial nature of embryonic and adult neural stem cells *Annu Rev Neurosci* **32**, 149–84.
57. Malatesta, P., Appolloni, I., and Calzolari, F. (2008) Radial glia and neural stem cells *Cell Tissue Res* **331**, 165–178.
58. Haubensak, W., Attardo, A., Denk, W., and Huttner, W.B. (2004) Neurons arise in the basal neuroepithelium of the early mammalian

telencephalon: a major site of neurogenesis *Proc Natl Acad Sci USA* **101**, 3196–201.

59. Miyata, T., Kawaguchi, A., Saito, K., Kawano, M., Muto, T., and Ogaw, M. (2004) Asymmetric production of surface-dividing and non-surface-dividing cortical progenitor cells *Development* **131**, 3133–45.
60. Noctor, S.C., Martinez-Cerdeno, V., Ivic, L., and Kriegstein, A.R. (2004) Cortical neurons arise in symmetric and asymmetric division zones and migrate through specific phases *Nat Neurosci* 7, 136–44.
61. Noble, M. (2000) Precursor cell transitions in oligodendrocyte development *J Cell Biol* **148**, 839–42.
62. Marin, O., and Rubenstein, J.L. (2003) Cell migration in the forebrain *Annu Rev Neurosci* **26**, 441–83.
63. Miyata, T., and Ogawa, M. (2007) Twisting of neocortical progenitor cells underlies a spring-like mechanism for daughter-cell migration *Curr Biol* **17**, 146–51.
64. Haubst N, Georges-Labouesse E, De Arcangelis A, Mayer U, and Gotz M. (2006) Basement membrane attachment is dispensable for radial glial cell fate and for proliferation, but affects positioning of neuronal subtypes *Development* **133**, 3245–54.
65. Nadarajah, B., Brunstrom, J.E., Grutzendler, J., Wong, R.O., and Pearlman, A.L. (2001) Two modes of radial migration in early development of the cerebral cortex *Nat Neurosci* **4**, 143–50.
66. Miyata, T., Kawaguchi, A., Okano, H., and Ogawa, M. (2001) Asymmetric inheritance of radial glial fibers by cortical neurons *Neuron* **31**, 727–41.
67. Adams, N.C., Tomoda, T., Cooper, M., Dietz, G., and Hatten, M.E. (2002) Mice that lack astrotactin have slowed neuronal migration *Development* **129**, 965–72.
68. Elias, L.A., Wang, D.D., and Kriegstein, A.R. (2007) Gap junction adhesion is necessary for radial migration in the neocortex *Nature* **448**, 901–7.
69. Cina, C., Maass, K., Theis, M., Willecke, K., Bechberger, J.F., and Naus, C.C. (2009) Involvement of the cytoplasmic C-terminal domain of connexin43 in neuronal migration *J Neurosci* **29**, 2009–21.
70. Rice, D.S., and Curran, T. (2001) Role of the reelin signaling pathway in central nervous system development *Annu Rev Neurosci* **24**, 1005–39.
71. Rio, C., Rieff, H.I., Qi, P., Khurana, T.S., and Corfas, G. (1997) Neuregulin and erbB receptors play a critical role in neuronal migration *Neuron* **19**, 39–50.
72. Anton, E.S., Marchionni, M.A., Lee, K.F., and Rakic, P. (1997) Role of GGF/neuregulin signaling in interactions between migrating neurons and radial glia in the developing cerebral cortex *Development* **124**, 3501–10.
73. Tsai, J.W., Bremner, K.H., and Vallee, R.B. (2007) Dual subcellular roles for LIS1 and dynein in radial neuronal migration in live brain tissue *Nat Neurosci* **10**, 970–9.
74. Gleeson, J.G., Allen, K.M., Fox, J.W., Lamperti, E.D., Berkovic, S., Scheffer, I., Cooper, E.C., Dobyns, W.B., Minnerath, S.R., Ross, M.E., and Walsh, C.A. (1998) Doublecortin, a brain-specific gene mutated in human X-linked lissencephaly and double cortex syndrome, encodes a putative signaling protein *Cell* **92**, 63–72.
75. Metin, C., Vallee, R.B., Rakic, P., and Bhide, P.G. (2008) Modes and mishaps of neuronal migration in the mammalian brain *J Neurosci* **28**, 11746–52.
76. Noctor, S.C., Martinez-Cerdeno, V., and Kriegstein, A.R. (2008) Distinct behaviors of neural stem and progenitor cells underlie cortical neurogenesis *J Comp Neurol* **508**, 28–44.
77. Wright, D.E., Wagers, A.J., Gulati, A.P., Johnson, F.L, and Weissman IL. (2001) Physiological migration of hematopoietic stem and progenitor cell. *Science* **294**, 1933–36.
78. Mendez-Ferrer, S., Lucas, D., Battista, M., and Frenette, P.S. (2008) Haematopoietic stem cell release is regulated by circadian oscillations *Nature* **452**, 442–7.
79. Jacobson, L.O., Marks, E.K., et al. (1949) The role of the spleen in radiation injury *Proc Soc Exp Biol Med* **70**, 740–2.
80. Barnes, D.W., Corp, M.J., Loutit, J.F., and Nea,l F.E. (1956) Treatment of murine leukaemia with X rays and homologous bone marrow; preliminary communication *Br Med J* **2**, 626–7.
81. Lorenz, E., Uphoff, D., Reid, T.R., and Shelton, E. (1951) Modification of irradiation injury in mice and guinea pigs by bone marrow injections *J Natl Cancer Inst* **12**, 197–201.
82. Frenette, P.S., and Wagner, D.D. (1996) Adhesion molecules--Part 1 *N Engl J Med* **334**, 1526–29.
83. Frenette, P.S., and Wagner, D.D. (1996) Adhesion molecules--Part II: Blood vessels and blood cells *N Engl J Med* **335**, 43–5.
84. Labow, M.A., Norton, C.R., Rumberger, J.M., Lombard-Gillooly, K.M., Shuster, D.J., Hubbard, J., Bertko, R., Knaack, P.A., Terry, R.W., and Harbison, M.L., et al. (1994) Characterization of E-selectin-deficient mice: demonstration of overlapping function of the endothelial selectins *Immunity* **1**, 709–20.

85. Yang, J., Hirata, T., Croce, K., Merrill-Skoloff, G., Tchernychev, B., Williams, E., Flaumenhaft, R., Furie, B.C., and Furie, B. (1999) Targeted gene disruption demonstrates that P-selectin glycoprotein ligand 1 (PSGL-1) is required for P-selectin-mediated but not E-selectin-mediated neutrophil rolling and migration *J Exp Med* **190**, 1769–82.
86. Frenette, P.S., Subbarao, S., Mazo, I.B., von Andrian, U.H., and Wagner, D.D. (1998) Endothelial selectins and vascular cell adhesion molecule-1 promote hematopoietic progenitor homing to bone marrow *Proc Natl Acad Sci USA* **95**, 14423–28.
87. Mazo, I.B., Gutierrez-Ramos, J.C., Frenette, P.S., Hynes, R.O., Wagner, D.D., and von Andrian, U.H. (1998) Hematopoietic progenitor cell rolling in bone marrow microvessels: parallel contributions by endothelial selectins and vascular cell adhesion molecule 1 *J Exp Med* **188**, 465–74.
88. Hidalgo, A., and Frenette, P.S. (2005) Enforced fucosylation of neonatal CD34+ cells generates selectin ligands that enhance the initial interactions with microvessels but not homing to bone marrow *Blood* **105**, 567–75.
89. Xia, L., McDaniel, J.M., Yago, T., Doeden, A., and McEver, R.P. (2004) Surface fucosylation of human cord blood cells augments binding to P-selectin and E-selectin and enhances engraftment in bone marrow *Blood* **104**, 3091–96.
90. Papayannopoulou, T., Craddock, C., Nakamoto, B., Priestley, G.V., and Wolf, N.S. (1995) The VLA4/VCAM-1 adhesion pathway defines contrasting mechanisms of lodgement of transplanted murine hemopoietic progenitors between bone marrow and spleen *Proc Natl Acad Sci USA* **92**, 9647–51.
91. Katayama, Y., Hidalgo, A., Peired, A., and Frenette, P.S. (2004) Integrin alpha4beta7 and its counterreceptor MAdCAM-1 contribute to hematopoietic progenitor recruitment into bone marrow following transplantation *Blood* **104**, 2020–26.
92. Qian, H., Tryggvason, K., Jacobsen, S.E., and Ekblom, M. (2006) Contribution of alpha6 integrins to hematopoietic stem and progenitor cell homing to bone marrow and collaboration with alpha4 integrins *Blood* **107**, 3503–10.
93. Peled, A., Petit, I., Kollet, O., Magid, M., Ponomaryov, T., Byk, T., Nagler, A., Ben-Hur, H., Many, A., Shultz, L., Lider, O., Alon, R., Zipori, D., and Lapidot, T. (1999) Dependence of human stem cell engraftment and repopulation of NOD/SCID mice on CXCR4 *Science* **283**, 845–8.
94. Peled, A., Kolle, O., Ponomaryov, T., Petit, I., Franitza, S., Grabovsky, V., Slav, M.M., Nagler, A., Lider, O., Alon, R., Zipori, D., and Lapidot, T. (2000) The chemokine SDF-1 activates the integrins LFA-1, VLA-4, and VLA-5 on immature human CD34(+) cells: role in transendothelial/stromal migration and engraftment of NOD/SCID mice *Blood* **95**, 3289–96.
95. Avigdor, A., Goichberg, P., Shivtiel, S., Dar, A., Peled, A., Samira, S., Kollet, O., Hershkoviz, R., Alon, R., Hardan, I., Ben-Hur, H., Naor, D., Nagler, A., and Lapidot, T. (2004) CD44 and hyaluronic acid cooperate with SDF-1 in the trafficking of human CD34+ stem/progenitor cells to bone marrow *Blood* **103**, 2981–89.
96. Fukuda, S., Broxmeyer, HE., and Pelus, L.M. (2005) Flt3 ligand and the Flt3 receptor regulate hematopoietic cell migration by modulating the SDF-1alpha(CXCL12)/CXCR4 axis *Blood* **105**, 3117–26.
97. Bonig, H., Priestley, G.V., Nilsson, L.M., Jiang, Y., and Papayannopoulou, T. (2004) PTX-sensitive signals in bone marrow homing of fetal and adult hematopoietic progenitor cells *Blood* **104**, 2299–306.
98. Nilsson, S.K., Johnston, H.M., and Coverdale, J.A. (2001) Spatial localization of transplanted hemopoietic stem cells: inferences for the localization of stem cell niches *Blood* **97**, 2293–99.
99. Calvi, L.M., Adams, G.B., Weibrecht, K.W., Weber, J.M., Olson, D.P., Knight, M.C., Martin, R.P., Schipani, E., Divieti, P., Bringhurst, F.R., Milner, L.A., Kronenberg, H.M., and Scadden, D.T. (2003) Osteoblastic cells regulate the haematopoietic stem cell niche *Nature* **425**, 841–46.
100. Zhang, J., Niu, C., Ye, L., Huang, H., He, X., Tong, W.G., Ross, J., Haug, J., Johnson, T., Feng, J.Q., Harris, S., Wiedemann, L.M., Mishina, Y., and Li, L. (2003) Identification of the haematopoietic stem cell niche and control of the niche size *Nature* **425**, 836–41.
101. Kopp, H.G., Avecilla, S.T., Hooper, A.T., Shmelkov, S.V., Ramos, C.A., Zhang, F., and Rafii, S. (2005) Tie2 activation contributes to hemangiogenic regeneration after myelosuppression *Blood* **106**, 505–13.
102. Lord, B.I., Testa, N.G., and Hendry, J.H. (1975) The relative spatial distributions of CFUs and CFUc in the normal mouse femur *Blood* **46**, 65–72.
103. Gong, J.K. (1978) Endosteal marrow: a rich source of hematopoietic stem cells *Science* **199**, 1443–45.

104. Visnjic, D., Kalajzic, Z., Rowe, D.W., Katavic, V., Lorenzo, J., and Aguila, H.L. (2004) Hematopoiesis is severely altered in mice with an induced osteoblast deficiency *Blood* **103**, 3258–64.

105. Adams, G.B., Chabner, K.T., Alley, I.R., Olson, D.P., Szczepiorkowski, Z.M., Poznansky, M.C., Kos, C.H., Pollak, M.R., Brown, E.M., and Scadden, D.T. (2006) Stem cell engraftment at the endosteal niche is specified by the calcium-sensing receptor *Nature* **439**, 599–603.

106. Nilsson, S.K., Johnston, H.M., Whitty, G.A., Williams, B., Webb, R.J., Denhardt, D.T., Bertoncello, I., Bendall, L.J., Simmons, P.J., and Haylock, D.N. (2005) Osteopontin, a key component of the hematopoietic stem cell niche and regulator of primitive hematopoietic progenitor cells *Blood* **106**, 1232–39.

107. Stier, S., Ko, Y., Forkert, R., Lutz, C., Neuhaus, T., Grunewald, E., Cheng, T., Dombkowski, D., Calvi, L.M., Rittling, S.R., and Scadden, D.T. (2005) Osteopontin is a hematopoietic stem cell niche component that negatively regulates stem cell pool size *J Exp Med* **201**, 1781–91.

108. Arai, F., Hirao, A., Ohmura, M., Sato, H., Matsuoka, S., Takubo, K., Ito, K., Koh, G.Y., and Suda, T. (2004) Tie2/angiopoietin-1 signaling regulates hematopoietic stem cell quiescence in the bone marrow niche *Cell* **118**, 149–61.

109. Nie, Y., Han, Y.C., and Zou, Y.R. (2008) CXCR4 is required for the quiescence of primitive hematopoietic cells *J Exp Med* **205**, 777–83.

110. Adams, G.B., Alley, I.R., Chung, U.I., Chabner, K.T., Jeanson, N.T., Lo Celso, C., Marsters, E.S., Chen, M., Weinstein, L.S., Lin, C.P., Kronenberg, H.M., and Scadden, D.T. (2009) Haematopoietic stem cells depend on Galpha(s)-mediated signalling to engraft bone marrow *Nature* **459**, 103–7.

111. Mendez-Ferrer, S., and Frenette, P.S. (2009) Galpha(s) uncouples hematopoietic stem cell homing and mobilization *Cell Stem Cell* **4**, 379–80.

112. Driessen, R.L., Johnston, H.M., and Nilsson, S.K. (2003) Membrane-bound stem cell factor is a key regulator in the initial lodgment of stem cells within the endosteal marrow region *Exp Hematol* **31**, 1284–91.

113. Kiel, M.J., Yilmaz, O.H., Iwashita, T., Yilmaz, O.H., Terhorst, C., and Morrison, S.J. (2005) SLAM family receptors distinguish hematopoietic stem and progenitor cells and reveal endothelial niches for stem cells *Cell* **121**, 1109–21.

114. Sugiyama, T., Kohara, H., Noda, M., and Nagasawa, T. (2006) Maintenance of the hematopoietic stem cell pool by CXCL12-CXCR4 chemokine signaling in bone marrow stromal cell niches *Immunity* **25**, 977–88.

115. Lo Celso, C., Fleming, H.E., Wu, J.W., Zhao, C.X., Miake-Lye, S., Fujisaki, J., Cote, D., Rowe, D.W., Lin, C.P., and Scadden, D.T. (2009) Live-animal tracking of individual haematopoietic stem/progenitor cells in their niche *Nature* **457**, 92–6.

116. Xie, Y., Yin, T., Wiegraebe, W., He, X.C., Miller, D., Stark, D., Perko, K., Alexander, R., Schwartz, J., Grindley, J.C., Park, J., Haug, J.S., Wunderlich, J.P., Li, H., Zhang, S., Johnson, T., Feldman, R.A., and Li, L. (2009) Detection of functional haematopoietic stem cell niche using real-time imaging *Nature* **457**, 97–101.

117. Brecher, G., and Cronkite EP. (1951) Post-radiation parabiosis and survival in rats *Proc Soc Exp Biol Med* 77, 292–4.

118. Warren, S., Chute, R.N., and Farrington, E.M. (1960) Protection of the hematopoietic system by parabiosis *Lab Invest* **9**, 191–8.

119. Massberg, S., Schaerli, P., Knezevic-Maramica, I., Kollnberger, M., Tubo, N., Moseman, E.A., Huff, I.V., Junt, T., Wagers, A.J., Mazo, I.B., and von Andrian, U.H. (2007) Immunosurveillance by hematopoietic progenitor cells trafficking through blood, lymph, and peripheral tissues *Cell* **131**, 994–1008.

120. Liu, F., Poursine-Laurent, J., and Link, D.C. (2000) Expression of the G-CSF receptor on hematopoietic progenitor cells is not required for their mobilization by G-CSF *Blood* **95**, 3025–31.

121. Levesque, J.P., Takamatsu, Y., Nilsson, S.K., Haylock, D.N., and Simmons, P.J. (2001) Vascular cell adhesion molecule-1 (CD106) is cleaved by neutrophil proteases in the bone marrow following hematopoietic progenitor cell mobilization by granulocyte colony-stimulating factor *Blood* **98**, 1289–97.

122. Petit, I., Szyper-Kravitz, M., Nagler, A., Lahav, M., Peled, A., Habler, L., Ponomaryov, T., Taichman, R.S., Arenzana-Seisdedos, F., Fujii, N., Sandbank, J., Zipori, D., and Lapidot, T. (2002) G-CSF induces stem cell mobilization by decreasing bone marrow SDF-1 and up-regulating CXCR4 *Nat Immunol* **3**, 687–94.

123. Heissig, B., Hattori, K., Dias, S., Friedrich, M., Ferris, B., Hackett, N.R., Crystal, R.G., Besmer, P., Lyden, D., Moore, M.A., Werb, Z., and Rafii, S. (2002) Recruitment of stem and progenitor cells from the bone marrow

niche requires MMP-9 mediated release of kit-ligand *Cell* **109**, 625–37.

124. Levesque, J.P., Liu, F., Simmons, P.J., Betsuyaku, T., Senior, R.M., Pham, C., and Link, D.C. (2004) Characterization of hematopoietic progenitor mobilization in protease-deficient mice *Blood* **104**, 65–72.
125. Katayama, Y., Battista, M., Kao, W.M., Hidalgo, A., Peired, A.J., Thomas, S.A., and Frenette, P.S. (2006) Signals from the sympathetic nervous system regulate hematopoietic stem cell egress from bone marrow *Cell* **124**, 407–21.
126. Schneider, A., Kruger, C., Steigleder, T., Weber, D., Pitzer, C., Laage, R., Aronowski, J., Maurer, M.H., Gassler, N., Mier, W., Hasselblatt, M., Kollmar, R., Schwab, S., Sommer, C., Bach, A., Kuhn, H.G., and Schabitz, W.R. (2005) The hematopoietic factor G-CSF is a neuronal ligand that counteracts programmed cell death and drives neurogenesis *J Clin Invest* **115**, 2083–98.
127. Meuer, K., Pitzer, C., Teismann, P., Kruger, C., Goricke, B., Laage, R., Lingor, P., Peters, K., Schlachetzki, J.C., Kobayashi, K., Dietz, G.P., Weber, D., Ferger, B., Schabitz, W.R., Bach, A., Schulz, J.B., Bahr, M., Schneider, A., and Weishaupt, J.H. (2006) Granulocyte-colony stimulating factor is neuroprotective in a model of Parkinson's disease *J Neurochem* **97**, 675–86.
128. Lois, C., Garcia-Verdugo, J.M., and Alvarez-Buylla, A. (1996) Chain migration of neuronal precursors *Science* **271**, 978–81.
129. Doetsch, F., and Alvarez-Buylla, A. (1996) Network of tangential pathways for neuronal migration in adult mammalian brain *Proc Natl Acad Sci USA* **93**, 14895–900.
130. Hu, H., Tomasiewicz, H., Magnuson, T., and Rutishauser, U. (1996) The role of polysialic acid in migration of olfactory bulb interneuron precursors in the subventricular zone *Neuron* **16**, 735–43.
131. Chazal, G., Durbec, P., Jankovski, A., Rougon, G., and Cremer, H. (2000) Consequences of neural cell adhesion molecule deficiency on cell migration in the rostral migratory stream of the mouse *J Neurosci* **20**, 1446–57.
132. Belvindrah, R., Hankel, S., Walker, J., Patton, B.L., and Muller, U. (2007) Beta1 integrins control the formation of cell chains in the adult rostral migratory stream *J Neurosci* **27**, 2704–17.
133. Wu, W., Wong, K., Chen, J., Jiang, Z., Dupuis, S., Wu, J.Y., and Rao, Y. (1999) Directional guidance of neuronal migration in the olfactory system by the protein Slit *Nature* **400**, 331–6.
134. Hu, H. (2001) Cell-surface heparan sulfate is involved in the repulsive guidance activities of Slit2 protein *Nat Neurosci* **4**, 695–701.
135. Nguyen-Ba-Charvet, K.T., Picard-Riera, N., Tessier-Lavigne, M., Baron-Van Evercooren, A., Sotelo, C., and Chedotal, A. (2004) Multiple roles for slits in the control of cell migration in the rostral migratory stream *J Neurosci* **24**, 1497–506.
136. Sawamoto, K., Wichterle, H., Gonzalez-Perez, O., Cholfin, J.A., Yamada, M., Spassky, N., Murcia, N.S., Garcia-Verdugo, J.M., Marin, O., Rubenstein, J.L., Tessier-Lavigne, M., Okano, H., and Alvarez-Buylla, A. (2006) New neurons follow the flow of cerebrospinal fluid in the adult brain. *Science* **311**, 629–32.
137. Murase, S., and Horwitz, A.F. (2002) Deleted in colorectal carcinoma and differentially expressed integrins mediate the directional migration of neural precursors in the rostral migratory stream *J Neurosci* **22**, 3568–79.
138. Anton, E.S., Ghashghaei, H.T., Weber, J.L., McCann, C., Fischer, T.M., Cheung, I.D., Gassmann, M., Messing, A., Klein, R., Schwab, M.H., Lloyd, K.C., and Lai ,C. (2004) Receptor tyrosine kinase ErbB4 modulates neuroblast migration and placement in the adult forebrain *Nat Neurosci* **7**, 1319–28.
139. Conover, J.C., Doetsch, F., Garcia-Verdugo, J.M., Gale, N.W., Yancopoulos, G.D., and Alvarez-Buylla, A. (2000) Disruption of Eph/ephrin signaling affects migration and proliferation in the adult subventricular zone *Nat Neurosci* **3**, 1091–97.
140. Chiaramello, S., Dalmasso, G., Bezin, L., Marcel, D., Jourdan, F., Peretto, P., Fasolo, A., and De Marchis, S. (2007) BDNF/ TrkB interaction regulates migration of SVZ precursor cells via PI3-K and MAP-K signalling pathways *Eur J Neurosci* **26**, 1780–90.
141. Bovetti, S., Hsieh, Y.C., Bovolin, P., Perroteau, I., Kazunori, T., and Puche, A.C. (2007) Blood vessels form a scaffold for neuroblast migration in the adult olfactory bulb *J Neurosci* **27**, 5976–80.
142. Snapyan, M., Lemasson, M., Brill, M.S., Blais, M., Massouh, M., Ninkovic, J., Gravel, C., Berthod, F., Gotz, M., Barker, P.A., Parent, A., and Saghatelyan, A. (2009) Vasculature guides migrating neuronal precursors in the adult mammalian forebrain via brain-derived neurotrophic factor signaling *J Neurosci* **29**, 4172–88.
143. Seki, T., Namba, T., Mochizuki, H., and Onodera, M. (2007) Clustering, migration, and neurite formation of neural precursor cells in the adult rat hippocampus *J Comp Neurol* **502**, 275–90.

144. Burgess, A., Wainwright, S.R., Shihabuddin, L.S., Rutishauser, U., Seki, T., and Aubert, I. (2008) Polysialic acid regulates the clustering, migration, and neuronal differentiation of progenitor cells in the adult hippocampus *Dev Neurobiol* **68**, 1580–90.

145. Reynolds, B.A., Weiss, S. (1992) Generation of neurons and astrocytes from isolated cells of the adult mammalian central nervous system *Science* **255**, 1707–10.

146. Snyder, E.Y., Deitcher, D.L., Walsh, C., Arnold-Aldea, S., Hartwieg, E.A., and Cepko, C.L. (1992) Multipotent neural cell lines can engraft and participate in development of mouse cerebellum *Cell* **68**, 33–51.

147. Flax, J.D., Aurora, S., Yang, C., Simonin, C., Wills, A.M., Billinghurst, L.L., Jendoubi, M., Sidman, R.L., Wolfe, J.H., Kim, S.U., and Snyder, E.Y. (1998) Engraftable human neural stem cells respond to developmental cues, replace neurons, and express foreign genes *Nat Biotechnol* **16**, 1033–39.

148. Snyder, E.Y., Taylor, R.M., and Wolfe, J.H. (1995) Neural progenitor cell engraftment corrects lysosomal storage throughout the MPS VII mouse brain *Nature* **374**, 367–70.

149. Snyder, E.Y., Yoon, C., Flax, J.D., and Macklis, J.D. (1997) Multipotent neural precursors can differentiate toward replacement of neurons undergoing targeted apoptotic degeneration in adult mouse neocortex *Proc Natl Acad Sci USA* **94**, 11663–8.

150. Aboody, K.S., Brown, A., Rainov, N.G., Bower, K.A., Liu, S., Yang, W., Small, J.E., Herrlinger, U., Ourednik, V., Black, P.M., Breakefield, X.O., and Snyder, E.Y. (2000) Neural stem cells display extensive tropism for pathology in adult brain: evidence from intracranial gliomas *Proc Natl Acad Sci USA* **97**, 12846–51.

151. Parent, J.M., Yu, T.W., Leibowitz, R.T., Geschwind, D.H., Sloviter, R.S., and Lowenstein, D.H. (1997) Dentate granule cell neurogenesis is increased by seizures and contributes to aberrant network reorganization in the adult rat hippocampus *J Neurosci* **17**, 3727–38.

152. Bengzon, J., Kokaia, Z., Elmer, E., Nanobashvili, A., Kokaia, M., and Lindvall, O. (1997) Apoptosis and proliferation of dentate gyrus neurons after single and intermittent limbic seizures *Proc Natl Acad Sci USA* **94**, 10432–7.

153. Zhang, R.L., Zhang, Z.G., Zhang, L., and Chopp, M. (2001) Proliferation and differentiation of progenitor cells in the cortex and the subventricular zone in the adult rat after focal cerebral ischemia *Neuroscience* **105**, 33–41.

154. Jin, K., Minami, M., Lan, J.Q., Mao, X.O., Batteur, S., Simon, R.P., and Greenberg, D.A. (2001) Neurogenesis in dentate subgranular zone and rostral subventricular zone after focal cerebral ischemia in the rat *Proc Natl Acad Sci USA* **98**, 4710–5.

155. Magavi, S.S., Leavitt, B.R., and Macklis, J.D. (2000) Induction of neurogenesis in the neocortex of adult mice *Nature* **405**, 951–5.

156. Arvidsson, A., Collin, T., Kirik, D., Kokaia, Z., and Lindvall, O. (2002) Neuronal replacement from endogenous precursors in the adult brain after stroke *Nat Med* **8**, 963–70.

157. Huang, D., Han, Y., Rani, M.R., Glabinski, A., Trebst, C., Sorensen, T., Tani, M., Wang, J., Chien, P., O'Bryan, S., Bielecki, B., Zhou, Z.L., Majumder, S., and Ransohoff, R.M. (2000) Chemokines and chemokine receptors in inflammation of the nervous system: manifold roles and exquisite regulation *Immunol Rev* **177**, 52–67.

158. Aarum, J., Sandberg, K., Haeberlein, S.L., and Persson, M.A. (2003) Migration and differentiation of neural precursor cells can be directed by microglia *Proc Natl Acad Sci USA* **100**, 15983–8.

159. Hayashi, T., Noshita, N., Sugawara, T., and Chan, P.H. (2003) Temporal profile of angiogenesis and expression of related genes in the brain after ischemia *J Cereb Blood Flow Metab* **23**, 166–80.

160. Belmadani, A., Tran, P.B., Ren, D., and Miller, R.J. (2006) Chemokines regulate the migration of neural progenitors to sites of neuroinflammation *J Neurosci* **26**, 3182–91.

161. Ohab, J.J., and Carmichael, S.T. (2008) Poststroke neurogenesis: emerging principles of migration and localization of immature neurons *Neuroscientist* **14**, 369–80.

162. Xu, Q., Wang, S., Jiang, X., Zhao, Y., Gao, M., Zhang, Y., Wang, X., Tano, K., Kanehara, M., Zhang, W., and Ishida, T. (2007) Hypoxia-induced astrocytes promote the migration of neural progenitor cells via vascular endothelial factor, stem cell factor, stromal-derived factor-1alpha and monocyte chemoattractant protein-1 upregulation in vitro *Clin Exp Pharmacol Physiol* **34**, 624–31.

163. Teramoto, T., Qiu, J., Plumier, J.C., and Moskowitz, M.A. (2003) EGF amplifies the replacement of parvalbumin-expressing striatal interneurons after ischemia *J Clin Invest* **111**, 1125–32.

164. Thored, P., Arvidsson, A., Cacci, E., Ahlenius, H., Kallur, T., Darsalia, V., Ekdahl, C.T., Kokaia, Z., and Lindvall, O. (2006) Persistent production of neurons from adult brain stem

cells during recovery after stroke *Stem Cells* **24**, 739–47.

165. Yamashita, T., Ninomiya, M., Hernandez Acosta, P., Garcia-Verdugo, J.M., Sunabori, T., Sakaguchi, M., Adachi, K., Kojima, T., Hirota, Y., Kawase,T., Araki, N., Abe, K., Okano, H., and Sawamoto, K. (2006) Subventricular zone-derived neuroblasts migrate and differentiate into mature neurons in the post-stroke adult striatum *J Neurosci* **26**, 6627–36.
166. Thored, P., Wood, J., Arvidsson, A., Cammenga, J., Kokaia, Z., and Lindvall, O. (2007) Long-term neuroblast migration along blood vessels in an area with transient angiogenesis and increased vascularization after stroke *Stroke* **38**, 3032–9.
167. Tran, P.B., Banisadr, G., Ren, D., Chenn, A., and Miller, R.J. (2007) Chemokine receptor expression by neural progenitor cells in neurogenic regions of mouse brain *J Comp Neurol* **500**, 1007–33.
168. Imitola, J., Raddassi, K., Park, K.I., Mueller, F.J., Nieto, M., Teng, Y.D., Frenkel, D., Li, J., Sidman, R.L., Walsh, C.A., Snyder, E.Y., and Khoury, S.J. (2004) Directed migration of neural stem cells to sites of CNS injury by the stromal cell-derived factor 1alpha/CXC chemokine receptor 4 pathway *Proc Natl Acad Sci USA* **101**, 18117–22.
169. Ohab, J.J., Fleming, S., Blesch, A., and Carmichael, S.T. (2006) A neurovascular niche for neurogenesis after stroke *J Neurosci* **26**, 13007–16.
170. Gordon, R.J., McGregor, A.L., and Connor, B. (2009) Chemokines direct neural progenitor cell migration following striatal cell loss *Mol Cell Neurosci* **41**, 219–32.
171. Sun, L., Lee, J., and Fine, H.A. (2004) Neuronally expressed stem cell factor induces neural stem cell migration to areas of brain injury *J Clin Invest* **113**, 1364–74.
172. Doitsidou, M., Reichman-Fried, M., Stebler, J., Koprunner, M., Dorries, J., Meyer, D., Esguerra, C.V., Leung, T., and Raz, E. (2002) Guidance of primordial germ cell migration by the chemokine SDF-1 *Cell* **111**, 647–59.
173. Knaut, H., Werz, C., Geisler, R., and Nusslein-Volhard, C. (2003) A zebrafish homologue of the chemokine receptor Cxcr4 is a germ-cell guidance receptor *Nature* **421**, 279–82.
174. Boldajipour, B., Mahabaleshwar, H., Kardash, E., Reichman-Fried, M., Blaser, H., Minina, S., Wilson, D., Xu, Q., and Raz, E. (2008) Control of chemokine-guided cell migration by ligand sequestration *Cell* **132**, 463–73.

Chapter 2

Migratory Strategies of Normal and Malignant Stem Cells

David A. Hess and Alison L. Allan

Abstract

The regulated migration of stem cells is critical for organogenesis during development and for tissue homeostasis and repair during adulthood. Human bone marrow (BM) represents an accessible reservoir containing regenerative cell types from hematopoietic, endothelial, and mesenchymal-stromal lineages that together coordinate hematopoiesis and promote the repair of damaged vasculature and tissues throughout the body. Thus, a detailed understanding of lineage-specific stem cell mobilization, homing, and subsequent engraftment in areas of injury or disease is of critical importance to the rational development of novel cell-mediated regenerative therapies. Stem cell trafficking via the circulation from site of origin to peripheral tissues requires fundamental molecular pathways governing (1) niche-specific deadhesion of progenitor cells; (2) chemoattraction to guide progenitor cell homing; and (3) interstitial navigation and adhesion/retention of recruited progenitor cells. This overview chapter summarizes the diversity of migratory strategies employed by hematopoietic, endothelial, and mesenchymal-stromal progenitor cells during repair and regeneration after tissue damage. Further elucidation of stem cell homing and migration pathways will allow greater application of stem cells for targeted cell therapy and/or drug delivery for tissue repair. Strikingly similar migratory mechanisms appear to govern the in vivo migration of recently characterized cancer stem cells (CSC) in leukemias and solid tumors, indicating that conserved principles of stem cell migration and niche specificity will provide new information to target CSC in anticancer therapy.

Key words: Adult stem cells, Niche specificity, Homing/chemotaxis, SDF-1/CXCR-4 axis, Adhesion, CD44, Cancer stem cells, Metastasis, Epithelial-to-mesenchymal transition

1. Introduction

Stem cells are rare, unspecialized precursor cells that are capable of multipotent differentiation to produce mature cells that carry out specific tissue functions and self-renewal to sustain or replenish the stem cell pool (for review, see ref. 1). These cells play a critical role in organ development in utero and during tissue maintenance and repair throughout adulthood, where they support the ongoing replacement of short-lived, diseased, or damaged

Marie-Dominique Filippi and Hartmut Geiger (eds.), *Stem Cell Migration: Methods and Protocols*,
Methods in Molecular Biology, vol. 750, DOI 10.1007/978-1-61779-145-1_2,

daughter cells. In addition, new studies provide increasing support for the notion that tissue-resident adult stem cells participate in the formulation of defined microenvironments or "niches" to support the endogenous regeneration or repair of diseased or damaged organs (2, 3). An important consideration for the development of cell-based strategies for regenerative medicine relies on understanding the endogenous capacity of stem and progenitor cells to migrate throughout the body in order to coordinate cell replacement and/or tissue repair in response to disease and injury. Recent evidence suggests that circulating and/or tissue-resident stem cells from hematopoietic (4), endothelial (5, 6), and/or mesenchymal lineages (7) participate directly in immune modulation and surveillance (4, 8, 9), new blood vessel formation (5, 6, 10), and endogenous tissue repair (11, 12). Implicit in the regenerative functions of tissue-specific stem cells is the proper localization of precursor cells for tissue homeostasis and repair that requires directed migration, engraftment, and retention within supportive stem cell niches. Thus, achieving targeted homing of stem cells to damaged organs is critical for efficient tissue regeneration from transplanted cells for the future development of alternative clinical therapies.

In this chapter, we review stem cell migration and homing during postnatal tissue homeostasis and regeneration, to specify the molecular mechanisms and fundamental themes governing the trafficking of various stem cells within the body (for review, see ref. 13). In addition, we will focus on the identification of conserved migratory strategies between these highly studied adult stem cell subtypes and newly characterized malignant cancer stem cells (CSC). Consistent with the notion that controlled mobilization, homing, and tissue engraftment of malignant stem cells to a pre-metastatic niche is important in the establishment of metastatic tumors (14), targeted interference in conserved stem cell migration or homing pathways may delay cancer progression and metastasis.

2. Mechanisms of Normal Stem Cell Migration

Optimal tissue function during normal cell turnover and after injury-induced repair relies on efficient stem cell homing. Homing can be defined as the process whereby stem cells are disseminated throughout the body passively via the bloodstream, and are ultimately directed to a supportive anatomical niche within the target organ (Fig. 1). Thus, stem cell trafficking via the circulation from site of origin to peripheral tissues requires fundamental molecular pathways governing (1) niche-specific deadhesion of progenitor cells; (2) chemoattraction to guide progenitor cell homing; and (3) interstitial navigation and adhesion/retention of recruited

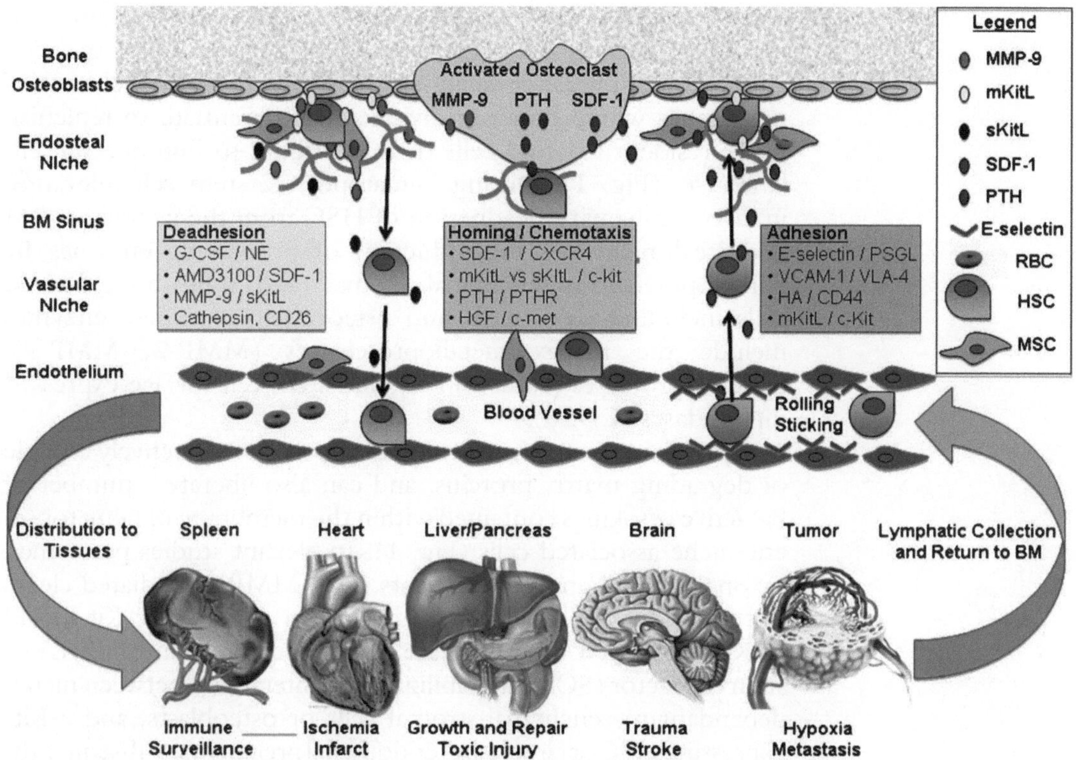

Fig. 1. Migratory strategies of normal and malignant stem cells. HSC reside in the bone marrow (BM) endosteal niche where interactions with mesenchymal-stromal cells, osteoblasts, and activated osteoclasts provide signals for proliferation, self-renewal and differentiation of mature hematopoietic cells. HSC are liberated from the endosteal niche through the release of metalloproteinases (MMPs) that convert stem cell factor (SCF) from the membrane bound (mKitL) to the soluble form (sKitL) of the ligand, resulting in deadhesion and migration to the vascular niche. Cathepsin K released from activated osteoclasts, and dipeptidyl peptidases (DPPIV), or CD26, on HSC also degrade stem cell supportive niche components resulting in deadhesion and mobilization of HSC to the peripheral circulation. Parathyroid hormone (PTH) binding to the PTH receptor (PTHR) leads to increased expression of the notch-ligand, Jagged-1 in osteoblasts, resulting in the overall expansion, and mobilization of BM-derived HSC pool via the Notch-signaling developmental pathway. Homing and chemotaxis between the BM, the circulation, and the peripheral tissues relies on the stromal-derived factor-1 (SDF-1 or CXCL12) and the CXC chemokine receptor, CXCR4. SDF-1 production by mesenchymal-stromal cells (MSC) in the BM in response to irradiation is involved in the recruitment of transplanted hematopoietic progenitor cells (HPC) to the BM. SDF-1 production by tissue-resident MSC and endothelial cells in hypoxic, damaged, or diseased tissues, including tumors, results in the directed homing of multiple stem cell types. Extravasation from the peripheral circulation into perivascular regions of tissues is mediated by adhesion and retention of HSC on the endothelium, through the activities of the adhesion moieties, including the α and β integrins, and the hyaluronic acid receptor, CD44.

progenitor cells. Blood to bone marrow homing is best described for transplanted hematopoietic stem cells (HSC), but these processes also apply to the trafficking of endogenous HSC and other nonhematopoietic stem cell types including metastasizing CSC.

2.1. Deadhesion Mechanisms and Stem Cell Mobilization

In the adult hematopoietic system, multipotent HSC or committed progenitors reside primarily in the bone marrow (BM) endosteum and/or vascular niche, respectively, (15–17), where they replenish circulating pools of short-lived, mature red blood cells and leukocytes (1). However, under both homeostatic and mobilization-induced states, these cells can leave the marrow, enter the peripheral circulation, and subsequently travel to various tissues where they can divide and differentiate to replenish tissue-resident myeloid cells that contribute to immune surveillance (4) (Fig. 1). During hematopoiesis, stem cell migration in vivo begins with deadhesion of HSC from the protective BM endosteal niche via the induction of proteolytic enzymes by hematopoietic, mesenchymal-stromal, and solid bone-derived cells including osteoblasts and osteoclasts (2). These enzymes include the matrix metalloproteinases (MMP-2, MMP-9), cysteine proteases (cathepsin K), and stem cell surface-expressed dipeptidases (CD26).

MMPs are zinc-dependent endopeptidases collectively capable of degrading matrix proteins, and can also liberate a number of bioactive cytokines contained within the membrane of hematopoietic-niche associated cells (Fig. 1). In elegant studies performed by Shahin Rafii and collaborators (16), MMP-9-mediated cleavage of membrane bound stem cell factor (SCF) or c-Kit ligand (mKitL) resulted in the release of soluble Kit ligand (sKitL) or stem cell factor (SCF), destabilizing the interaction between niche-dependant mesenchymal-stromal cells or osteoblasts, and c-Kit-expressing HSC or circulating endothelial precursors. Subsequently, sKitL conferred signals resulted in the translocation of these cells from the endosteal surface into a vascular-enriched niche favoring differentiation and release into the peripheral circulation (16). In a similar manner, Cathepsin K production by activated osteoclasts during bone resorption (Fig. 1), has been shown to degrade stem cell supportive niche components including stromal-derived factor-1 (SDF-1 or CXCL12), SCF, and osteopontin (OPN) resulting in HSC and progenitor cell mobilization (18). HSC also control the expression of the cell surface dipeptidase IV, or CD26, that contributes to the silencing of retention signals in part by cleavage-mediated inactivation of SDF-1 (Fig. 1) (19–22). An interesting role for parathyroid hormone (PTH) has also been demonstrated in the activation of osteoblasts, leading to increased expression of the notch-ligand, Jagged 1, and resulting in the overall expansion, and mobilization of BM-derived HSC pool via Notch-signaling developmental pathways (Fig. 1) (23, 24).

Mobilization of hematopoietic progenitor cells (HPC), and subsequent harvest from the peripheral circulation has emerged as the preferred strategy for the treatment of many hematological malignancies. Previously, granulocyte colony stimulating factor (G-CSF) was considered the gold standard for the mobilization of HSC or HPC for leukaphoresis and transplantation. Following G-CSF treatment, serine proteases including neutrophil elastase (NE) accumulate in the BM, and their broad substrates include vascular cell adhesion molecule-1 (VCAM-1), c-kit, CXCR4 and its ligand SDF-1 (25, 26). However, G-CSF mobilization is relatively inefficient, requiring 4–5 days treatment, and demonstrates broad interindividual variation in circulating HPC numbers resulting in reduced CD34+ cell harvests. More recently, AMD3100 or plerixafor, a potent CXCR4 antagonist, has been shown to efficiently mobilize CD34+ HPC into the periphery within 4 h (27) by selective blockade of CXCR4 binding to SDF-1 in the BM. AMD3100 in combination with G-CSF has been shown to be superior to G-CSF alone in mobilizing CD34+ non-obese diabetic/severe combined immune deficient (NOD/SCID) repopulating cells (SRC) in mice and humans (28, 29). Furthermore, interference in the CXCR4/SDF-1 axis has also been shown to mobilize nonhematopoietic endothelial and mesenchymal-stromal progenitor cells (30–32), underscoring the fundamental importance of this pathway in regulating stem and progenitor cell migration.

2.2. Guidance Factors that Mediate Chemoattraction and Chemoretention

Chemokines are cytokines that are best known for their ability to direct the migration of distinct subsets of leukocytes to sites of tissue inflammation. For HSC, functional interactions between the chemokine receptor CXCR4 and its ligand SDF-1 (CXCL12) have been implicated as the principle axis regulating survival (33–35), directed chemotaxis (36–38), and BM engraftment (39–44) (for review, see ref. 45). SDF-1 is expressed by vascular endothelium, osteoblasts, and stromal cells (46). Total body irradiation prior to transplantation into mice increases SDF-1 and SCF secretion within 24–28 h in the murine BM and spleen (46, 47). Human cell engraftment after transplantation of human CD34+ HSC into immunodeficient NOD/SCID recipients is efficiently blocked using CXCR4 antibodies (43), indicating the importance of this pathway in coordinated chemotaxis. However, murine CXCR4$^{-/-}$ cells can home to the BM, suggesting that other chemokine signaling mechanisms likely contribute to BM cell migration (48, 49). Interestingly, the SDF-1/CXCR4 axis is also active in the recruitment of HSC and other cell types to damaged tissues in response to hypoxia. Hypoxia inducible factor-1 (HIF-1) transcriptional activity is supported by low-oxygen conditions, and results in the increased expression of SDF-1 and vascular endothelial growth factor (VEGF) by endothelial cells in the liver (50), heart (51), and brain (52) during hypoxic stress (Fig. 1). SDF-1 secreted into the circulation from the injured liver can cross the endothelium in the BM and further recruit proangiogenic hematopoietic and progenitors into the circulation (50). These stimuli, together with secretion of other cytokines, chemokines, proteases, and adhesion molecules mediate stem cell migration between bone marrow and tissue-specific stem cell microenvironments (Fig. 1).

CXCR4 expression on stem cells is dynamically controlled by several well-characterized hematopoietic cytokines. IL-6, SCF, and/or hepatocyte growth factor (HGF) rapidly increase CXCR4 surface expression by externalization of intracellular stores on human CD34+ cells, leading to increased SDF-1-mediated migration and homing (43, 50). G-protein-coupled receptor signaling via the activation of CXCR4 has pleiotropic effects on stem cell function. Many of these effects are dependent on the activation of the atypical protein kinase, PKCz, which co-localizes with CXCR4 upon SDF-1-binding. PKCz-activation induces chemotaxis, polarization, MMP-9 secretion, and upregulation of cell surface adhesion molecules (53). CXCR4 signaling via the Rho family of GTPases (Rac1 and Rac2) is required for HSC retention and BM homing after transplantation, as deletion of Rac1 and Rac2 or administration of small molecule inhibitors induces immediate mobilization of progenitors in to the circulation (54, 55). A novel class of G-protein-coupled receptors, the lysophospholipid shingosine-1 phosphate (S1P) and its receptors (S1PR) act synergistically with SDF-1 via the Rho family kinases to increase HSC migration from the tissues into the lymphatic system, enabling their return to the blood via the thoracic duct (4).

Multipotent stromal/stem cells (MSC) are undifferentiated cells of mesenchymal lineage that can be isolated from many adult tissues including BM, kidney, liver, pancreas, adipose, and placenta (7, 56). In addition to their ability to differentiate into effector cells of mesodermal lineages (fat, bone, cartilage, and muscle), MSC have been localized to the perivascular niche and are precursors of smooth muscle cells (7). Their ubiquitous distribution, high expansion potential ex vivo, and potential immunomodulatory properties make MSC ideal candidates in cellular therapies for the repair and regeneration of a large variety of tissues (for review, see ref. 57). Although resting MSC show variable CXCR4 expression, exposure to low oxygen results in externalization of intracellular CXCR4 (58) and increased migration in response to SDF-1 (56).

2.3. Cell Adhesion and Retention

Once released into the bloodstream, circulating stem cells respond to guidance factors through the upregulation of adherence molecules that mediate the multistep process of adhesion to microvascular endothelial cells, similar to that initially described for the adhesion and homing of mature blood leukocytes. Initially, circulating HSCs are tethered to the vessel wall by the action of primary adhesion molecules (vascular selectins) that bind to sialyl-Lewis^x-like carbohydrate moieties associated with p-selectin glycoprotein ligand (PSGL-1) and the hyaluronic acid (HA) receptor CD44. Selectin binding together with the interaction of vascular cell adhesion molecule-1 (VCAM-1) and $\alpha4\beta1$ integrin (very late antigen 4, VLA-4), mediates further cell rolling and arrest in the microvasculature. In the presence of chemoattractive factors such as SDF-1 or CXCL12, G-protein-coupled receptor (CXCR4) signaling on HSC upregulates integrin (VLA-4) expression and coordinates the extravasation of the HSC into the tissue mesenchyme in a VLA-4-dependent fashion (41, 42). SDF-1 and other cytokines also induce HSC expression of proteolytic enzymes and including MMP-2 and MMP-9, allowing tissue infiltration and localization toward the SDF-1 gradient (Fig. 1).

Similar to HSC, MSC express a variety of cell surface adhesion molecules including the β1 integrins and CD44 (56). mRNA microarray analyses have shown that MSC express the cell surface adhesion molecules CD54 (ICAM-1), CD56 (NCAM), CD106 (VCAM-1), CD49a, b, c, e, f (integrins α1, 2, 3, 4, 6), and E-cadherin (56). However, MSC do not express significant levels of ligands to endothelial selectins such as PSGL-1 or sialyl Lewis^x carbohydrates. Thus, MSC likely coordinate rolling and adhesion to endothelial cells in a P-selectin and VCAM-1-dependent manner, respectively (59). Accordingly, MSC migrate in response to SDF-1/CXCR4 and HGF/c-met axes, and upregulate metalloproteinases (MMPs) that allow subsequent extravasation (60, 61).

In summary, given the adult HSC and MSC migration pathways described above, it is clear that a number of conserved adhesive and chemotactic signaling pathways act in concert to allow the trafficking of normal stem cells in the adult organism. Many of these systems not only support stem cell migration in adult tissues, but are utilized by stem cells during embryonic development (for review, see ref. 2). This suggests that much can be learned about the regenerative activities of adult stem cells through the study of organogenesis from embryonic precursors. As discussed in the remainder of this chapter, the majority of these pathways are also shared by malignant CSC, providing potential insights into metastatic progression.

3. Mechanisms of Malignant Stem Cell Migration

Recently, there has been increasing support for the cancer stem cell hypothesis, which postulates that cancer arises from a subpopulation of tumor-initiating cells or "cancer stem cells" (CSC) (62–72). Similar to a normal stem cell, the term "cancer stem cell" is an operational term defined as a cancer cell that has the ability to both self-renew to give rise to another tumorigenic cell, as well as undergo differentiation to give rise to the phenotypically diverse (and nontumorigenic) cell population that makes up the rest of the tumor (73). However, the definitive cellular origin of the CSC has remained elusive and is currently the topic of intense debate and experimental investigation. If these cells arise from mature, differentiated cells, oncogenic mutations are required for dedifferentiation and self-renewal (73–77). On the other hand, if CSC arise from adult stem cells, then cancer cells could appropriate the existing stem cell regulatory pathways for self-renewal, migration, and protection from cytotoxic drugs (78). The fact that multiple mutations are necessary for a cell to become tumorigenic and metastatic (79) has implications for the cellular origin of CSC. It can be argued that mature cells have a very limited lifespan, and thus it is unlikely that all the necessary mutations could occur during the relatively short life of these cells. In contrast, the infinite self-renewal capacity of adult stem cells means that these cells may be the only cells that are around long enough to accumulate the necessary mutations (74–77, 80). There is some evidence to suggest that many leukemias arise from mutation of normal hematopoietic stem/progenitor cells (81, 82), although the same has not yet been definitively proven in solid cancers.

Although the majority of studies investigating CSC have focused on the role of these cells in initiation and maintenance of primary cancers, growing evidence from our group and others (67, 78, 83–87) indicates that CSC may also be the cells responsible for metastasis, the process whereby tumor cells disseminate or migrate from the site of the primary tumor and establish themselves as secondary tumors in distant organs. Metastasis involves a series of sequential steps, including escape from the primary tumor (intravasation), migration and survival within the circulation, homing to secondary organs, arrest and extravasation into these organs, initiation of micrometastatic growth, and maintenance of growth into clinically detectable macrometastases (88–90). Given the onerous nature of this process, it is not surprising that metastasis is highly inefficient, with the main rate-limiting steps being initiation and maintenance of growth at secondary sites, a process called colonization (89, 91–94). Taken together with the heterogeneous nature of solid tumors, this metastatic inefficiency suggests that only a small subset of cells (i.e., CSC) can successfully navigate the metastatic cascade and eventually re-initiate tumor growth to form metastases. The successful metastatic CSC must therefore possess several key functional properties, including the ability to adhere, migrate, invade, stimulate angiogenesis, and grow. Interestingly, many of these properties mirror those used by adult stem cells for mobilization and homing to sites of tissue damage, as presented in the first section of this chapter and in Fig. 1. These similarities suggest that parallel and highly conserved migratory mechanisms may be operational in both adult stem cells and metastatic cancer stem cells, and this is discussed in greater detail below.

3.1. The Metastatic Niche

Adult stem cells require a specific niche or microenvironment in order to grow and survive (95–97). The stem cell niche is an anatomically defined space that has been identified in many different tissue types, and it serves to regulate stem cell number and function as well as to modulate stem cells under conditions of physiologic change. The niche cells and the microenvironment they create allow the niche to maintain the stem cell pool and prevent its differentiation, while at the same time also directing tissue growth and repair through more differentiated daughter progenitor cells (95). Furthermore, the niche provides protection to stem cells through provision of nourishment and exclusion from molecules that may cause differentiation, mutation, and/or apoptosis (97, 98).

Metastatic cells, like adult stem cells, require a particular microenvironment or niche in which to grow. This has been elegantly demonstrated by Kaplan et al., who showed that bone marrow-derived hematopoietic progenitor cells (HPC) expressing vascular endothelial growth factor receptor 1 (VEGFR1) can home to tumor-specific pre-metastatic sites and form cellular clusters before the mobilization and arrival of metastatic tumor cells (99). At these sites, $VEGFR1^+$ HPC express several factors such as CD34, CD116, c-kit, and Sca-1, which help in maintaining their progenitor cell status within the tissue parenchyma in the pre-metastatic niche. Together with fibronectin, these $VEGFR1^+$ HPCs alter the local microenvironment, which leads to activation of integrins and chemokines that promote attachment, survival, and growth of tumor cells. When treated with an anti-$VEGFR1^+$ antibody, the supportive pre-metastatic cell clusters are abolished and metastasis can be prevented, indicating that these clusters play an important role in the metastatic process (14, 99, 100). Additional supportive studies have also demonstrated that the pre-metastatic niche is "primed" by expression of chemoattractant factors and proteases in order to assist with mobilization of tumor cells and delivery to the secondary site (101, 102).

There has been some controversy surrounding whether or not CSC require specific niches for metastatic homing and/or growth. However, studies have shown that leukemic stem cells actually occupy a similar region as normal adult HSC (the endosteal region), suggesting that the stem cell niche may in fact be required in order to protect and maintain the pool of tumor-initiating/tumor-sustaining CSC (95). In solid cancers, it has been hypothesized that CSC in secondary organs are similarly regulated by signaling from the metastatic niche (103, 104). Currently, the relationship between stem cell niches in different tissues remains poorly understood, in particular with regard to whether tissue-specific stem cells can be regulated by stem cell niches in other organs. This knowledge will have important implications for understanding metastatic recruitment and growth in secondary sites, including the possibility that CSC in some cancer types (i.e., breast, prostate) may favor metastasis to the bone marrow because it provides a particularly rich stem cell niche (78, 94).

3.2. Guidance Factors that Mediate Metastatic Homing to Secondary Organs

Another migration strategy that is conserved between adult stem cells and metastatic CSC is the use of chemokine pathways, in particular the SDF-1/CXCR4 axis (105). SDF-1 is an ideal candidate for aiding in metastasis because its major biological effects involve induction of motility and chemotactic responses, as well as secretion of MMPs and angiopoietic factors such as VEGF by cells that express CXCR4. SDF-1 also increases adhesion of cells to VCAM-1, fibronectin, and fibrinogen by activating/modulating the function of several cell surface integrins (106). SDF-1 can promote tumor progression through recruitment of endothelial progenitor cells for angiogenesis (107). In addition, stromal cell expression of SDF-1 and tumor cell expression of CXCR4 is often increased within hypoxic areas of the tumor, subsequently triggering tumor cell growth, motility, and invasiveness. Secretion of SDF-1 by mesenchymal-stromal cells in the tumor microenvironment allows adhesion of CXCR4-expressing tumor cells and can confer resistance to apoptosis (98). For example, activation of CXCR4 can induce leukemia cell trafficking and homing to the bone marrow, where interaction with SDF-1 facilitates leukemia cell adhesion to BM stromal cells that provide growth and drug resistance signals (108). Therapies targeting CXCR4 (such as AMD3100) can disrupt these cell–cell adhesions and release leukemia cells from their protective stromal microenvironment, thus making them more accessible for targeting using conventional chemotherapy (98). Similarly, solid tumor types such as glioblastoma and medulloblastoma that are treated with AMD3100 show reduced cell growth and increased tumor cell apoptosis (109).

From a mobilization point of view, many CXCR4-expressing metastatic cells use the SDF-1/CXCR4 axis to migrate through the body according to an SDF-1 gradient, homing to organs that express high levels of SDF-1 (106, 110). In support of this, breast cancer has been shown to metastasize experimentally using the SDF-1/CXCR4 axis, with CXCR4-expressing breast cancer cells preferentially metastasizing to SDF-1-expressing organs such as bone, lymph node, and liver (88, 90, 111). Additional studies have demonstrated that breast cancer cells treated with a CXCR4 inhibitor show significantly inhibited metastatic ability (112). Expression of CXCR4 in many cancer types is indicative of poor prognosis (98, 106, 110, 111), and growing evidence suggests that CXCR4 expression correlates with the CSC content (and thus the aggressiveness) of tumors and cancer cell lines. For example, relative to nonmetastatic MCF-7 breast cancer cells, highly metastatic MDA-MB-231 cells have a larger proportion of CSC and express higher levels of CXCR4 (84, 110). Furthermore, in pancreatic cancer, it has been shown that the $CD133^+$ CSC population is heterogeneous with regard to CXCR4 expression, and that only the $CD133^+CXCR4^+$ CSC are able to metastasize (67). Thus, the SDF-1/CXCR4 axis appears to be a key migratory strategy of malignant/metastatic CSC.

3.3. Cell Adhesion Factors

Similar to adult stem cells, CSC also express a broad range of cell surface adhesion molecules that aid in metastatic migration and seeding. In particular, the adhesion molecule CD44 is a cell surface receptor for hyaluronan (HA) and osteopontin (OPN), and has a well-established role in cell adhesion, migration, and metastasis of cancer cells (113–115). CD44 has been shown to enhance the activity of proteases such as MMP9 and MMP2 in order to facilitate degradation of the extracellular matrix and subsequently promote cell migration and invasion (116). Direct binding of CD44 to adhesion molecules on the vascular endothelium can also promote transendothelial migration and tumor cell invasion (117). Finally, CD44 expression has been shown to be associated with the expression of proteins that mediate the epithelial-to-mesenchymal transition, an important contributor to metastasis (118) that is discussed further below.

From a functional perspective, it is therefore not surprising that CD44 would select for highly migratory and aggressive tumor cells. Interestingly, in solid cancers, CSC were first prospectively isolated from primary tumors and pleural effusions from breast cancer patients based on a $CD44^{+}CD24^{-}$ phenotype (62). Subsequent experimental studies have shown that $CD44^{+}CD24^{-}$ breast cancer cells demonstrate increased expression of stem cell markers; an enhanced capacity for in vitro mammosphere formation, invasion, and self-renewal; and the ability to recapitulate a heterogeneous tumor population (62, 119–121). Furthermore, clinical studies indicate that $CD44^{+}CD24^{-}$ tumor-initiating cells express an invasive gene signature (122) and may be associated with aggressive basal-like (triple-negative) disease (123) and tumor cell dissemination to secondary organs (86, 87). In addition to breast cancer, CD44 has also been shown to be an important functional marker for identifying CSC in prostate (64), colon (65), head and neck (70), pancreatic (124), and ovarian cancers (125). Taken together, these studies indicate that CD44 is not simply a marker for highly aggressive CSC, but instead that this adhesion molecule plays an important functional and mechanistic role in regulating malignant/metastatic behavior.

3.4. Epithelial-to-Mesenchymal Transition

Finally, the change in cell phenotype between epithelial and mesenchymal states (called the epithelial-to-mesenchymal transition, or EMT) has been identified to have a key role in migration related to both development and cancer. EMT is characterized by loss of planar and apical-basal polarity, loss of cell–cell adhesion, and dramatic cytoskeletal remodeling. Cells undergoing EMT also acquire expression of mesenchymal proteins and develop an enhanced ability to migrate, thus assisting in cell distribution throughout the embryo and organ development (126–128) (Fig. 2a). The earliest occurrence of EMT in embryogenesis involves the formation of the mesoderm during gastrulation. EMT has also been implicated in the formation of the placenta, neural crest, and urogenital tract,

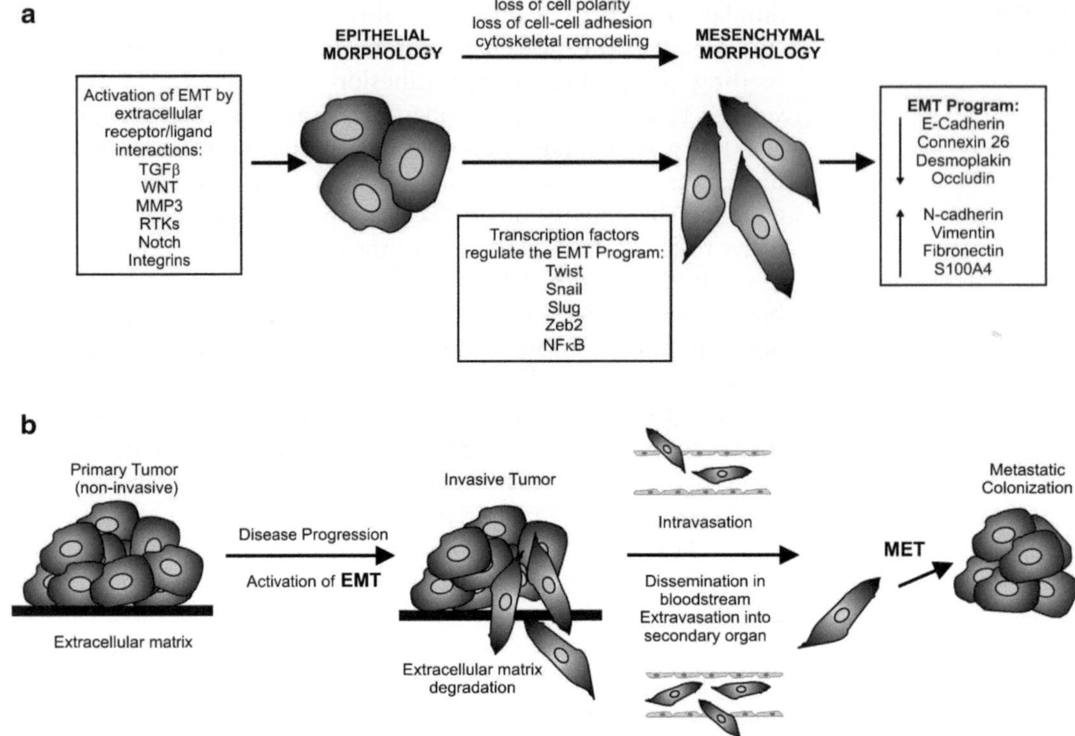

Fig. 2. The epithelial-to-mesenchymal transition. The change in cell phenotype between epithelial and mesenchymal states (called the epithelial-to-mesenchymal transition, or EMT) has been identified to have a key role in migration related to both development and cancer. (**a**) EMT can be activated by extracellular receptor/ligand interactions which activate downstream signaling including transforming growth factor β, Wnt, receptor tyrosine kinase (RTK), Notch, and integrin pathways. Subsequent regulation of EMT by Twist, Snail, Slug, ZEB2, NFκB, and other transcription factors leads to loss of planar and apical-basal polarity, loss of cell–cell adhesion, and dramatic cytoskeletal remodeling. Cells undergoing EMT also acquire expression of mesenchymal proteins and develop an enhanced ability to migrate, thus assisting in cell distribution throughout the embryo and organ development. (**b**) The EMT phenotypic transition is reversible, and it is hypothesized that once normal or cancer stem cells have migrated to their destination, they may transform back into an epithelial phenotype (so-called mesenchymal-to-epithelial transition, or MET) in order to facilitate growth in the secondary site.

as well as mediating branching morphogenesis in multiple organ types (128, 129). In cancer, it is believed that epithelial tumor cells (or CSC) may be able to somehow activate this primitive developmental program, thus converting differentiated epithelial cancer cells into de-differentiated cells that possess more primitive stem cell characteristics, the central of which is enhanced migration and metastasis (130, 131).

The EMT phenotypic transition is reversible, and it is hypothesized that once normal or cancer stem cells have migrated to their destination, they may transform back into an epithelial phenotype (so-called mesenchymal-to-epithelial transition, or MET) in order to facilitate growth in the secondary site (132) (Fig. 2b). EMT and MET are characterized by the expression of various factors responsible for mediating these processes at the molecular level.

Transforming growth factor β (TGF-β) has been shown to induce reversible EMT, along with Wnt pathway proteins (in particular β-catenin), Notch, and Hedgehog signaling pathways, which often act in a sequential manner to induce EMT (127). Additionally, transcription factors such as Twist, Snail, Slug, Zeb2, and NFκB have been shown to activate EMT programs in both development and cancer (126, 133, 134). Interestingly, recent studies have demonstrated that induction of EMT via overexpression of Twist or Snail (118) or activation of the Ras-MAPK pathway (135) in normal or malignant human mammary epithelial cells can lead to the generation of cells with CSC properties, including increased expression of stem cell proteins and an enhanced ability to grow as mammospheres (118, 135). These findings therefore implicate EMT not only in the migratory aspects of metastasis, but also in the last rate-limiting colonization step via EMT-mediated self-renewal capacity (118).

4. Conclusions

In summary, normal and malignant stem cells share many parallel and highly conserved migratory mechanisms related to (1) niche specificity, (2) chemoattraction to guide cell homing to target tissues, and (3) adhesion/retention of recruited stem cells (Fig. 1). In normal stem cells, many of these strategies are utilized during embryonic development, as well as supporting stem cell migration in adult tissues. Further elucidation of stem cell homing and migration pathways will allow greater application of stem cells for targeted cell therapy and/or drug delivery for tissue repair. The strikingly similar migratory mechanisms that govern malignant CSC migration in leukemias and solid tumors indicates that conserved principles of stem cell migration and niche specificity will provide new information to target CSC in anticancer therapy.

Acknowledgments

We thank members of our laboratory and our collaborators for their research work and helpful discussions. The authors' research on adult and malignant stem cells is supported by grants from the Canadian Institutes of Health Research (CIHR) (#MOP86759, MOP86702 to D.A.H.), The Krembil Foundation (to D.A.H), Canada Foundation for Innovation (#13199 to A.L.A.), and the Ontario Institute for Cancer Research (#08NOV-230 to A.L.A. and D.A.H). A.L.A. is supported by a CIHR New Investigator Award and an Early Researcher Award from the Ontario Ministry of Research and Innovation.

References

1. Weissman, I. L. (2000) Stem cells: units of development, units of regeneration, and units in evolution *Cell* **100**, 157–68.
2. Laird, D. J., von Andrian, U. H., and Wagers, A. J. (2008) Stem cell trafficking in tissue development, growth, and disease *Cell* **132**, 612–30.
3. Morrison, S. J. and Spradling, A. C. (2008) Stem cells and niches: mechanisms that promote stem cell maintenance throughout life *Cell* **132**, 598–611.
4. Massberg, S., Schaerli, P., Knezevic-Maramica, I., Kollnberger, M., Tubo, N., Moseman, E. A., Huff, I. V., Junt, T., Wagers, A. J., Mazo, I. B., and von Andrian, U. H. (2007) Immunosurveillance by hematopoietic progenitor cells trafficking through blood, lymph, and peripheral tissues *Cell* **131**, 994–1008.
5. Asahara, T., Murohara, T., Sullivan, A., Silver, M., van der Zee, R., Li, T., Witzenbichler, B., Schatteman, G., and Isner, J. M. (1997) Isolation of putative progenitor endothelial cells for angiogenesis *Science* **275**, 964–7.
6. Yoder, M. C., Mead, L. E., Prater, D., Krier, T. R., Mroueh, K. N., Li, F., Krasich, R., Temm, C. J., Prchal, J. T., and Ingram, D. A. (2007) Redefining endothelial progenitor cells via clonal analysis and hematopoietic stem/progenitor cell principals *Blood* **109**, 1801–9.
7. Crisan, M., Yap, S., Casteilla, L., Chen, C. W., Corselli, M., Park, T. S., Andriolo, G., Sun, B., Zheng, B., Zhang, L., Norotte, C., Teng, P. N., Traas, J., Schugar, R., Deasy, B. M., Badylak, S., Buhring, H. J., Giacobino, J. P., Lazzari, L., Huard, J., and Peault, B. (2008) A perivascular origin for mesenchymal stem cells in multiple human organs *Cell Stem Cell* **3**, 301–13.
8. Uccelli, A., Moretta, L., and Pistoia, V. (2008) Mesenchymal stem cells in health and disease *Nat Rev Immunol.*
9. Spaggiari, G. M., Capobianco, A., Abdelrazik, H., Becchetti, F., Mingari, M. C., and Moretta, L. (2008) Mesenchymal stem cells inhibit natural killer-cell proliferation, cytotoxicity, and cytokine production: role of indoleamine 2,3-dioxygenase and prostaglandin E2 *Blood* **111**, 1327–33.
10. Au, P., Tam, J., Fukumura, D., and Jain, R. K. (2008) Bone marrow-derived mesenchymal stem cells facilitate engineering of long-lasting functional vasculature *Blood* **111**, 4551–8.
11. Hess, D., Li, L., Martin, M., Sakano, S., Hill, D., Strutt, B., Thyssen, S., Gray, D. A., and Bhatia, M. (2003) Bone marrow-derived stem cells initiate pancreatic regeneration *Nat Biotechnol* **21**, 763–70.
12. Lee, R. H., Seo, M. J., Reger, R. L., Spees, J. L., Pulin, A. A., Olson, S. D., and Prockop, D. J. (2006) Multipotent stromal cells from human marrow home to and promote repair of pancreatic islets and renal glomeruli in diabetic NOD/scid mice *Proc Natl Acad Sci USA* **103**, 17438–43.
13. Zhou, P., Hohm, S., Capoccia, B., Wirthlin, L., Hess, D., Link, D., and Nolta, J. (2008) Immunodeficient mouse models to study human stem cell-mediated tissue repair *Methods Mol Biol* **430**, 213–25.
14. Kaplan, R. N., Rafii, S., and Lyden, D. (2006) Preparing the "soil": the premetastatic niche *Cancer Res* **66**, 11089–93.
15. Avecilla, S. T., Hattori, K., Heissig, B., Tejada, R., Liao, F., Shido, K., Jin, D. K., Dias, S., Zhang, F., Hartman, T. E., Hackett, N. R., Crystal, R. G., Witte, L., Hicklin, D. J., Bohlen, P., Eaton, D., Lyden, D., de Sauvage, F., and Rafii, S. (2004) Chemokine-mediated interaction of hematopoietic progenitors with the bone marrow vascular niche is required for thrombopoiesis *Nat Med* **10**, 64–71.
16. Heissig, B., Hattori, K., Dias, S., Friedrich, M., Ferris, B., Hackett, N. R., Crystal, R. G., Besmer, P., Lyden, D., Moore, M. A., Werb, Z., and Rafii, S. (2002) Recruitment of stem and progenitor cells from the bone marrow niche requires MMP-9 mediated release of kit-ligand *Cell* **109**, 625–37.
17. Heissig, B., Ohki, Y., Sato, Y., Rafii, S., Werb, Z., and Hattori, K. (2005) A role for niches in hematopoietic cell development *Hematology* **10**, 247–53.
18. Kollet, O., Dar, A., Shivtiel, S., Kalinkovich, A., Lapid, K., Sztainberg, Y., Tesio, M., Samstein, R. M., Goichberg, P., Spiegel, A., Elson, A., and Lapidot, T. (2006) Osteoclasts degrade endosteal components and promote mobilization of hematopoietic progenitor cells *Nat Med* **12**, 657–64.
19. Broxmeyer, H. E., Hangoc, G., Cooper, S., Campbell, T., Ito, S., and Mantel, C. (2007) AMD3100 and CD26 modulate mobilization, engraftment, and survival of hematopoietic stem and progenitor cells mediated by the SDF-1/CXCL12-CXCR4 axis *Ann N Y Acad Sci* **1106**, 1–19.
20. Campbell, T. B., Hangoc, G., Liu, Y., Pollok, K., and Broxmeyer, H. E. (2007) Inhibition of CD26 in human cord blood CD34+ cells enhances their engraftment of nonobese diabetic/severe combined immunodeficiency mice *Stem Cells Dev* **16**, 347–54.
21. Christopherson, K. W., 2nd, Hangoc, G., and Broxmeyer, H. E. (2002) Cell surface

peptidase CD26/dipeptidylpeptidase IV regulates CXCL12/stromal cell-derived factor-1 alpha-mediated chemotaxis of human cord blood CD34+ progenitor cells *J Immunol* **169**, 7000–8.

22. Christopherson, K. W., 2nd, Hangoc, G., Mantel, C. R., and Broxmeyer, H. E. (2004) Modulation of hematopoietic stem cell homing and engraftment by CD26 *Science* **305**, 1000–3.
23. Stier, S., Ko, Y., Forkert, R., Lutz, C., Neuhaus, T., Grunewald, E., Cheng, T., Dombkowski, D., Calvi, L. M., Rittling, S. R., and Scadden, D. T. (2005) Osteopontin is a hematopoietic stem cell niche component that negatively regulates stem cell pool size *J Exp Med* **201**, 1781–91.
24. Calvi, L. M., Adams, G. B., Weibrecht, K. W., Weber, J. M., Olson, D. P., Knight, M. C., Martin, R. P., Schipani, E., Divieti, P., Bringhurst, F. R., Milner, L. A., Kronenberg, H. M., and Scadden, D. T. (2003) Osteoblastic cells regulate the haematopoietic stem cell niche *Nature* **425**, 841–6.
25. Levesque, J. P., Hendy, J., Winkler, I. G., Takamatsu, Y., and Simmons, P. J. (2003) Granulocyte colony-stimulating factor induces the release in the bone marrow of proteases that cleave c-KIT receptor (CD117) from the surface of hematopoietic progenitor cells *Exp Hematol* **31**, 109–17.
26. Levesque, J. P., Hendy, J., Takamatsu, Y., Simmons, P. J., and Bendall, L. J. (2003) Disruption of the CXCR4/CXCL12 chemotactic interaction during hematopoietic stem cell mobilization induced by GCSF or cyclophosphamide *J Clin Invest* **111**, 187–96.
27. Liles, W. C., Broxmeyer, H. E., Rodger, E., Wood, B., Hubel, K., Cooper, S., Hangoc, G., Bridger, G. J., Henson, G. W., Calandra, G., and Dale, D. C. (2003) Mobilization of hematopoietic progenitor cells in healthy volunteers by AMD3100, a CXCR4 antagonist *Blood* **102**, 2728–30.
28. Broxmeyer, H. E., Orschell, C. M., Clapp, D. W., Hangoc, G., Cooper, S., Plett, P. A., Liles, W. C., Li, X., Graham-Evans, B., Campbell, T. B., Calandra, G., Bridger, G., Dale, D. C., and Srour, E. F. (2005) Rapid mobilization of murine and human hematopoietic stem and progenitor cells with AMD3100, a CXCR4 antagonist *J Exp Med* **201**, 1307–18.
29. Hess, D. A., Bonde, J., Craft, T. P., Wirthlin, L., Hohm, S., Lahey, R., Todt, L. M., Dipersio, J. F., Devine, S. M., and Nolta, J. A. (2007) Human progenitor cells rapidly mobilized by AMD3100 repopulate NOD/SCID mice with increased frequency in comparison to cells from the same donor mobilized by granulocyte colony stimulating factor *Biol Blood Marrow Transplant* **13**, 398–411.
30. Shepherd, R. M., Capoccia, B. J., Devine, S. M., Dipersio, J., Trinkaus, K. M., Ingram, D., and Link, D. C. (2006) Angiogenic cells can be rapidly mobilized and efficiently harvested from the blood following treatment with AMD3100 *Blood* **108**, 3662–7.
31. Capoccia, B. J., Shepherd, R. M., and Link, D. C. (2006) G-CSF and AMD3100 mobilize monocytes into the blood that stimulate angiogenesis in vivo through a paracrine mechanism *Blood* **108**, 2438–45.
32. Pitchford, S. C., Furze, R. C., Jones, C. P., Wengner, A. M., and Rankin, S. M. (2009) Differential mobilization of subsets of progenitor cells from the bone marrow *Cell Stem Cell* **4**, 62–72.
33. Broxmeyer, H. E., Kohli, L., Kim, C. H., Lee, Y., Mantel, C., Cooper, S., Hangoc, G., Shaheen, M., Li, X., and Clapp, D. W. (2003) Stromal cell-derived factor-1/CXCL12 directly enhances survival/antiapoptosis of myeloid progenitor cells through CXCR4 and G(alpha)i proteins and enhances engraftment of competitive, repopulating stem cells *J Leukoc Biol* **73**, 630–8.
34. Broxmeyer, H. E., Cooper, S., Kohli, L., Hangoc, G., Lee, Y., Mantel, C., Clapp, D. W., and Kim, C. H. (2003) Transgenic expression of stromal cell-derived factor-1/CXC chemokine ligand 12 enhances myeloid progenitor cell survival/antiapoptosis in vitro in response to growth factor withdrawal and enhances myelopoiesis in vivo *J Immunol* **170**, 421–9.
35. Lee, Y., Gotoh, A., Kwon, H. J., You, M., Kohli, L., Mantel, C., Cooper, S., Hangoc, G., Miyazawa, K., Ohyashiki, K., and Broxmeyer, H. E. (2002) Enhancement of intracellular signaling associated with hematopoietic progenitor cell survival in response to SDF-1/CXCL12 in synergy with other cytokines *Blood* **99**, 4307–17.
36. Jo, D. Y., Rafii, S., Hamada, T., and Moore, M. A. (2000) Chemotaxis of primitive hematopoietic cells in response to stromal cell-derived factor-1 *J Clin Invest* **105**, 101–11.
37. Naiyer, A. J., Jo, D. Y., Ahn, J., Mohle, R., Peichev, M., Lam, G., Silverstein, R. L., Moore, M. A., and Rafii, S. (1999) Stromal derived factor-1-induced chemokinesis of cord blood CD34(+) cells (long-term culture-initiating cells) through endothelial cells is mediated by E-selectin *Blood* **94**, 4011–9.
38. Mohle, R., Bautz, F., Rafii, S., Moore, M. A., Brugger, W., and Kanz, L. (1998) The chemokine receptor CXCR-4 is expressed on

CD34+ hematopoietic progenitors and leukemic cells and mediates transendothelial migration induced by stromal cell-derived factor-1 *Blood* **91**, 4523–30.

39. Kollet, O., Petit, I., Kahn, J., Samira, S., Dar, A., Peled, A., Deutsch, V., Gunetti, M., Piacibello, W., Nagler, A., and Lapidot, T. (2002) Human CD34(+)CXCR4(−) sorted cells harbor intracellular CXCR4, which can be functionally expressed and provide NOD/SCID repopulation *Blood* **100**, 2778–86.
40. Kollet, O., Spiegel, A., Peled, A., Petit, I., Byk, T., Hershkoviz, R., Guetta, E., Barkai, G., Nagler, A., and Lapidot, T. (2001) Rapid and efficient homing of human CD34(+) CD38(−/low)CXCR4(+) stem and progenitor cells to the bone marrow and spleen of NOD/SCID and NOD/SCID/B2m(null) mice *Blood* **97**, 3283–91.
41. Peled, A., Kollet, O., Ponomaryov, T., Petit, I., Franitza, S., Grabovsky, V., Slav, M. M., Nagler, A., Lider, O., Alon, R., Zipori, D., and Lapidot, T. (2000) The chemokine SDF-1 activates the integrins LFA-1, VLA-4, and VLA-5 on immature human CD34(+) cells: role in transendothelial/stromal migration and engraftment of NOD/SCID mice *Blood* **95**, 3289–96.
42. Peled, A., Grabovsky, V., Habler, L., Sandbank, J., Arenzana-Seisdedos, F., Petit, I., Ben-Hur, H., Lapidot, T., and Alon, R. (1999) The chemokine SDF-1 stimulates integrin-mediated arrest of CD34(+) cells on vascular endothelium under shear flow *J Clin Invest* **104**, 1199–211.
43. Peled, A., Petit, I., Kollet, O., Magid, M., Ponomaryov, T., Byk, T., Nagler, A., Ben-Hur, H., Many, A., Shultz, L., Lider, O., Alon, R., Zipori, D., and Lapidot, T. (1999) Dependence of human stem cell engraftment and repopulation of NOD/SCID mice on CXCR4 *Science* **283**, 845–8.
44. Hattori, K., Heissig, B., Tashiro, K., Honjo, T., Tateno, M., Shieh, J. H., Hackett, N. R., Quitoriano, M. S., Crystal, R. G., Rafii, S., and Moore, M. A. (2001) Plasma elevation of stromal cell-derived factor-1 induces mobilization of mature and immature hematopoietic progenitor and stem cells *Blood* **97**, 3354–60.
45. Lapidot, T., Dar, A., and Kollet, O. (2005) How do stem cells find their way home? *Blood* **106**, 1901–10.
46. Ponomaryov, T., Peled, A., Petit, I., Taichman, R. S., Habler, L., Sandbank, J., Arenzana-Seisdedos, F., Magerus, A., Caruz, A., Fujii, N., Nagler, A., Lahav, M., Szyper-Kravitz, M., Zipori, D., and Lapidot, T. (2000) Induction of the chemokine stromal-derived factor-1 following DNA damage improves human stem cell function *J Clin Invest* **106**, 1331–9.
47. Zhao, Y., Zhan, Y., Burke, K. A., and Anderson, W. F. (2005) Soluble factor(s) from bone marrow cells can rescue lethally irradiated mice by protecting endogenous hematopoietic stem cells *Exp Hematol* **33**, 428–34.
48. Kawabata, K., Ujikawa, M., Egawa, T., Kawamoto, H., Tachibana, K., Iizasa, H., Katsura, Y., Kishimoto, T., and Nagasawa, T. (1999) A cell-autonomous requirement for CXCR4 in long-term lymphoid and myeloid reconstitution *Proc Natl Acad Sci USA* **96**, 5663–7.
49. Ma, Q., Jones, D., and Springer, T. A. (1999) The chemokine receptor CXCR4 is required for the retention of B lineage and granulocytic precursors within the bone marrow microenvironment *Immunity* **10**, 463–71.
50. Kollet, O., Shivtiel, S., Chen, Y. Q., Suriawinata, J., Thung, S. N., Dabeva, M. D., Kahn, J., Spiegel, A., Dar, A., Samira, S., Goichberg, P., Kalinkovich, A., Arenzana-Seisdedos, F., Nagler, A., Hardan, I., Revel, M., Shafritz, D. A., and Lapidot, T. (2003) HGF, SDF-1, and MMP-9 are involved in stress-induced human CD34+ stem cell recruitment to the liver *J Clin Invest* **112**, 160–9.
51. Ratajczak, M. Z., Kucia, M., Reca, R., Majka, M., Janowska-Wieczorek, A., and Ratajczak, J. (2004) Stem cell plasticity revisited: CXCR4-positive cells expressing mRNA for early muscle, liver and neural cells 'hide out' in the bone marrow *Leukemia* **18**, 29–40.
52. Stumm, R. K., Rummel, J., Junker, V., Culmsee, C., Pfeiffer, M., Krieglstein, J., Hollt, V., and Schulz, S. (2002) A dual role for the SDF-1/CXCR4 chemokine receptor system in adult brain: isoform-selective regulation of SDF-1 expression modulates CXCR4-dependent neuronal plasticity and cerebral leukocyte recruitment after focal ischemia *J Neurosci* **22**, 5865–78.
53. Petit, I., Goichberg, P., Spiegel, A., Peled, A., Brodie, C., Seger, R., Nagler, A., Alon, R., and Lapidot, T. (2005) Atypical PKC-zeta regulates SDF-1-mediated migration and development of human CD34+ progenitor cells *J Clin Invest* **115**, 168–76.
54. Cancelas, J. A., Lee, A. W., Prabhakar, R., Stringer, K. F., Zheng, Y., and Williams, D. A. (2005) Rac GTPases differentially integrate signals regulating hematopoietic stem cell localization *Nat Med* **11**, 886–91.
55. Gu, Y., Filippi, M. D., Cancelas, J. A., Siefring, J. E., Williams, E. P., Jasti, A. C., Harris, C. E., Lee, A. W., Prabhakar, R., Atkinson, S. J., Kwiatkowski, D. J., and Williams, D. A. (2003) Hematopoietic cell

regulation by Rac1 and Rac2 guanosine triphosphatases *Science* **302**, 445–9.

56. Brooke, G., Tong, H., Levesque, J. P., and Atkinson, K. (2008) Molecular trafficking mechanisms of multipotent mesenchymal stem cells derived from human bone marrow and placenta *Stem Cells Dev* **17**, 929–40.
57. Uccelli, A., Moretta, L., and Pistoia, V. (2008) Mesenchymal stem cells in health and disease *Nat Rev Immunol* **8**, 726–36.
58. Hung, S. C., Pochampally, R. R., Hsu, S. C., Sanchez, C., Chen, S. C., Spees, J., and Prockop, D. J. (2007) Short-term exposure of multipotent stromal cells to low oxygen increases their expression of CX3CR1 and CXCR4 and their engraftment in vivo *PLoS ONE* **2**, e416.
59. Ruster, B., Gottig, S., Ludwig, R. J., Bistrian, R., Muller, S., Seifried, E., Gille, J., and Henschler, R. (2006) Mesenchymal stem cells display coordinated rolling and adhesion behavior on endothelial cells *Blood* **108**, 3938–44.
60. Sordi, V., Malosio, M. L., Marchesi, F., Mercalli, A., Melzi, R., Giordano, T., Belmonte, N., Ferrari, G., Leone, B. E., Bertuzzi, F., Zerbini, G., Allavena, P., Bonifacio, E., and Piemonti, L. (2005) Bone marrow mesenchymal stem cells express a restricted set of functionally active chemokine receptors capable of promoting migration to pancreatic islets *Blood* **106**, 419–27.
61. Son, B. R., Marquez-Curtis, L. A., Kucia, M., Wysoczynski, M., Turner, A. R., Ratajczak, J., Ratajczak, M. Z., and Janowska-Wieczorek, A. (2006) Migration of bone marrow and cord blood mesenchymal stem cells in vitro is regulated by stromal-derived factor-1-CXCR4 and hepatocyte growth factor-c-met axes and involves matrix metalloproteinases *Stem Cells* **24**, 1254–64.
62. Al-Hajj, M., Wicha, M. S., Benito-Hernandez, A., Morrison, S. J., and Clarke, M. F. (2003) Prospective identification of tumorigenic breast cancer cells *Proc Natl Acad Sci USA* **100**, 3983–8.
63. Bonnet, D. and Dick, J. E. (1997) Human acute myeloid leukemia is organized as a hierarchy that originates from a primitive hematopoietic cell *Nat Med* **3**, 730–7.
64. Collins, A. T., Berry, P. A., Hyde, C., Stower, M. J., and Maitland, N. J. (2005) Prospective identification of tumorigenic prostate cancer stem cells *Cancer Res* **65**, 10946–51.
65. Dalerba, P., Dylla, S. J., Park, I. K., Liu, R., Wang, X., Cho, R. W., Hoey, T., Gurney, A., Huang, E. H., Simeone, D. M., Shelton, A. A., Parmiani, G., Castelli, C., and Clarke, M. F. (2007) Phenotypic characterization of human colorectal cancer stem cells *Proc Natl Acad Sci USA* **104**, 10158–63.
66. Ginestier, C., Hur, M. H., Charafe-Jauffret, E., Monville, F., Dutcher, J., Brown, M. J. J., Viens, P., Kleer, C. G., Liu, S., Schott, A., Hayes, D., Birnbaum, D., Wicha, M. S., and Dontu, G. (2007) ALDH1 is a marker of normal and malignant human mammary stem cells and a predictor of poor clinical outcome *Cell Stem Cell* **1**, 555–567.
67. Hermann, P. C., Huber, S. L., Herrler, T., Aicher, A., Ellwart, J. W., Guba, M., Bruns, C. J., and Heeschen, C. (2007) Distinct populations of cancer stem cells determine tumor growth and metastatic activity in human pancreatic cancer *Cell Stem Cell* **1**, 313–323.
68. Hope, K. J., Jin, L., and Dick, J. E. (2004) Acute myeloid leukemia originates from a hierarchy of leukemic stem cell classes that differ in self-renewal capacity *Nat Immunol* **5**, 738–43.
69. O'Brien, C. A., Pollett, A., Gallinger, S., and Dick, J. E. (2007) A human colon cancer cell capable of initiating tumour growth in immunodeficient mice *Nature* **445**, 106–10.
70. Prince, M. E., Sivanandan, R., Kaczorowski, A., Wolf, G. T., Kaplan, M. J., Dalerba, P., Weissman, I. L., Clarke, M. F., and Ailles, L. E. (2007) Identification of a subpopulation of cells with cancer stem cell properties in head and neck squamous cell carcinoma *Proc Natl Acad Sci USA* **104**, 973–8.
71. Ricci-Vitiani, L., Lombardi, D. G., Pilozzi, E., Biffoni, M., Todaro, M., Peschle, C., and De Maria, R. (2007) Identification and expansion of human colon-cancer-initiating cells *Nature* **445**, 111–5.
72. Schatton, T., Murphy, G. F., Frank, N. Y., Yamaura, K., Waaga-Gasser, A. M., Gasser, M., Zhan, Q., Jordan, S., Duncan, L. M., Weishaupt, C., Fuhlbrigge, R. C., Kupper, T. S., Sayegh, M. H., and Frank, M. H. (2008) Identification of cells initiating human melanomas *Nature* **451**, 345–9.
73. Clarke, M. F., Dick, J. E., Dirks, P. B., Eaves, C. J., Jamieson, C. H., Jones, D. L., Visvader, J., Weissman, I. L., and Wahl, G. M. (2006) Cancer Stem Cells--Perspectives on Current Status and Future Directions: AACR Workshop on Cancer Stem Cells *Cancer Res* **66**, 9339–44.
74. Pardal, R., Clarke, M. F., and Morrison, S. J. (2003) Applying the principles of stem-cell biology to cancer *Nat Rev Cancer* **3**, 895–902.
75. Bjerkvig, R., Tysnes, B. B., Aboody, K. S., Najbauer, J., and Terzis, A. J. (2005) Opinion: the origin of the cancer stem cell: current

controversies and new insights *Nat Rev Cancer* **5**, 899–904.

76. Al-Hajj, M. and Clarke, M. F. (2004) Self-renewal and solid tumor stem cells *Oncogene* **23**, 7274–82.
77. Reya, T., Morrison, S. J., Clarke, M. F., and Weissman, I. L. (2001) Stem cells, cancer, and cancer stem cells *Nature* **414**, 105–11.
78. Croker, A. K. and Allan, A. L. (2008) Cancer stem cells: implications for the progression and treatment of metastatic disease *J Cell Mol Med* **12**, 374–90.
79. Hanahan, D. and Weinberg, R. A. (2000) The hallmarks of cancer *Cell* **100**, 57–70.
80. Marx, J. (2003) Cancer research. Mutant stem cells may seed cancer *Science* **301**, 1308–10.
81. Pereira, D. S., Dorrell, C., Ito, C. Y., Gan, O. I., Murdoch, B., Rao, V. N., Zou, J. P., Reddy, E. S., and Dick, J. E. (1998) Retroviral transduction of TLS-ERG initiates a leukemogenic program in normal human hematopoietic cells *Proc Natl Acad Sci USA* **95**, 8239–44.
82. Kelly, L. M. and Gilliland, D. G. (2002) Genetics of myeloid leukemias *Annu Rev Genomics Hum Genet* **3**, 179–98.
83. Charafe-Jauffret, E., Ginestier, C., Iovino, F., Wicinski, J., Cervera, N., Finetti, P., Hur, M. H., Diebel, M. E., Monville, F., Dutcher, J., Brown, M., Viens, P., Xerri, L., Bertucci, F., Stassi, G., Dontu, G., Birnbaum, D., and Wicha, M. S. (2009) Breast cancer cell lines contain functional cancer stem cells with metastatic capacity and a distinct molecular signature *Cancer Res* **69**, 1302–13.
84. Croker, A. K., Goodale, D., Chu, J., Postenka, C., Hedley, B. D., Hess, D. A., and Allan, A. L. (2008) High aldehyde dehydrogenase and expression of cancer stem cell markers selects for breast cancer cells with enhanced malignant and metastatic ability *J Cell Mol Med.*
85. Yu, F., Yao, H., Zhu, P., Zhang, X., Pan, Q., Gong, C., Huang, Y., Hu, X., Su, F., Lieberman, J., and Song, E. (2007) let-7 regulates self renewal and tumorigenicity of breast cancer cells *Cell* **131**, 1109–23.
86. Abraham, B. K., Fritz, P., McClellan, M., Hauptvogel, P., Athelogou, M., and Brauch, H. (2005) Prevalence of CD44+/CD24-/low cells in breast cancer may not be associated with clinical outcome but may favor distant metastasis *Clin Cancer Res* **11**, 1154–9.
87. Balic, M., Lin, H., Young, L., Hawes, D., Giuliano, A., McNamara, G., Datar, R. H., and Cote, R. J. (2006) Most early disseminated cancer cells detected in bone marrow of breast cancer patients have a putative breast cancer stem cell phenotype *Clin Cancer Res* **12**, 5615–21.
88. Chambers, A. F., Groom, A. C., and MacDonald, I. C. (2002) Dissemination and growth of cancer cells in metastatic sites *Nat Rev Cancer* **2**, 563–72.
89. Chambers, A. F., Naumov, G. N., Varghese, H. J., Nadkarni, K. V., MacDonald, I. C., and Groom, A. C. (2001) Critical steps in hematogenous metastasis: an overview *Surg Oncol Clin N Am* **10**, 243–55, vii.
90. Pantel, K. and Brakenhoff, R. H. (2004) Dissecting the metastatic cascade *Nat Rev Cancer* **4**, 448–56.
91. Cameron, M. D., Schmidt, E. E., Kerkvliet, N., Nadkarni, K. V., Morris, V. L., Groom, A. C., Chambers, A. F., and MacDonald, I. C. (2000) Temporal progression of metastasis in lung: cell survival, dormancy, and location dependence of metastatic inefficiency *Cancer Res* **60**, 2541–6.
92. Luzzi, K. J., MacDonald, I. C., Schmidt, E. E., Kerkvliet, N., Morris, V. L., Chambers, A. F., and Groom, A. C. (1998) Multistep nature of metastatic inefficiency: dormancy of solitary cells after successful extravasation and limited survival of early micrometastases *Am J Pathol* **153**, 865–73.
93. Weiss, L. (1990) Metastatic inefficiency *Adv Cancer Res* **54**, 159–211.
94. Allan, A. L., Vantyghem, S. A., Tuck, A. B., and Chambers, A. F. (2006) Tumor dormancy and cancer stem cells: implications for the biology and treatment of breast cancer metastasis *Breast Dis* **26**, 87–98.
95. Scadden, D. T. (2007) The stem cell niche in health and leukemic disease *Best Pract Res Clin Haematol* **20**, 19–27.
96. Hendrix, M. J., Seftor, E. A., Seftor, R. E., Kasemeier-Kulesa, J., Kulesa, P. M., and Postovit, L. M. (2007) Reprogramming metastatic tumour cells with embryonic microenvironments *Nat Rev Cancer* **7**, 246–55.
97. Chepko, G. and Dickson, R. B. (2003) Ultrastructure of the putative stem cell niche in rat mammary epithelium *Tissue Cell* **35**, 83–93.
98. Burger, J. A. and Peled, A. (2009) CXCR4 antagonists: targeting the microenvironment in leukemia and other cancers *Leukemia* **23**, 43–52.
99. Kaplan, R. N., Riba, R. D., Zacharoulis, S., Bramley, A. H., Vincent, L., Costa, C., MacDonald, D. D., Jin, D. K., Shido, K., Kerns, S. A., Zhu, Z., Hicklin, D., Wu, Y., Port, J. L., Altorki, N., Port, E. R., Ruggero, D., Shmelkov, S. V., Jensen, K. K., Rafii, S., and Lyden, D. (2005) VEGFR1-positive haematopoietic bone marrow progenitors initiate the pre-metastatic niche *Nature* **438**, 820–7.

100. Psaila, B., Kaplan, R. N., Port, E. R., and Lyden, D. (2006) Priming the 'soil' for breast cancer metastasis: the pre-metastatic niche *Breast Dis* **26**, 65–74.
101. Hiratsuka, S., Nakamura, K., Iwai, S., Murakami, M., Itoh, T., Kijima, H., Shipley, J. M., Senior, R. M., and Shibuya, M. (2002) MMP9 induction by vascular endothelial growth factor receptor-1 is involved in lung-specific metastasis *Cancer Cell* **2**, 289–300.
102. Hiratsuka, S., Watanabe, A., Sakurai, Y., Akashi-Takamura, S., Ishibashi, S., Miyake, K., Shibuya, M., Akira, S., Aburatani, H., and Maru, Y. (2008) The S100A8-serum amyloid A3-TLR4 paracrine cascade establishes a pre-metastatic phase *Nat Cell Biol* **10**, 1349–55.
103. Wicha, M. S., Liu, S., and Dontu, G. (2006) Cancer stem cells: an old idea--a paradigm shift *Cancer Res* **66**, 1883–90; discussion 1895–6.
104. Li, L. and Neaves, W. B. (2006) Normal stem cells and cancer stem cells: the niche matters *Cancer Res* **66**, 4553–7.
105. Kucia, M., Reca, R., Miekus, K., Wanzeck, J., Wojakowski, W., Janowska-Wieczorek, A., Ratajczak, J., and Ratajczak, M. Z. (2005) Trafficking of normal stem cells and metastasis of cancer stem cells involve similar mechanisms: pivotal role of the SDF-1-CXCR4 axis *Stem Cells* **23**, 879–94.
106. Ratajczak, M. Z., Zuba-Surma, E., Kucia, M., Reca, R., Wojakowski, W., and Ratajczak, J. (2006) The pleiotropic effects of the SDF-1-CXCR4 axis in organogenesis, regeneration and tumorigenesis *Leukemia* **20**, 1915–24.
107. Orimo, A., Gupta, P. B., Sgroi, D. C., Arenzana-Seisdedos, F., Delaunay, T., Naeem, R., Carey, V. J., Richardson, A. L., and Weinberg, R. A. (2005) Stromal fibroblasts present in invasive human breast carcinomas promote tumor growth and angiogenesis through elevated SDF-1/CXCL12 secretion *Cell* **121**, 335–48.
108. Spoo, A. C., Lubbert, M., Wierda, W. G., and Burger, J. A. (2007) CXCR4 is a prognostic marker in acute myelogenous leukemia *Blood* **109**, 786–91.
109. Rubin, J. B., Kung, A. L., Klein, R. S., Chan, J. A., Sun, Y., Schmidt, K., Kieran, M. W., Luster, A. D., and Segal, R. A. (2003) A small-molecule antagonist of CXCR4 inhibits intracranial growth of primary brain tumors *Proc Natl Acad Sci USA* **100**, 13513–8.
110. Dewan, M. Z., Ahmed, S., Iwasaki, Y., Ohba, K., Toi, M., and Yamamoto, N. (2006) Stromal cell-derived factor-1 and CXCR4 receptor interaction in tumor growth and metastasis of breast cancer *Biomed Pharmacother* **60**, 273–6.
111. Muller, A., Homey, B., Soto, H., Ge, N., Catron, D., Buchanan, M. E., McClanahan, T., Murphy, E., Yuan, W., Wagner, S. N., Barrera, J. L., Mohar, A., Verastegui, E., and Zlotnik, A. (2001) Involvement of chemokine receptors in breast cancer metastasis *Nature* **410**, 50–6.
112. Liang, Z., Wu, T., Lou, H., Yu, X., Taichman, R. S., Lau, S. K., Nie, S., Umbreit, J., and Shim, H. (2004) Inhibition of breast cancer metastasis by selective synthetic polypeptide against CXCR4 *Cancer Res* **64**, 4302–8.
113. Shipitsin, M., Campbell, L. L., Argani, P., Weremowicz, S., Bloushtain-Qimron, N., Yao, J., Nikolskaya, T., Serebryiskaya, T., Beroukhim, R., Hu, M., Halushka, M. K., Sukumar, S., Parker, L. M., Anderson, K. S., Harris, L. N., Garber, J. E., Richardson, A. L., Schnitt, S. J., Nikolsky, Y., Gelman, R. S., and Polyak, K. (2007) Molecular definition of breast tumor heterogeneity *Cancer Cell* **11**, 259–73.
114. Tuck, A. B., Chambers, A. F., and Allan, A. L. (2007) Osteopontin overexpression in breast cancer: knowledge gained and possible implications for clinical management *J Cell Biochem* **102**, 859–68.
115. Marhaba, R. and Zoller, M. (2004) CD44 in cancer progression: adhesion, migration and growth regulation *J Mol Histol* **35**, 211–31.
116. Ponta, H., Sherman, L., and Herrlich, P. A. (2003) CD44: from adhesion molecules to signalling regulators *Nat Rev Mol Cell Biol* **4**, 33–45.
117. Zen, K., Liu, D. Q., Guo, Y. L., Wang, C., Shan, J., Fang, M., Zhang, C. Y., and Liu, Y. (2008) CD44v4 is a major E-selectin ligand that mediates breast cancer cell transendothelial migration *PLoS ONE* **3**, e1826.
118. Mani, S. A., Guo, W., Liao, M. J., Eaton, E. N., Ayyanan, A., Zhou, A. Y., Brooks, M., Reinhard, F., Zhang, C. C., Shipitsin, M., Campbell, L. L., Polyak, K., Brisken, C., Yang, J., and Weinberg, R. A. (2008) The epithelial-mesenchymal transition generates cells with properties of stem cells *Cell* **133**, 704–15.
119. Fillmore, C. M. and Kuperwasser, C. (2008) Human breast cancer cell lines contain stem-like cells that self-renew, give rise to phenotypically diverse progeny and survive chemotherapy *Breast Cancer Res* **10**, R25.
120. Ponti, D., Costa, A., Zaffaroni, N., Pratesi, G., Petrangolini, G., Coradini, D., Pilotti, S., Pierotti, M. A., and Daidone, M. G. (2005) Isolation and in vitro propagation of

tumorigenic breast cancer cells with stem/progenitor cell properties *Cancer Res* **65**, 5506–11.

121. Sheridan, C., Kishimoto, H., Fuchs, R. K., Mehrotra, S., Bhat-Nakshatri, P., Turner, C. H., Goulet, R., Jr., Badve, S., and Nakshatri, H. (2006) CD44+/CD24- breast cancer cells exhibit enhanced invasive properties: an early step necessary for metastasis *Breast Cancer Res* **8**, R59.
122. Liu, R., Wang, X., Chen, G. Y., Dalerba, P., Gurney, A., Hoey, T., Sherlock, G., Lewicki, J., Shedden, K., and Clarke, M. F. (2007) The prognostic role of a gene signature from tumorigenic breast-cancer cells *N Engl J Med* **356**, 217–26.
123. Honeth, G., Bendahl, P. O., Ringner, M., Saal, L. H., Gruvberger-Saal, S. K., Lovgren, K., Grabau, D., Ferno, M., Borg, A., and Hegardt, C. (2008) The CD44+/CD24- phenotype is enriched in basal-like breast tumors *Breast Cancer Res* **10**, R53.
124. Li, C., Heidt, D. G., Dalerba, P., Burant, C. F., Zhang, L., Adsay, V., Wicha, M., Clarke, M. F., and Simeone, D. M. (2007) Identification of pancreatic cancer stem cells *Cancer Res* **67**, 1030–7.
125. Zhang, S., Balch, C., Chan, M. W., Lai, H. C., Matei, D., Schilder, J. M., Yan, P. S., Huang, T. H., and Nephew, K. P. (2008) Identification and characterization of ovarian cancer-initiating cells from primary human tumors *Cancer Res* **68**, 4311–20.
126. Polyak, K. and Weinberg, R. A. (2009) Transitions between epithelial and mesenchymal states: acquisition of malignant and stem cell traits *Nat Rev Cancer*.
127. Thiery, J. P. and Sleeman, J. P. (2006) Complex networks orchestrate epithelial-mesenchymal transitions *Nat Rev Mol Cell Biol* **7**, 131–42.
128. Yang, J. and Weinberg, R. A. (2008) Epithelial-mesenchymal transition: at the crossroads of development and tumor metastasis *Dev Cell* **14**, 818–29.
129. Hugo, H., Ackland, M. L., Blick, T., Lawrence, M. G., Clements, J. A., Williams, E. D., and Thompson, E. W. (2007) Epithelial--mesenchymal and mesenchymal--epithelial transitions in carcinoma progression *J Cell Physiol* **213**, 374–83.
130. Eccles, S. A. and Welch, D. R. (2007) Metastasis: recent discoveries and novel treatment strategies *Lancet* **369**, 1742–57.
131. Kang, Y. and Massague, J. (2004) Epithelial-mesenchymal transitions: twist in development and metastasis *Cell* **118**, 277–9.
132. Stover, D. G., Bierie, B., and Moses, H. L. (2007) A delicate balance: TGF-beta and the tumor microenvironment *J Cell Biochem* **101**, 851–61.
133. Moreno-Bueno, G., Portillo, F., and Cano, A. (2008) Transcriptional regulation of cell polarity in EMT and cancer *Oncogene* **27**, 6958–69.
134. Peinado, H., Olmeda, D., and Cano, A. (2007) Snail, Zeb and bHLH factors in tumour progression: an alliance against the epithelial phenotype? *Nat Rev Cancer* **7**, 415–28.
135. Morel, A. P., Lievre, M., Thomas, C., Hinkal, G., Ansieau, S., and Puisieux, A. (2008) Generation of breast cancer stem cells through epithelial-mesenchymal transition *PLoS ONE* **3**, e2888.

Part II

Stem Cell Identification and Microscopic Technologies to Track Transplanted Stem Cells In Vivo

Chapter 3

Hematopoietic Stem Cell Characterization and Isolation

Lara Rossi, Grant A. Challen, Olga Sirin, Karen Kuan-Yin Lin, and Margaret A. Goodell

Abstract

Hematopoietic stem cells (HSCs) are defined by the capabilities of multi-lineage differentiation and long-term self-renewal. Both these characteristics contribute to maintain the homeostasis of the system and allow the restoration of hematopoiesis after insults, such as infections or therapeutic ablation. Reconstitution after lethal irradiation strictly depends on a third, fundamental property of HSCs: the capability to migrate under the influence of specific chemokines. Directed by a chemotactic compass, after transplant HSCs find their way to the bone marrow, where they eventually home and engraft.

HSCs represent a rare population that primarily resides in the bone marrow with an estimated frequency of 0.01% of total nucleated cells. Separating HSCs from differentiated cells that reside in the bone marrow has been the focus of intense investigation for years. In this chapter, we will describe in detail the strategy routinely used by our laboratory to purify murine HSCs, by exploiting their antigenic phenotype (KSL), combined with the physiological capability to efficiently efflux the vital dye Hoechst 33342, generating the so-called Side Population, or SP.

Key words: Hematopoietic stem cells, Side population, Hoechst 33342, c-Kit$^+$ Sca-1$^+$ Lineage$^-$ cells (KSL)

1. Introduction

HSCs represents by far the most extensively studied population of stem cells in the adult. In particular, the murine model represents an excellent investigation system, where putative HSCs can be tested for long-term reconstitution of the lympho-hematopoietic system in lethally irradiated recipients. As demonstrated by the first transplantation assays performed decades ago, the hematopoietic activity resides primarily in the bone marrow. However, the cellular composition of the bone marrow is extremely

Marie-Dominique Filippi and Hartmut Geiger (eds.), *Stem Cell Migration: Methods and Protocols*, Methods in Molecular Biology, vol. 750, DOI 10.1007/978-1-61779-145-1_3, © Springer Science+Business Media, LLC 2011

heterogeneous and includes different populations of progenitors that can be hierarchically organized according to their self-renewing and differentiation potential. Long-term HSCs (LT-HSCs) represent the foundation pillars of hematopoiesis: their ability to self-renew indefinitely guarantees the homeostatic and continuous turn-over of blood cells that organisms require throughout life. LT-HSCs can also give rise to short-term HSCs (ST-HSCs), whose extensive proliferation and differentiation contributes to generate multipotent progenitors (MMPs) and all the downstream progenitors that will eventually produce terminally differentiated blood cells. Conversely to the subset of quiescent LT-HSCs, the highly proliferative ST-HSCs and MMPs, when transplanted, can only sustain hematopoiesis in the short-term and rapidly exhaust. Furthermore, deeper investigations have shown that the hematopoietic hierarchy might be more complicated than originally thought. Dykstra et al. (1) assessed single HSCs by serial transplantation and retrospectively classified them based on their pattern of peripheral blood reconstitution. Their analysis proved that even the LT-HSC compartment is a heterogeneous and multifaceted entity, comprising cells that are partly biased toward myeloid or lymphoid phenotypes. Identifying the rare cell population, on which the hematopoietic homeostasis is elegantly built, represents therefore one of the major challenges in the field (2–4). Nonetheless, despite the numerous efforts, a single specific marker, that can be employed alone to isolate HSCs, has yet to be discovered. Hence, investigators must turn to combinations of different markers or physiological properties. Benefiting from the advances in multicolor flow cytometry and monoclonal antibody development, several laboratories have proposed over the last two decades different isolation schemes that, however, lead to extremely similar HSC populations (5, 6).

Among the principal criteria utilized for HSC identification and isolation is the expression, or lack of expression, of specific cell surface markers. The isolation of one of the most thoroughly characterized populations of HSCs relies on the positive expression of the tyrosine kinase receptor c-Kit (CD117) and the membrane glycoprotein Sca-1 (7), concomitantly with the lack of markers of terminal differentiation (Ter119, Gr-1, Mac-1, B220, CD4, and CD8), collectively known as Lineage markers. The resulting c-Kit^{+} Sca-1^{+} Lineage- population, commonly referred to as KSL cells, contains cells capable of hematopoietic reconstitution. However, different studies showed that the KSL fraction contains a variety of progenitors, including ST-HSCs. Thanks to the contribution of different groups, schemes to further enrich the KSL fraction in HSCs have been developed over time. These strategies are based on either the combination with other surface markers, such as Thy1.1 (KSL Thy^{low} or KTSL), CD34 (KSL CD34$^{neg/low}$), and Flk2 (KSL CD34^{-} Flk2^{-}) (8), or on physiological properties, such as the capability to efflux Hoechst observed in SP cells (SP^{KSL} or SPKLS, pronounced *SParKLeS*) (4, 7, 9, 10).

More recently, alternative methods to identify HSCs have been described, that do not rely on the KSL scheme. These strategies include the use of markers such as Tie-2 (11), Endoglin (12), or endothelial protein C receptor (EPCR) (13). Morrison and colleagues recently described an alternative method based on markers from the signaling lymphocytic activation molecule (SLAM) family (CD150+ CD244− CD48−) (14). However, in order to obtain high purity, this strategy should be used in conjunction with other purification schemes.

In this chapter, we will focus on the purification of murine SP^{KLS} cells, based on the peculiar pattern that bone marrow cells acquire after Hoechst 33342 staining.

Hoechst 33342 fluorescent dye is a bisbenzimidazole derivative, capable of permeating through cell membranes and binding to nucleic acids. The emission of fluorescence is highly affected by DNA properties, such as chromatine rearrangements, DNA conformation, and nucleic acid composition. In particular, Hoechst dyes bind in a stoichiometric manner to AT-rich regions of the minor groove of double-stranded DNA (this property has been extensively used by genetists to develop the Q-bands staining for chromosomes).

Interestingly, when Hoechst dyes bind to DNA, their fluorescence undergoes a small spectral shift, that can be detected and used as a measurement of the amount of cellular DNA. This property has been exploited in flow cytometry to study ploidy and distribution in the different cell-cycle stages of a heterogeneous population, such as bone marrow samples. Traditionally, cell cycle studies have been performed by analyzing Hoechst emission at a short wavelength (450 nm), through a "blue" bandpass on a fluorescence-activated cell-sorter. However, Hoechst fluorescence can be detected with "red" (650 nm) bandpass optics as well. When Hoechst blue and red fluorescence signals are simultaneously collected and plotted against each other, a characteristic tail-shaped population, displaying low fluorescence, can be observed and distinguished from the main bulk that conversely emits high levels of fluorescence. This "tail" is the so-called Side Population, or SP, and comprises cells that display low Hoechst fluorescence. Conversely to the main bulk of bone marrow cells (whose Hoechst fluorescence is directly proportional to the DNA content), the atypical cytometric morphology of SP cells is a direct consequence of their capability to efflux with high efficiency the vital dye Hoechst 33342. However, what makes this peculiar bone marrow population so interesting for the stem cell field is the fact that SP cells are highly enriched in HSCs, capable of sustaining multilineage and long-term engraftment in the murine model. Since the first description of SP cells in 1996 (10), follow-up studies also proved that the SP fraction encompasses entirely the hematopoietic activity that resides in the murine bone marrow, thus making Hoechst staining a unique experimental tool in stem cell biology (2, 4–6, 15).

The capability of SP cells to efflux vital dyes at a higher rate than other bone marrow cells is believed to reside in the activity of membrane pumps belonging to the superfamily of ATP-binding cassette (ABC) transporters. Members of this family are, for instance, multidrug resistance 1 (murine Mdr 1a/1b; human MDR1) and breast cancer resistance protein 1 (Bcrp1)/ABC, superfamily G, member2 (ABCG2). Interestingly, drugs such as verapamil block the activity of these transporters and concomitantly cause the SP profile to disappear.

Knock-out and retroviral-driven overexpression models helped shed some light onto the role ABC transporters play in HSC biology. MDR1 overexpression only slightly increases the SP fraction; on the other hand, Mdr 1a/1b$^{-/-}$ bone marrow shows numbers of SP cells comparable to the *wild type*, thus indicating that this membrane transporter only plays a marginal role in the SP phenotype (16, 17). Conversely, the enforced expression of ABCG2 significantly expands SP cells, while loss of ABCG2 expression has been shown to drastically reduce the size of the SP fraction. Nonetheless, since HSC numbers and function in these mice are preserved, it is not yet clear whether the efflux plays a functional role in HSCs. Furthermore, ABCG2 knock-out mice still contain in their bone marrow a few residual SP cells, suggesting that multiple drug transporters are likely to be involved in the appearance of this phenotype (18–21).

However, if ABC membrane pumps are not crucial determinants of stem cell activity, why are they expressed at high levels in stem cells? This observation could be teleologically interpreted as a mechanism that biological systems adopt to protect from the environment crucial subsets of cells, like HSCs. Also, membrane pumps could play a role in extruding differentiation factors from HSCs, thus helping maintaining their stemness throughout the life of an organism.

2. Materials

2.1. Sample Preparation: Isolation of Murine Bone Marrow Cells

1. Murine bone marrow cells obtained from C57Bl/6 mice, 5–8 weeks old (see Note 1).
2. HBSS. Hanks Balanced Salt Solution, supplemented with 2% Fetal Bovine Serum and 10 mM HEPES buffer.

The solution so prepared will be hereafter referred to as HBSS+.

3. Needles (27 Gauge and 18 Gauge).
4. Cell strainer (70 μm).
5. Red Blood Cells (RBC) lysis buffer. 0.17 M TrisCl, pH 7.6:0.16 M $NH_4Cl = 1:9$.

2.2. Staining of Murine Bone Marrow Cells with Hoechst 33342

1. DMEM. Dulbecco's Modified Eagle's Medium with High Glucose, supplemented with 2% Fetal Bovine Serum and 10 mM HEPES buffer. The solution so prepared will be referred to as DMEM+.
2. Hoechst 33342, bisBenzimide H33342 trihydrochloride (Sigma-Aldrich). To make concentrated stock solutions of Hoechst 33342, dissolve the powder in water (recommended concentration: 1 mg/mL, 200× solution) and filter-sterilize (see Note 2).
3. Verapamil (Sigma-Aldrich). Prepare a concentrated stock (100×) in 95% Ethanol and use at the final concentration of 50 μM in the staining buffer (HBSS+ and Hoechst 33342) (see Note 3).
4. Circulating water bath at exactly 37°C (see Note 4).
5. Refrigerated centrifuge at 4°C (see Note 5).

2.3. Isolation of SP Sca-1$^+$ c-Kit$^+$ Lineage$^-$ Cells

1. HBSS+ (as described in Subheading 2.1).
2. Anti-Sca-1 antibodies either biotinylated or FITC-conjugated (BD Pharmingen).
3. Anti-Biotin magnetic microbeads (Miltenyi Biotech).
4. AutoMACS separator (Miltenyi Biotech).
5. Anti-c-Kit antibody. We use a PE-conjugated antibody.
6. Anti-Lineage antibody cocktail. The cocktail comprises a mixture of the following PE-Cy5-conjugated antibodies (all from eBioscience): anti-B220, anti-CD4, anti-CD8, anti-Gr-1, anti-Mac-1, and anti-TER119.
7. Propidium Iodide (PI, Sigma-Aldrich). Prepare a stock solution at 10 mg/mL in water and store at −20°C. From this solution, prepare a working solution at 200 μg/mL and keep it at 4°C, protected from light. The final concentration of PI in the sample should be 2 μg/mL (100× dilution of the working solution).

2.4. Identification and Sorting of SP^{KLS} Cells

1. Flow cytometer equipped with a UV laser, such as a MoFlo sorter (Dako) or a FACSAria (BD Biosciences) (see Note 6).

3. Methods

Because the method relies on detecting dye efflux from a cell, which is a dynamic biological process, a successful SP staining is highly dependent on cell and Hoechst concentration, as well as temperature and time of staining. Even small variations in any of these parameters can affect significantly the composition and purity of the SP. Here we illustrate the protocol as it was originally established for the staining of C57Bl/6 bone marrow and we recommend the protocol to be followed exactly as we describe before attempting the use in different species, tissues, or mouse strains.

3.1. Sample Preparation: Isolation of Murine Bone Marrow Cells

1. Anesthetize the mouse and sacrifice it by cervical dislocation. Lay the mouse on its back and profusely spray with 70% Ethanol to sterilize.
2. Make a horizontal abdominal incision at the level of the knees and pull the skin until the legs are exposed completely.
3. Proceed to remove the tibias by cutting through the ankles and the knees. Clean the muscle off the tibias and place them in a Petri dish containing HBSS+ (5 mL) on ice.
4. Proceed now to remove the femurs, by cutting at the level of the hips. Carefully remove the muscle from the femurs and put them into the Petri dish with the tibias. Femurs are extremely rich in bone marrow, so we recommend to cut off the bone as close to the hip as possible.
5. Load a 10cc syringe with HBSS+ buffer and, holding a bone over a new Petri dish, insert the needle (27 Gauge) into one of the extremities and proceed to flush the bone marrow out of the bone. As the bone marrow is expelled, the bones will appear clearer. Repeat the same by inserting the needle into the second extremity of the same bone and flush thoroughly (see Note 7).
6. Using a syringe with an 18-Gauge needle, proceed to resuspend the bone marrow in the Petri dish. Repeat several times (four to five times), until the clusters of bone marrow will convert into a homogeneous single-cell suspension. Pay special attention to avoid the formation of air bubbles while resuspending cells, because of their detrimental effect on cell survival. Transfer the cell suspension into a 50 mL-conical tube and filter through a 70 μm cell-strainer to remove from the sample cell clumps or bone fragments.
7. Carefully count the bone marrow cells, paying particular attention to exclude red blood cells (RBCs) (see Note 8). To do so, prepare a 1:20 dilution of an aliquot of bone marrow cell suspension (e.g., 5 μL) in RBC-lysis buffer (95 μL) for

counting. One C57Bl/6 mouse (5–8 weeks old) will averagely yield 5–7 × 10^7 nucleated cells. Note that, in order to proceed to the following staining, no Ficoll separation or lysis of red blood cells of the whole sample is necessary.

3.2. Staining of Murine Bone Marrow Cells with Hoechst 33342

1. Pre-warm the staining medium (DMEM+) in a circulating water bath at 37°C.
2. Spin down bone marrow cells and resuspend in pre-warmed DMEM+ at the concentration of 10^6 cells/mL (see Notes 8 and 9).
3. Add Hoechst 33342 to the cell suspension to a final concentration of 5 μg/mL (from the 200× working solution).
4. Incubate the sample for *exactly* 90 min at 37°C in a circulating water bath. During the incubation, periodically mix the tubes and always ensure that the tubes are fully immersed in the water.
5. Once the 90-min staining is completed, *always* keep your sample at 4°C and *always* use a refrigerated centrifuge to spin cells down, in order to prevent continuous Hoechst expulsion from the stained cells (see Note 10).
6. Spin down the Hoechst-stained cells in a refrigerated centrifuge and resuspend in iced HBSS+ buffer at the concentration of 10^8 cells/mL. Bone marrow cells are now stained with Hoechst and ready for the following staining procedures with monoclonal antibodies. Any further handling of the sample must be performed at 4°C or on ice (see Note 11).

3.3. Isolation of SP Sca-1⁺ c-Kit⁺ Lineage⁻ Cells

1. Sca-1 enrichment (see Note 12). Incubate cells on ice in the presence of anti-Sca-1 biotinylated antibody (0.5 μg/10^6 cells, 1:100 dilution) (see Note 13). After 10 min, wash out the unbound antibody by adding a tenfold volume of iced HBSS+. Spin cells down at 4°C and resuspend in HBSS+ buffer.
2. Label bone marrow cells with anti-biotin magnetic microbeads (1:5 dilution). Incubate for 15 min at 4°C.
3. Wash the sample with a tenfold volume of HBSS+ buffer and spin cells down at 4°C.
4. Resuspend at 2 × 10^8 cells/mL in HBSS+. The sample is now ready to be processed by AutoMACS (choose the program for stringent positive selections) (see Note 14).
5. Spin down at 4°C the Sca-1-enriched cells and resuspend in iced HBSS+ buffer.
6. Label the cells with anti-c-Kit antibody and with an anti-Lineage cocktail, comprising anti-B220, anti-CD4, anti-CD8, anti-Gr-1, anti-Mac-1, and anti-TER119 antibodies. Although the sample has been previously enriched for Sca-1+ cells,

we recommend staining the sample with an anti-Sca-1 antibody as well, as a control during the sorting. Incubate for 15 min on ice.

7. Wash the sample with a tenfold volume of HBSS+ buffer and spin cells down at 4°C. Resuspend HBSS+ buffer containing PI. The sample is now ready for sorting of SP c-Kit$^+$ Lin$^-$ Sca-1$^+$ cells.

3.4. Identification and Sorting of SPKLS Cells

1. Excitation of Hoechst 33342. In order to view the SP, the flow cytometer must be equipped with a high power ultraviolet laser (35–100 mW), which is capable to excite both Hoechst 33342 and Propidium Iodide (PI) at 350 nm (see Note 15). A second laser is necessary to excite additional fluorochromes involved in the staining, such as a 488 nm laser for FITC and Phycoerthrin.
2. Detection of Hoechst 33342 emission. The emission of Hoechst 33342 is measured bimodally and commonly referred to as *Hoechst Blue* and *Hoechst Red*. Hoechst Blue is measured with a 450BP filter, whereas Hoechst Red is measured with a 675LP filter. In order to separate the different emission wavelength, a dichroic mirror is used (we use a 610 DMSP). PI emission is also measured with the 675LP filter, but its signal is significantly brighter than the one captured for Hoechst Red, so that PI-positive cells line up to the very far right side of the SP profile (Fig. 1).
3. FACS Analysis. The characteristic SP profile can be visualized by plotting Hoechst Blue emission (on the vertical axis) vs. Hoechst Red emission (on the horizontal axis). The detectors for both parameters must be set on linear mode. The voltage must be adjusted so that the PI-positive dead cells will appear at the far right vertical line. Also, if the voltage is set correctly, red blood cells should group together in the lower left corner. The majority of the bone marrow cells will be displayed in the central area or in the upper right quadrant of the plot. If the cytometer settings are arranged correctly, the SP profile should appear as displayed in Fig. 1.
4. Identification and gating of SP^{KLS} cells. Once the instrument set-up has been performed, follow the gating strategy described in Fig. 2. Briefly, start by drawing the first gate around the SP population. Proceed by checking the morphological phenotype of SP cells (FSC vs. SSC plot) and gate out all the events not compatible with stem cell morphology (low granulosity and small/medium size). Finally, proceed to analyze the KSL phenotype: first, gate Lineage$^-$ cells and then display these events as shown in the last panel of Fig. 2. The events that simultaneously fulfill the criteria of both c-Kit and Sca-1 positivity represent the desired SP^{KLS} population (see Note 16).

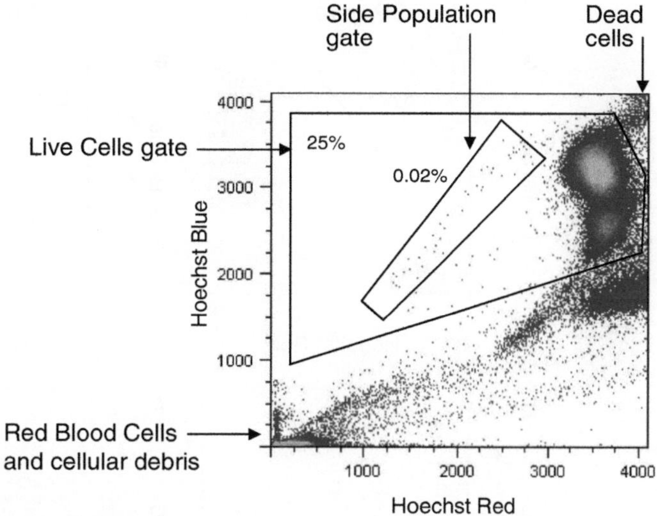

Fig. 1. Example of an SP population from an unenriched whole bone marrow sample. In order to visualize the characteristic SP pattern of bone marrow cells, the emission of Hoechst 33342 must be displayed bimodally as Hoechst Red vs. Hoechst Blue, both in a linear scale. The cells concentrated at the lower left corner represent red blood cells and cellular debris, while the rest of the sample is mainly grouped on the upper right side of the acquisition window. The SP gate is drawn around the tail that diagonally emerges from the main population and usually represents 0.02–0.05% of whole bone marrow cells.

4. Notes

1. This protocol was originally established and optimized for murine bone marrow cells, derived from normal C57Bl/6 mice. Because of the high sensitivity of Hoechst efflux to multiple parameters, we strongly recommend investigators, who are attempting this procedure for the first time, to follow the protocol *exactly* as we describe, until proficiency in SP staining and identification is achieved. In order to optimize the protocol for different species, we suggest to change one parameter at a time (for instance, duration of the staining or Hoechst concentration).
2. For long-term storage, prepare aliquots of the stock solution (e.g., 1 mL aliquots) and store them at -20°C, protected from light. Avoid, when possible, repeated thawing/freezing cycles. We strongly recommend using a new Hoechst aliquot for each experiment.
3. Verapamil is a drug that blocks the activity of the membrane transporters responsible for the efflux of Hoechst 33342. When Verapamil (50 μM) is included in the Hoechst staining solution and in the washing buffers, the SP fraction is no longer detectable and becomes part of the main population.

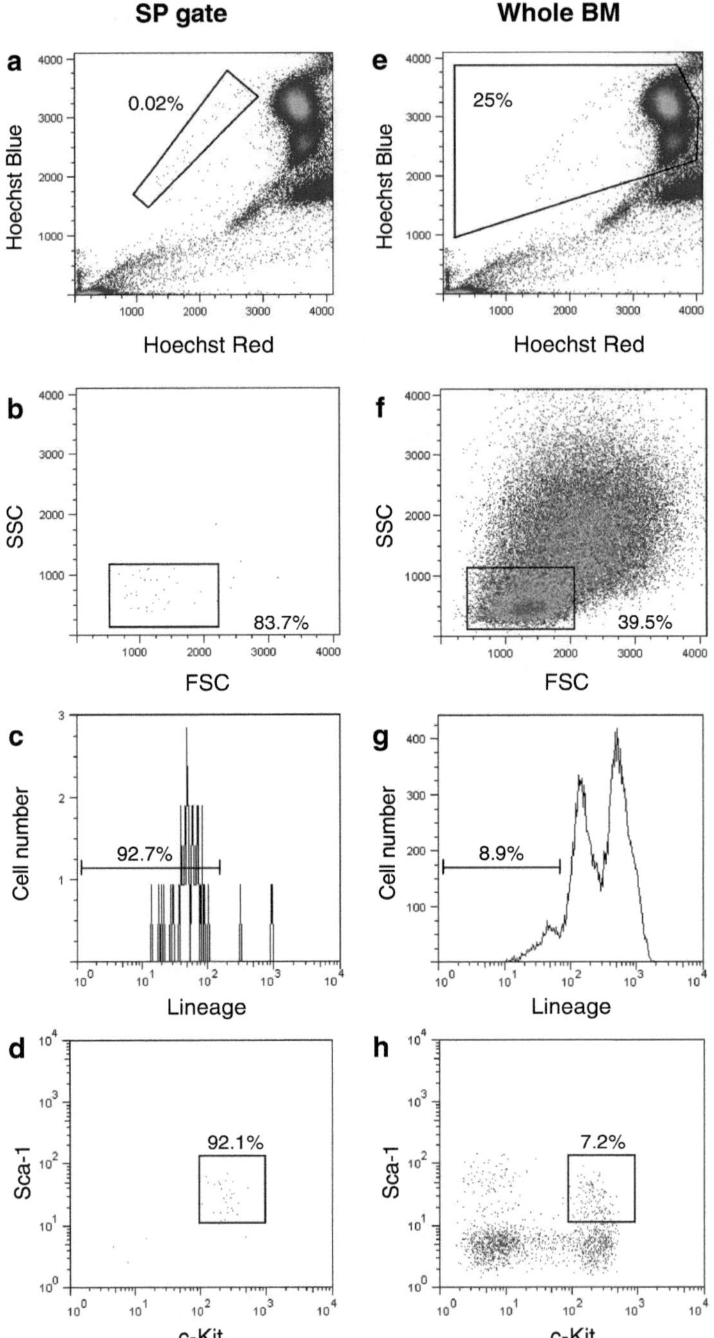

Fig. 2. Sorting strategy for SP[KLS] (SP c-Kit+ Lineage- Sca-1+) cells. (**a**) SP gate: the first step consists in displaying the Hoechst 33342 efflux pattern is linear mode (as Hoechst Red vs. Hoechst Blue) and gating the SP population. (**b**) Morphological characteristics: display the SP cells gated in the first panel as FSC (*forward scatter*) vs. SSC (*side scatter*) and draw a second gate as shown in figure. (**c**) Lineage staining (PE-Cy5): gate out cells that express markers of hematopoietic terminal differentiation and select Lineage-negative cells. (**d**) c-Kit vs. Sca-1: the last panel shows the expression of the stem cell markers c-Kit (PE) and Sca-1 (FITC) in SP/Lineage-negative cells. This is the sorting gate, comprising the SP[KLS] population. (**e**–**h**) The *panels on the right* show, by comparison, how unenriched bone marrow cells (gated only on the live population from (**e**)) distribute on the same parameters.

We highly recommend the use of Verapamil-treated cells as negative control to help investigators identify the "true" SP population and draw the sorting gate. However, once the method has been routinely established, Verapamil treatment can be left out.

4. Hoechst staining is highly sensitive to temperature. Therefore, the water bath must be set at *precisely* 37°C. Avoid using water baths whose temperature fluctuates (we recommend using a circulating water bath) and avoid immersing ice-cold or frozen reagents into the water during the staining.
5. Use a refrigerated centrifuge for spinning cells down and *always* keep the sample at 4°C or on ice. In the case the stained cells are exposed to higher temperatures, they might expel Hoechst to the point they will become undistinguishable from the "true" SP cells. This will eventually affect the composition and decrease the purity of the SP.
6. Although it is possible to detect SP using a violet laser, in order to obtain optimal results, we recommend using a UV laser.
7. When isolating tibias and femurs, it is important to remove as much muscle as possible in order to prevent the bone marrow from sticking to it once it is flushed out of the bone.
8. Cell dilution, Hoechst concentration, and staining time are all critical factors in determining an optimal staining. In particular, dye concentration and number of nucleated cells should be carefully determined.
9. In order to prevent cells from sticking to the plastic, we recommend using polypropylene tubes while staining with Hoechst.
10. Because of the aforementioned sensitivity of the procedure to temperature, even when the staining process is over, the samples must be maintained at 4°C, in order to prohibit efflux of the dye from the cells. Therefore, whether you are going to directly sort SP cells or you are going to perform antibody staining, *always* keep your sample at 4°C.
11. If interested in SP isolation only, disregard the following KSL staining. Resuspend the sample in HBSS+ buffer and PI and proceed to sort. However, keep in mind that combination of SP staining with KSL markers significantly increases HSC purity, other than being an internal diagnostic parameter for optimal staining conditions. Likewise, if this protocol is used to isolate stem cells from other tissues, SP staining should be combined, whenever possible, with tissue-specific stem cell markers.

12. Enrichment of the bone marrow sample before sorting is not strictly necessary, but strongly recommended. Enrichment helps increase purity and yield after sorting and sensibly decreases sort time.
13. The antibody concentration of 0.5 μg/10^6 cells reflects the optimal staining conditions that have been identified in our laboratory and is consistently used for each antibody mentioned throughout this protocol. However, especially for samples different from murine bone marrow cells, we recommend to adjust the antibody titration ad hoc.
14. Alternatively, the Sca-1 enrichment can be performed manually using Miltenyi MS/LS columns for positive selection.
15. In the case that the sorting strategy relies also on conjugated antibodies (as in the case of SP^{KLS} purification), the flow cytometer must have the corresponding additional lasers (e.g., a 488 nm laser, if cells are stained with FITC and PE).
16. Despite the unique pattern of SP cells, uninitiated investigators usually are challenged by deciding where to draw the SP gate, especially when it comes to deciding how far toward the top of the tail it is possible to go, without including cells that are not "true" HSCs. In our laboratory, we tend to use a conservative gate, while attempting to maximize cell yield and minimize contamination from non-HSCs. An excellent internal quality control for drawing the SP gate in the correct position is provided by the KSL staining itself. Since SP cells are highly enriched in HSCs, the SP gate should not contain more than 25% $Lineage^+$ cells. Also, approximately 85% of SP should be KSL. If these criteria are not matched, it generally means that a more restricted gate should be drawn. Another possible reason is that the protocol has been poorly performed and consequently a high percentage of non-SP cells are contaminating the SP gate.

References

1. Dykstra B., Kent D., et al. (2007) Long-term propagation of hematopoietic differentiation programs in vivo *Cell Stem Cells* **1**, 218–29.
2. Camargo, F. D., Chambers S. M., et al. (2006) Hematopoietic stem cells do not engraft with absolute efficiencies *Blood* **107**, 501–7.
3. Morrison, S. J. and Weissman I. L. (1994) The long-term repopulating subset of hematopoietic stem cells is deterministic and isolatable by phenotype *Immunity* **1**, 661–73.
4. Goodell, M. A., Rosenzweig M. et al. (1997) Dye efflux studies suggest that hematopoietic stem cells expressing low or undetectable levels of CD34 antigen exist in multiple species *Nat Med* **3**, 1337–45.
5. Weksberg, D. C., Chambers S. M. et al. (2008). CD150- side population cells represent a functionally distinct population of long-term hematopoietic stem cells. *Blood* **111**, 2444–51.
6. Pearce, D. J., Ridler C. M. et al. (2004) Multiparameter analysis of murine bone marrow side population cells *Blood* **103**, 2541–6.
7. Okada, S., Nakauchi H. et al. (1992) In vivo and in vitro stem cell function of c-kit- and Sca-1-positive murine hematopoietic cells *Blood* **80**, 3044–50.

8. Christensen, J. L. and Weissman I. L. (2001) Flk-2 is a marker in hematopoietic stem cell differentiation: a simple method to isolate long-term stem cells *Proc Natl Acad Sci USA* **98**, 14541–6.
9. Challen, G. A., Boles N. et al. (2009) Mouse hematopoietic stem cell identification and analysis. *Cytometry A* **75**, 14–24.
10. Goodell, M. A., Brose K. et al. (1996) Isolation and functional properties of murine hematopoietic stem cells that are replicating in vivo *J Exp Med* **183**, 1797–806.
11. Arai, F., Hirao A. et al. (2004) Tie2/angiopoietin-1 signaling regulates hematopoietic stem cell quiescence in the bone marrow niche *Cell* **118**, 149–61.
12. Chen, C. Z., Li M. et al. (2002) Identification of endoglin as a functional marker that defines long-term repopulating hematopoietic stem cells *Proc Natl Acad Sci USA* **99**, 15468–73.
13. Balazs, A. B., Fabian A. J., et al. (2006) Endothelial protein C receptor (CD201) explicitly identifies hematopoietic stem cells in murine bone marrow Blood **107**, 2317–21.
14. Kiel, M. J., Yilmaz O. H. et al. (2005) SLAM family receptors distinguish hematopoietic stem and progenitor cells and reveal endothelial niches for stem cells *Cell* **121**, 1109–21.
15. Challen, G. A. and Little M. H. (2006) A side order of stem cells: the SP phenotype *Stem Cells* **24**, 3–12.
16. Bunting, K. D., Galipeau J. et al. (1998) Transduction of murine bone marrow cells with an MDR1 vector enables ex vivo stem cell expansion, but these expanded grafts cause a myeloproliferative syndrome in transplanted mice *Blood* **92**, 2269–79.
17. Bunting, K. D., Galipeau J. et al. (1999) Effects of retroviral-mediated MDR1 expression on hematopoietic stem cell self-renewal and differentiation in culture *Ann N Y Acad Sci* **872**, 125–40; discussion 140–1.
18. Bunting, K. D., Zhou S. et al. (2000) Enforced P-glycoprotein pump function in murine bone marrow cells results in expansion of side population stem cells in vitro and repopulating cells in vivo *Blood* **96**, 902–9.
19. Scharenberg, C. W., Harkey M. A. et al. (2002) The ABCG2 transporter is an efficient Hoechst 33342 efflux pump and is preferentially expressed by immature human hematopoietic progenitors *Blood* **99**, 507–12.
20. Zhou, S., Morris J. J. et al. (2002) Bcrp1 gene expression is required for normal numbers of side population stem cells in mice, and confers relative protection to mitoxantrone in hematopoietic cells in vivo *Proc Natl Acad Sci USA* **99**, 12339–44.
21. Zhou, S., Schuetz J. D. et al. (2001) The ABC transporter Bcrp1/ABCG2 is expressed in a wide variety of stem cells and is a molecular determinant of the side-population phenotype *Nat Med* 7, 1028–34.

Chapter 4

Isolation and Characterization of Adult Neural Stem Cells

Florian A. Siebzehnrubl, Vinata Vedam-Mai, Hassan Azari, Brent A. Reynolds, and Loic P. Deleyrolle

Abstract

It has been thought for a long time that the adult brain is incapable of generating new neurons, or that neurons cannot be added to its complex circuitry. However, recent technology has resulted in an explosion of research demonstrating that neurogenesis, or the birth of new neurons from adult stem cells constitutively occurs in two specific regions of the mammalian brain; namely the subventricular zone and hippocampal dentate gyrus. Adult CNS stem cells exhibit three main characteristics: (1) they are "self-renewing," i.e., they possess a theoretically unlimited ability to produce progeny indistinguishable from themselves, (2) they are proliferative (undergoing mitosis) and (3) they are multipotent for the different neuroectodermal lineages of the CNS, including the different neuronal, and glial subtypes. CNS stem cells and all progenitor cell types are broadly termed "precursors."

In this chapter, we describe methods to identify, isolate and experimentally manipulate stem cells of the adult brain. We outline how to prepare a precursor cell culture from naive brain tissue and how to test the "stemness" potential of different cell types present in that culture, which is achieved in a three-step paradigm. Following their isolation, stem/progenitor cells are expanded in neurosphere culture. Single cells obtained from these neurospheres are sorted for the expression of surface markers by flow cytometry. Finally, putative stem cells from cell sorting will be subjected to the so-called neural colony-forming cell assay, which allows discrimination between stem and progenitor cells. At the end of this chapter we will also describe how to identify neural stem cells in vivo.

Key words: Neural stem cell, Neurosphere assay, Flow cytometry, Neural colony-forming cell assay, Immunohistochemistry

1. Introduction

With the identification of multipotent stem cells in the adult brain, an assay that allowed the propagation of these cells – the neurosphere assay (NSA) was developed and described (1). The NSA has become the method of choice not only for the expansion of stem/progenitor cells, but is also widely used to determine

Marie-Dominique Filippi and Hartmut Geiger (eds.), *Stem Cell Migration: Methods and Protocols*,
Methods in Molecular Biology, vol. 750, DOI 10.1007/978-1-61779-145-1_4,

stem cell activity in vitro. However, several cell types other than stem cells can also form neurospheres, including neural progenitor cells, O2A cells, oligodendrocyte precursors, and possibly even some types of astrocytes (2, 3). This "promiscuity" of sphere formation results in an overestimation of stem cell numbers when calculating sphere-forming frequency from all plated cells. While the NSA is an appropriate tool to expand stem/progenitor cells for experimental manipulation, it is insufficient to discriminate stem cells from other sphere-forming cell types. Even though the NSA is the most popular method to detect neural stem cell activity, it has caveats and cannot be used as an accurate assay to measure neural stem cell (NSC) frequency. As the formation of an individual neurosphere does not reflect the presence of a single stem cell, and because progenitors can generate spheres, the one-to-one relationship between neurospheres and neural stem cells is incorrect. Therefore, quantification of the neurosphere-forming frequency is not an accurate measurement of stem cell enumeration. To address this issue, the neural colony-forming cell assay (N-CFCA) was designed (4). This assay discriminates between stem and progenitor cells on the basis of their proliferative potential. The N-CFCA is based on the observation that stem cells present higher proliferative capability compared to progenitor cells; therefore, the size of the clonally derived colonies (i.e., diameter) can be used to differentiate its founder cell type. Colonies were generated with a distinct size range, and subsequently four categories of colonies are identified based on their diameter 0.5, 0.5–1, 1–2, and >2 mm (Fig. 3). Only the large colonies (>2 mm) are derived from a cell exhibiting all of the stem cell features. Therefore, the frequency of large colony can be used as a read-out of NSCs frequency. Cogency and validity of the assay has been established with embryonic and adult stem/precursor cells (4).

Flow cytometry is a very powerful technology that allows for the purification of cell populations according to size, granularity, and antigens expressed on the cell surface. Unfortunately, adult neural stem/progenitor cells do not differ very much in size and granularity, and it is nearly impossible to purify one or the other population based on any of these characteristics. Over the last decade, a variety of such antigens constituting putative stem cell markers have been identified (e.g., CD133, LeX, EGFR, Nestin, Musashi and Sox2 (5–10). In addition, assays have been developed to reveal putative stem cell populations based on internal cell characteristics such as the side population (11) or ALDH1 activity (12). As flow cytometry can be viewed as live cell immunostaining and sorting of stained (or unstained) cells, the technology is only as good as the markers (i.e., antibodies) targeting the desired cell populations. Herein also lies the greatest pitfall of stem cell purification. However, in conjunction with functional stemness assays, flow cytometry becomes an indispensable tool in modern stem cell research. Neurosphere culture is arguably not the best method of stem cell enrichment (4, 13), but it is a very practicable culture method allowing for fast expansion of stem and progenitor cells. Cellular subpopulations of this heterogeneous mixture can be further purified based on their expression of certain antigens (we will use CD133 as exemplary marker) and then tested for their stemness in more complex assays, i.e., the N-CFCA.

Ideally, the identification of stem cells in vivo is based on the fact that the cells can be labeled as dividing in combination with the expression of several markers. In this chapter, we will use Sox2 expression as an example to identify in situ stem/precursor cells.

2. Materials

2.1. Culture Medium for Neurosphere Assay

To avoid inconsistency in experiments due to possible batch-to-batch differences of the in-laboratory prepared medium, optimized reagents and medium are available from Stem Cell Technologies (http://www.stemcell.com), Gibco or Sigma. Here we provide an example list of commercially available reagents that can be used to prepare the Neurosphere assay media.

1. Basal medium (NeuroCult NSC basal medium, Stem-Cell Technologies) supplemented or not with Bovine Serum Albumin (BSA).
2. 10× hormone mix (NeuroCult NSC proliferation supplement, StemCell Technologies).
3. Differentiation medium (NeuroCult differentiation supplement, StemCell Technologies).
4. Solution of trypsin (0.05%) and ethylenediamine tetraacetic acid (EDTA).
5. Fetal bovine serum.
6. Trypsin inhibitor solution: add 0.14 g of Trypsin Inhibitor to 10 ml of DNase Solution (100 mg DNase dissolved in 100 ml of HEM), then make the volume up to 1 l using HEM. Use ratio 1:1 of Trypsin inhibitor solution: Trypsin/EDTA 0.05% or tissue dissociation medium.

To prepare complete NSC medium, combine 450 ml of NeuroCult NSC basal medium with 50 ml of NeuroCult NSC proliferation supplement and then add required amount of growth factors (20 ng/ml EGF, 10 ng/ml bFGF, and 0.679 U/ml heparin).

2.2. Flow Cytometry

1. BSA is dissolved at 2% BSA in phosphate-buffered saline (PBS) (20 g of BSA in 1 l of PBS). Filter sterilize.
2. 200 mM EDTA solution: dissolve 584.5 mg of EDTA in 10 ml of PBS; then filter sterilize.
3. Rat anti-mouse CD133 monoclonal antibody, conjugated to Allophycocyanin (APC).
4. Fluorescence-activated Cell Sorter (e.g., BD FACSAria II).
5. Propidium Iodide is dissolve at 1 mg/ml in sterile water.

2.3. Neural Colony-Forming Cell Assay

1. Complete NeuroCult® Proliferation Medium (StemCell Technologies).
2. NeuroCult® NCFC Serum-Free Medium without Cytokines (StemCell Technologies).
3. Mouse NeuroCult® NSC Proliferation Supplements (StemCell Technologies).
4. Collagen Solution (StemCell Technologies).
5. Light microscope with 5× and 10× objectives.
6. 40 μm cell strainer.

2.4. Immunohistochemistry

1. Paraformaldehyde (PFA) is made up at 4% in PBS; pH 7.4.
2. Sucrose (20% in PBS, 30% in PBS); BP-220-1.
3. OCT tissue embedding compound (Tissue-Tek).
4. Triton-X 100.
5. Blocking solution (5–15% normal serum in PBS. The blocker varies with the source of antibodies to be used).
6. Positively charged Superfrost glass slides.
7. Forceps.
8. Primary and secondary antibodies.

3. Methods

3.1. Establishment of Primary Adult Neural Stem Cells Using the Neurosphere Assay

In the following section, we describe the isolation and expansion method for adult murine neural stem cells by means of growth factor stimulation. Sacrifice of animals, removal and dissection of brain are performed outside the laminar flow hood. Particular caution and sterile techniques should be exercised to avoid contamination.

3.1.1. Dissection

1. Warm the culture medium and tissue dissociation medium to 37°C in a water bath.
2. Anesthetize mice using 3–4% Isoflurane and sacrifice them by cervical dislocation. Pool tissues from two or four mice (4–8 weeks old) to start a culture.

3. Using large scissors excise the head just above the cervical spinal cord region. Rinse the head with 70% ethanol. Then using small pointed scissors make a median caudal–rostral cut and expose the skull.
4. Using the skin to hold the head in place, place each blade of small scissors in orbital cavity, so as to make a coronal cut between the orbits.
5. Using the foramen magnum as an entry point, make a longitudinal cut through the skull along the sagittal suture. Be cautious not to damage the underlying brain by making small cuts ensuring the angle of the blades is as shallow as possible. Cut the entire length of the skull to the coronal cut between the orbits.
6. Using curved, pointed forceps grasp and peel the skull of the each hemisphere outward to expose the brain, then using a small wetted curved spatula, scoop the brain into a 50 ml tube containing HEM.
7. Wash brains three times using HEM to remove blood and/or hairs and transfer them to 100-mm Petri dishes containing HEM.
8. To dissect the forebrain subventricular region, place the dish containing the brain under the dissecting microscope (×10) magnification. Position the brain flat on its ventral surface and hold it from the caudal side using fine curved forceps placed on either side of the cerebellum. Use scalpel to make a coronal cut just behind the olfactory bulbs.
9. After the removal of the olfactory bulbs, rotate the brain to expose the ventral aspect. Make a 90° coronal cut at the level of the optic chiasm, discarding the caudal aspect of the brain.
10. Switch to a (×25) magnification. Rotate the rostral aspect of the brain with the presumptive olfactory bulb facing downward. Using fine curved microscissors, first remove the septum and discard and then cut the thin layer of tissue surrounding the ventricles, excluding the striatal parenchyma and the corpus callosum. Pool dissected tissue in a newly labeled 35-mm Petri dish.
11. Upon harvesting the periventricular regions from all brains, transfer dish to tissue culture laminar flow hood. Continue to use strict sterile technique.

3.1.2. Tissue Dissociation

1. Mince tissue for ~ 1–2 min using a scalpel blade until only very small pieces remain (see Note 1). Add a total volume of 3 ml of tissue dissociation medium (Trypsin–EDTA); transfer all of the minced tissues into of a 15-ml tube. 3 ml dissociation medium is enough for good digestion of tissues harvested from up to eight mice. Then incubate the tube for 7 min in a 37°C water bath.

2. At the end of the enzymatic incubation, add an equal volume of trypsin inhibitor (3 ml).
3. Avoid generation of air bubbles, mix well and pellet the suspension by centrifugation at 100 × *g* for 5 min.
4. Aspirate the supernatant and discard it, then resuspend the cells in 150 μl of sterile basal medium containing BSA, reset the pipettor to 200 μl. Pipette up and down gently to break the clumps up until a milky single cell suspension is achieved (see Note 2).
5. Add medium for a total volume of 1 ml and pass the suspension through a 40-μm pore size strainer into a 15-ml tube, so as to remove debris or undissociated pieces, and then pellet the cells by centrifugation at 100 × *g* for 5 min (see Note 3).
6. Transfer one brain into a T25 flask (containing 5 ml of complete media). The cells are then incubated at 37°C, 5% CO_2 for 7–10 days by which time neurospheres should have formed. Tissue harvested from one brain usually can generate 400–600 spheres but a count of over 300 spheres is acceptable from a T 25 flask.

3.1.3. Passaging Neurospheres

Usually, neurospheres with a variety of diameters are apparent in the culture. To determine if spheres are ready to be passaged, the mean neurospheres diameter should be about 100–150 μm (see Notes 4 and 5). If neurospheres are allowed to grow too large, they become dark colored because of cell death at the center of the spheres, difficult to dissociate and eventually begin to differentiate in situ (attaching to the substrate and migrating toward the periphery).

1. If the neurospheres are ready to be passaged, remove medium with suspended spheres and place in an appropriate size sterile tissue culture tube. Wash the flasks out with 2 ml of warm basal medium (to prevent the cells from being shocked) and add that to the centrifuge tube(s). Centrifuge cells at 100 × *g* for 5 min at room temperature.
2. Remove supernatant and resuspend the spheres in 1 ml of dissociation medium, then incubate at 37°C in the water bath for 2–3 min, then inactivate the trypsin using an equal volume of trypsin inhibitor.
3. Mix well to ensure that all the trypsin has been completely inactivated, then spin at 100 × *g* for 5 min.
4. Remove by aspiration the supernatant down to the actual pellet and resuspend the cells in 1 ml of basal medium and mix well, but gently.
5. Transfer 10 μl of this suspension into a 0.6-ml tube that contains 90 μl of trypan blue. Perform a cell count. At this stage it is easy to see if the cells are single or are still aggregated.

If they are not "a single cell suspension" then it will be necessary to resuspend the cells a little more vigorously. Perform the cell count again.

6. Cells are seeded at a concentration of 2.5×10^5 cells in 5 ml of complete medium in a T25 Flask.

3.1.4. Undifferentiated Whole Neurosphere Preparation for Cellular Characterization

1. Transfer the contents of a primary or passaged neurosphere culture to an appropriately sized sterile tissue culture tube. Centrifuge at $30 \times g$ for 5 min.
2. Aspirate essentially the entire growth medium, then gently resuspend (so as not to dissociate any neurospheres) with an appropriate volume of basal medium.
3. Transfer neurosphere suspension (~500 spheres/ml) into individual wells of 24- or 96-well tissue culture plate with a poly-L-ornithine coated surface in neurosphere medium.
4. Centrifuge plate at 700 rpm for 10 min.
5. Leave the plate in the 37°C incubator for 20 min.
6. Carefully aspirate the entire medium (so as not to dislodge any neurospheres), then gently add appropriate volume of 4% paraformaldehyde (in PBS, pH 7.2) and leave it for 20 min at room temperature.
7. Remove the paraformaldehyde solution by aspiration.
8. Add PBS (pH 7.2) to the samples and incubate for 5 min. Aspirate PBS and repeat this washing procedure two more times for a total of three wash steps before immunolabeling.

3.1.5. Neural Stem Cell Differentiation

When cultured in the presence of EGF and/or bFGF, neural stem cells and progenitor cells proliferate and give rise to neurospheres which, when harvested at the appropriate time-point and using the appropriate methods as described here, can be passed practically indefinitely, demonstrating long-term self renewal, and can generate a large number of progenies. However, upon removal of growth factors, neurosphere-derived cells are induced to differentiate into neurons, astrocytes, and oligodendrocytes indicative of multipotency (Fig. 1e, f). In general, two methods have been described for the differentiation of neurospheres: as whole spheres (typically used to demonstrate individual spheres are multipotent) or as dissociated cells (used to determine the relative percentage of differentiated cell types generated).

3.1.5.1. Whole Neurosphere Differentiation

1. Once primary or passaged neurospheres reach 150 μm in diameter, transfer the contents of the flask to an appropriate size sterile tissue culture tube. Centrifuge at $30 \times g$ for 5 min.
2. Aspirate essentially the entire growth medium, then gently resuspend (so as not to dissociate any neurospheres) with an appropriate volume of basal medium.

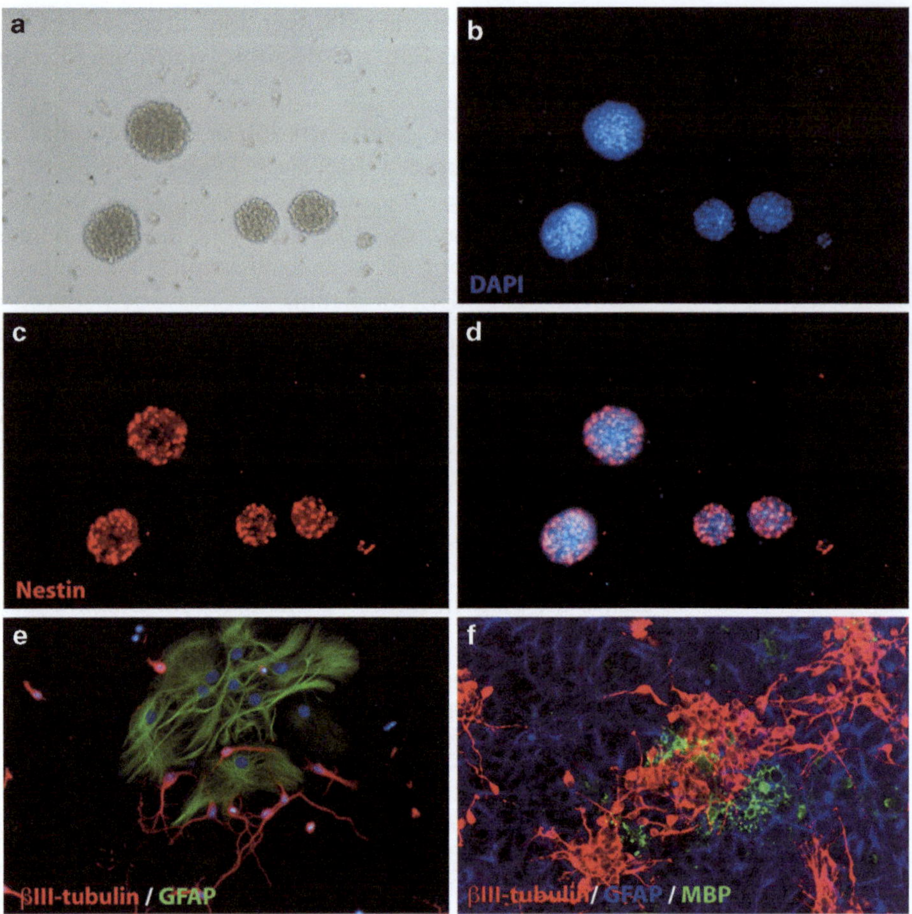

Fig. 1. Immunocytochemistry on undissociated neurospheres: (**a**) Phase contrast, (**b**) DAPI, (**c**) Nestin, and (**d**) Merged Nestin-DAPI. Differentiated neural stem cells: (**e**) Double labeling showing astrocyte (GFAP-*green*) and neurons (βIII-tubulin, *red*), (**f**) Triple labeling showing astrocytes (GFAP, *blue*), neurons (βIII-tubulin, *red*) and oligodendrocytes (MBP, *green*).

3. Transfer neurosphere suspension to a 60-mm dish (or other sized vessel) to enable the harvesting/plucking of individual neurospheres with a disposable plastic pipette.
4. Transfer approximately ten neurospheres into individual wells of 24- or 96-well tissue culture plate with a poly-L-ornithine coated surface in neurosphere medium containing 1% sterile fetal calf serum.
5. After 5–8 days in vitro, individual neurospheres should have attached to the substrate and dispersed in such a manner so as to appear as a flattened monolayer of cells.
6. Proceed to fix cells with the addition of 4% paraformaldehyde (in PBS, pH 7.2) for 20 min at room temperature and then process the adherent cells for immunocytochemistry as required.

3.1.5.2. Dissociated Neurosphere Differentiation

1. Once primary or passaged neurospheres reach 150 μm, transfer the contents of the flask to an appropriate size sterile tissue culture tube. Spin at $30 \times g$ for 5 min.
2. Remove supernatant, resuspend the cells in 1 ml of trypsin/EDTA and incubate for 3–4 min.
3. Add 1 ml of trypsin inhibitor to each tube, mix well, centrifuge at $100 \times g$ for 5 min and remove the supernatant before to resuspend the cells by the addition of 1 ml of basal medium. Dissociate the cells until suspension appears milky and no spheres can be seen (~ five to seven times pipetting).
4. Combine a 10 μl aliquot from the cell suspension with 90 μl of Trypan blue in a microcentrifuge tube, mix, and then transfer 10 μl to a hemocytometer so as to perform a cell count.
5. Seed individual wells of 24-well tissue culture plate containing a poly-L-ornithine coated glass coverslip with 5×10^5 cells.
6. After 4–6 days in vitro, neurosphere-derived cells will have differentiated sufficiently. Proceed to fix the cells with the addition of 4% paraformaldehyde (in PBS, pH 7.2) for 20 min at room temperature.
7. Remove the paraformaldehyde, add PBS (pH 7.2) to the samples and incubate for 5 min. Aspirate PBS and repeat this washing procedure two more times for a total of three wash steps before to process the cells for immunocytochemistry as required.

3.1.6. Immunocytochemistry

1. Block and permeabilize (if the antigen is intracellular) for 60 min in PBS-0.1% Triton-X100 + 10% Normal Goat Serum at 37°C.
2. Incubate the cells for 60–90 min at room temperature with the primary antibodies diluted in blocking solution (or overnight at 4°C) (see Table 1).
3. Wash the cells three times with PBS and incubate 45–60 min at 37°C with fluorochrome-conjugated secondary antibody diluted in blocking buffer at 1:700.
4. Wash the cells three times with PBS; include DAPI (1:1,000) in second wash for nuclear counter-stain.
5. Mount on slides using DAKO fluorescent mounting media (S3023).
6. Visualize the immunostaining using a fluorescent microscope using appropriate filters.

Figure 1a–d shows undifferentiated/undissociated neurospheres stained for nestin (marker to identify neural stem cells). Figure 1e shows the differentiation in neurons (βIII-tubulin) and

Table 1
Suggested primary antibodies and targeted antigens for the different neural lineages

	Antigen	Working dilution	Source
Neurons	βIII-tubulin	1:2,000	Promega#G7121
	Microtubule-associated protein-2 (MAP-2)	1:300	Chemicon # MAB3418
	Doublecortin	1:1,000	Chemicon # AB5910
	PSA-NCAM	1:300	Chemicon # MAB5324
Astrocytes	Glial fibrillary acidic protein (GFAP)	1:700	Dako Cytomation # Z0334
Oligodendrocytes	O4	1:300	Chemicon # MAB345
	Gal-c	1:300	Chemicon # MAB342
	Myelin basic protein (MBP)	1:300	Chemicon # AB980

astrocytes (GFAP) and Fig. 1f shows neuronal (βIII-tubulin), astrocytic (GFAP), and oligodendrocytic (MBP) triple-labeling in differentiated-dissociated neurosphere culture.

3.2. Neural Stem Cell Enrichment Using Flow Cytometry

1. Prepare a single cell solution from a neurosphere culture (see Note 6).
2. Wash the cell suspension once with PBS, count, and pellet the cells.
3. Adjust the cell suspension with PBS/2 % BSA to $1–5 \times 10^6$ cells/ml. Add 2.5 μl of 200 mM EDTA per ml suspension (final conc. 0.5 mM). Split cell suspension into a smaller negative control (approx. 2×10^5 cells) and the proper sample.
4. Add appropriate volume of primary antibody to the cell suspension (CD133) and incubate for 30 min on ice. Incubate the negative control with isotype control antibody (see Note 7).
5. Wash with PBS (resuspend the pellet in PBS and spin down again).
6. Resuspend the final pellet in an appropriate volume of PBS containing 0.5 mM EDTA (cell concentration should be about 1×10^7 cells/ml for faster sorting) and 1 μl/ml propidium iodide (PI) solution.
7. Run the samples on sorter; use the negative control to adjust voltage for forward/side scatter so the cells form a cloud that is roughly centered in the dot plot. Set the first gate (P1) to include the cloud (Fig. 2a). Adjust voltage for specific

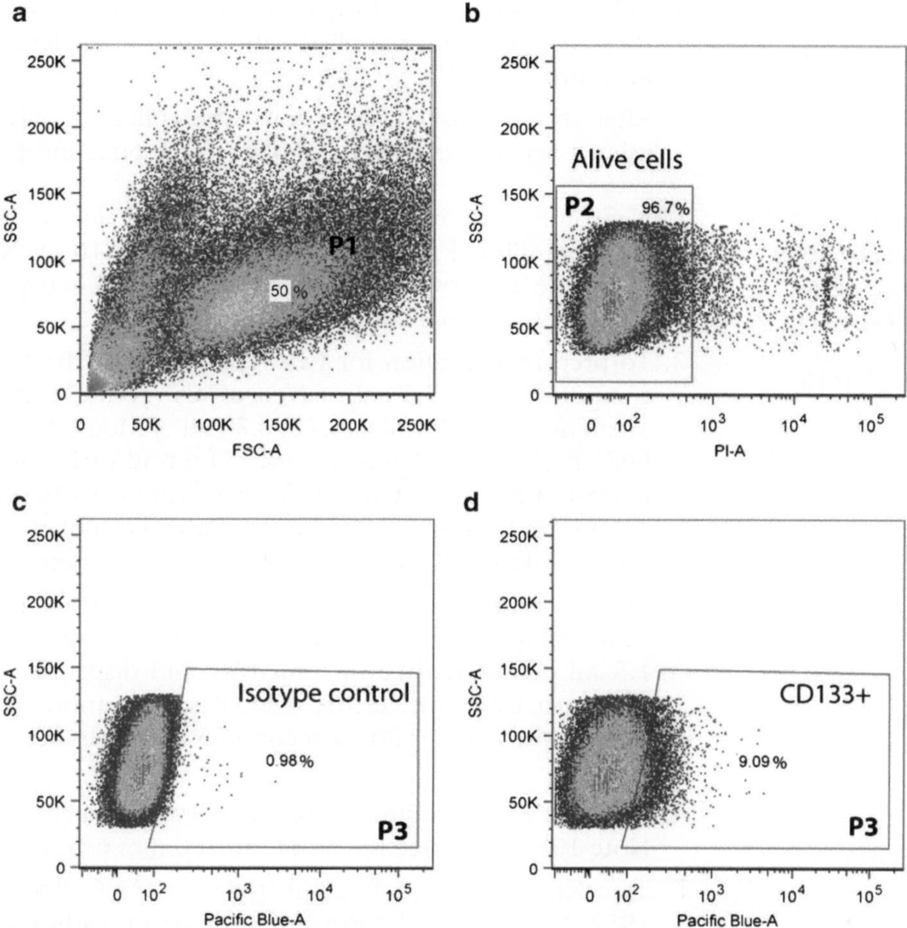

Fig. 2. Neural stem cell isolation. (**a**) Representative dot plot scatter of cells from neurosphere culture derived from adult periventricular area. Gating for cells in population 1 (P1) exclude the debris. (**b**) Representative dot plot comparing Side scatter and Propidium Iodide (PI) staining of the P1. A gate is determined around the PI negative population (P2) to exclude PI positive dead cells for further analysis. (**c, d**) Dot plot distribution of viable cells based on side scatter and CD133 staining intensity. CD133 positive gate is set on the dot plot using the background level of fluorescence of the unstained negative control (containing only the fluorochrome-conjugated secondary antibody without the primary or with isotype control).

antibody (depending on the fluorochrome) so that the events in gate 1 do not exceed a fluorescence intensity of greater than 10^2. Set second gate (P2, Fig. 2b) for all events of gate 1 that are negative for propidium iodide (i.e., live cells). Set third gate (P3) for all events with fluorescence intensities greater than the negative control (Fig. 2c, d).

8. After adjusting all voltages and acquiring 10,000 events of negative control (Fig. 2c), run proper sample (Fig. 2d). Acquire 10,000 events and check the gates. Cells should form a cloud on the FSC/SSC blot that falls into gate 1. A significant portion of events from gate 1 should be measurable in P2.

9. Before beginning the sort, set sorter to sort all events from P3 into collection tube filled with 2 ml of complete growth medium.
10. After sort is finished, spin down the collected cells; count using a hematocytometer and plate in the NSA and NCFCA.

3.3. Neural Stem Cell Quantification Using the Neural Colony-Forming Cell Assay

3.3.1. Culture Set-Up

1. Neurosphere-derived sorted cells (CD133 immunoreactive cells) are diluted to a concentration of 2.2×10^5 cells/ml in Complete NeuroCult® Proliferation Medium and plated at 2,500 cells/35 mm culture dish with 1.5 ml.
2. To prepare a solution for two replicates, mix the following components: (1) 1.7 ml of NeuroCult® NCFC Serum-Free Medium without Cytokines, (2) 0.33 ml of Mouse NeuroCult® NSC Proliferation Supplements, (3) 6.6 μl of Recombinant Human Epidermal Growth Factor (rhEGF) (10 μg/ml), 3.3 μl of Recombinant Human Basic Fibroblast Growth Factor (10 μg/ml) and 6.6 μl of Heparin Solution (0.2%).
3. Mix the medium containing the cells and transfer 1.3 ml of cold Collagen Solution to the tube and mix again. Remove 1.5 ml of the final culture mixture and dispense this volume into a 35 mm culture dish. Dispense another 1.5 ml in the same manner into a second 35 mm dish (see Notes 8 and 9).
4. Place the 35 mm culture dishes in a 100 mm petri dish (see Note 10) and replace the lid of the 100 mm petri dish.
5. Transfer the plates to an incubator set at 37°C, 5% CO_2 and >95% humidity. Gel formation will occur within approximately 1 h. Incubate the cultures for 21–28 days.
6. Due to the prolonged culture period, the medium need to be replenished by depositing 60 μl of complete liquid medium supplemented with concentrated EGF (0.5 μg/ml) plus fibroblast growth factor (0.25 μg/ml) and heparin (0.01%) in the center of the dish once every week for the total of 3–4 weeks.
7. Visually assess the cultures regularly for overall colony growth and morphology using an inverted microscope (see Note 11).

3.3.2. Neural Stem Cell Frequency Measurement

A number of the colonies stop growing after approximately 10–14 days while other colonies continue to expand. By day 21–28, four categories of colony size can be classified: (1) less than 0.5 mm in diameter, (2) 0.5–1 mm in diameter, (3) 1–2 mm in diameter, and (4) 2.0 or >2 mm in diameter.

The original cell that forms a large colony (2.0 or >2 mm in diameter) is referred to as a Neural Stem Cell, while colonies <2 mm are likely produced by a progenitor cell. Dividing the number of large colonies by the number of cells originally plated, and multiplying by 100 gives the neural stem cell frequency (Fig. 3).

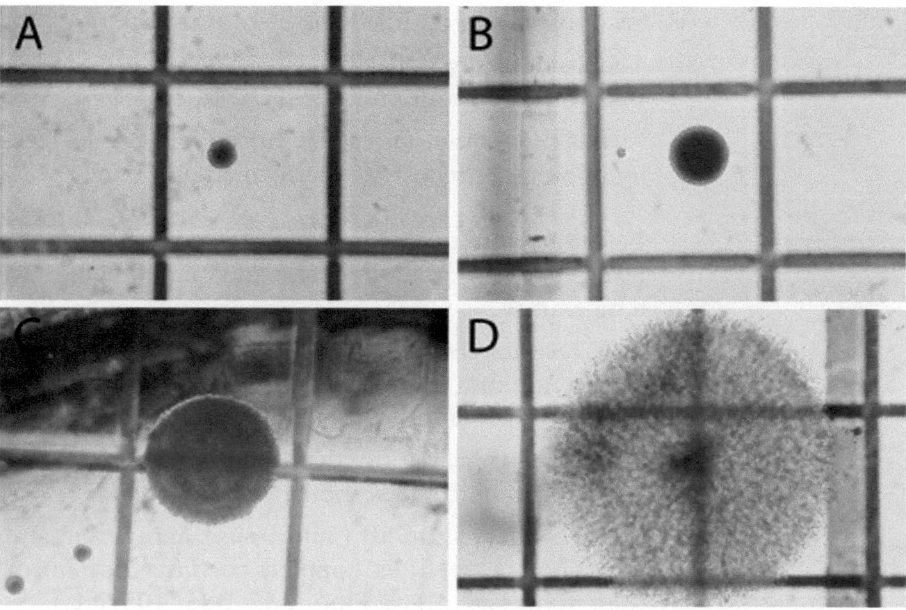

Fig. 3. Neural colony-forming cell assay. Four size categories are identified based on the diameter of the colonies: (**a**) <0.5 mm, (**b**) 0.5–1 mm, (**c**) 1–2 mm, and (**d**) >2 mm. Photos reproduced with permission from STEMCELL Technologies Inc.

3.4. In Vivo Identification of Neural Stem Cells Using Immunohistochemistry

Immunohistochemistry (IHC) is a commonly used method that definitively demonstrates not only the presence, but also the location of proteins in tissue sections. IHC is a less sensitive method than some immunoassays; however, it allows for examination of intact tissue. For optimal immunohistochemical studies, quick processing of brain tissue is very important.

3.4.1. Brain Perfusion

1. Animals are deeply anesthetized and then transcardially perfused with saline (0.9%) followed by ice-cold 4% paraformaldehyde (PFA). Volumes will have to be adjusted depending on the size of the animal.
2. The brain of the animal is removed from the skull, and is post-fixed in 4% PFA overnight, and is then placed in a phosphate-buffered sucrose solution (usually 20–30%) for 24 h.
3. The brain is now ready to be trimmed and frozen

3.4.2. Tissue Sectioning

1. Blocks of frozen tissue are sectioned in a coronal plane on a cryostat using OCT compound as an adhesive. The cryostat chuck should be placed in dry ice till it is chilled thoroughly. OCT compound can then be used to cover the surface of the chuck. Once the OCT has frozen and turned white, the block of frozen tissue can be attached to it.
2. It is important to let the temperature of the chuck equilibrate with that of the cryostat in order to avoid fracturing the tissue.

3. Sections should optimally be 5–14 μm thick.
4. The sections are collected with a fine brush, and stored in PBS with 0.1% sodium azide.
5. At this point, sections can be stored for several weeks before immunohistochemistry is performed on them.

3.4.3. Immunolabeling

1. Before beginning this procedure frozen sections need to be taken out of the freezer, and equilibrated to room temperature.
2. It is recommended to block sections with serum (Bovine Serum Albumin BSA, or Normal Donkey Serum NDS); 5% in PBS for about an hour.
3. Then, the primary antibody (Table 2) is appropriately diluted and applied onto the sections.
4. The secondary antibody (anti-mouse, anti-rabbit, anti-guinea pig, or anti-sheep) is coupled to different fluorochromes (e.g., Cy3, Cy5, Alexa Fluor 355, 385, 405, etc.).
5. Incubate the sections in primary antibody (appropriately diluted) at 4°C overnight. Figure 4 shows an example of a section of a mouse brain stained for Sox-2.
6. After the overnight incubation wash off the primary antibody using PBS 3 × 5 min.

Table 2
Suggested primary antibodies and targeted antigens to identify neural stem cells

Antibody	Properties	Source	Working dilution
Nestin	Intermediate filament protein; expressed by neuronal precursor cells of the SVZ	Millipore, MAB353	1:500
Sox2	Transcription factor essential for maintenance of self-renewal of stem cells	R&D, MAB2018	1:250
Ki67	Cellular marker for proliferation; is present during all active phases of the cell cycle	NovoCastra, NCL Ki67p	1:500
PCNA	Protein synthesized in early G1 and S phases; detectable in nuclei of proliferating cells	DAKO Cytomation, M0879	1:1,000
MCM2	Is involved in the initiation of eukaryotic genome replication	Cell Signaling Technologies, D7G11	1:500
GFAP	Intermediate filament protein; thought to be specific for astrocytes in the CNS	DAKO Cytomation, Z0334	1:500
Musashi-1	RNA-binding protein expressed in neural progenitor and stem cells	Chemicon	1:250

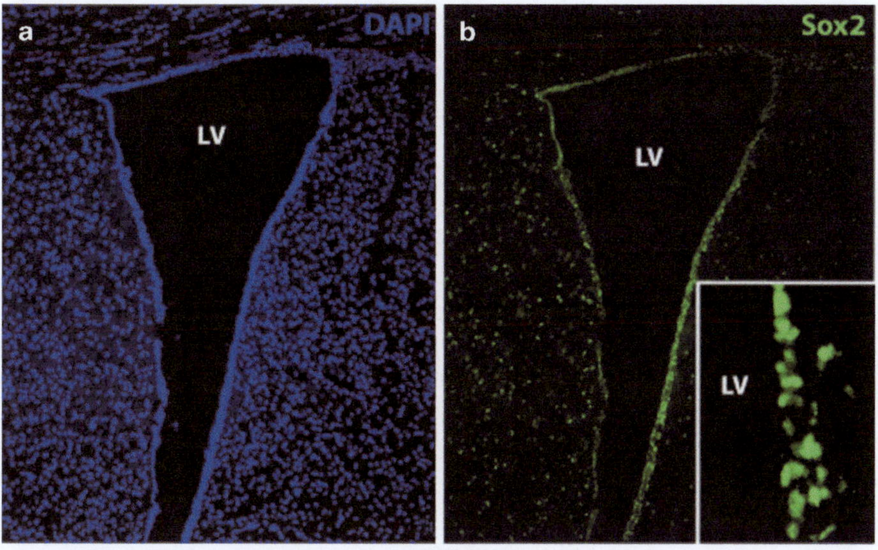

Fig. 4. Low magnification of a coronal section of adult mouse brain showing the lateral ventricle, (**a**) nuclei are stained *blue* with DAPI, (**b**) expression of Sox2, a key transcription factor required in pluripotent stem cells. *Inset* shows a higher magnification of the same along the wall of the LV.

7. Incubate the sections in species-specific fluorescent secondary antibody for 2–3 h, at room temperature, in a dark box.
8. Wash off the secondary antibody using PBS 3 × 5 min.
9. Keep the sections wet by mounting onto slides (if not already mounted) and coverslipping.
10. Once the slides have dried, the sections are ready to be viewed under either a fluorescent microscope, or a laser scanning confocal microscope. It is recommended that the imaging is done systematically in order to eliminate any errors or omissions in data collection. Images are captured and stored in digital format for analysis.

4. Notes

1. Ensure that the scalpel blade does not cut into the plastic of the petri dish, as this could be toxic to the cells. It would be preferable to use a 10 cm glass petri dish.
2. Make sure to avoid generating air bubbles, as this reduces the number of viable cells and makes for inefficient dissociation. Also, the expulsion of cells during the dissociation should not be too forceful, as this will also significantly reduce viability.
3. In primary cultures from adult brain significant debris is normally present, together with adherent cells. In general, debris and adherent cells are eliminated after about two passages.

4. Spheres must be rounded but not compacted; they should measure between 100 and 150 μm (Fig. 1a).
5. Primary neurospheres are often associated with cellular debris; however, subculturing will effectively select for proliferating precursor cells and remove cell aggregates, debris, and dead cells.
6. The spheres should preferably be of early passage primary cells, between 2 and 8.
7. Direct conjugated primary antibodies are preferred for flow cytometry, but a primary/secondary staining is possible as well. In case secondary staining is necessary, wash twice with PBS after the primary and incubate with the secondary for 30 min on ice. Proceed with step 7 of the FACS-sorting protocol. A negative control for the secondary antibody is recommended.
8. Remove any air bubbles by gently touching bubble with the end of the pipette.
9. The collagen starts to polymerize within several minutes after to be added to the cell suspension. Gently tip each culture dish using a circular motion to allow the mixture in the dishes to spread evenly over the surface.
10. The 100 mm petri dish should contain an open 35 mm culture dish filled with 3 ml of sterile water to maintain optimal humidity during the incubation period.
11. Do not leave cultures at room temperature for extended periods of time as the collagen gel will begin to liquefy.

Acknowledgments

The authors would like to thank Dr. Mohammad G. Golmohammadi for kindly providing the neurosphere pictures. This work was supported by the Overstreet foundation.

References

1. Reynolds, B.A., and Weiss, S. (1992) Generation of neurons and astrocytes from isolated cells of the adult mammalian central nervous system *Science* **255**, 1707–10.
2. Engstrom, C.M., Demers, D., Dooner, M., McAuliffe, C., Benoit, B.O., Stencel, K., Joly, M., Hulspas, R., Reilly, J.L., Savarese, T., Recht, L.D., Ross, A.H., and Quesenberry, P.J. (2002) A method for clonal analysis of epidermal growth factor-responsive neural progenitors *J Neurosci Methods* **117**, 111–21.
3. Gritti, A., Parati, E.A., Cova, L., Frolichsthal, P., Galli, R., Wanke, E., Faravelli, L., Morassutti, D.J., Roisen, F., Nickel, D.D., Vescovi, A.L. (1996) Multipotential stem cells from the adult mouse brain proliferate and self-renew in response to basic fibroblast growth factor *J Neurosci* **16**, 1091–100.
4. Louis, S.A., Rietze, R.L., Deleyrolle, L., Wagey, R.E., Thomas, T.E., Eaves, A.C., Reynolds, B.A. (2008) Enumeration of neural stem and progenitor cells in the neural colony-forming cell assay *Stem Cells* **26**, 988–96.
5. Pastrana, E., Cheng, L.C., Doetsch, F. (2009) Simultaneous prospective purification of adult

subventricular zone neural stem cells and their progeny *Proc Natl Acad Sci USA* **106**, 6387–92.

6. Capela, A., Temple, S. (2002)LeX/ssea-1 is expressed by adult mouse CNS stem cells, identifying them as nonependymal *Neuron* **35**, 865–75.
7. Uchida, N., Buck, D.W., He, D., Reitsma, M.J., Masek, M., Phan, T.V., Tsukamoto, A.S., Gage, F.H., Weissman, I.L. (2000) Direct isolation of human central nervous system stem cells *Proc Natl Acad Sci USA* **97**, 14720–5.
8. Gilyarov, A.V. (2008) Nestin in central nervous system cells *Neurosci Behav Physiol* **38**, 165–9.
9. Sakakibara, S., Nakamura, Y., Satoh, H., Okano, H. (2001) Rna-binding protein Musashi2: developmentally regulated expression in neural precursor cells and subpopulations of neurons in mammalian CNS *J Neurosci* **21**, 8091–107.
10. Graham, V., Khudyakov, J., Ellis, P., Pevny, L. (2003) SOX2 functions to maintain neural progenitor identity *Neuron* **39**, 749–65.
11. Kim, M., Morshead, C.M. (2003) Distinct populations of forebrain neural stem and progenitor cells can be isolated using side-population analysis *J Neurosci* **23**, 10703–9.
12. Corti, S., Locatelli, F., Papadimitriou, D., Donadoni, C., Salani, S., Del Bo, R., Strazzer, S., Bresolin, N., Comi, G.P. (2006) Identification of a primitive brain-derived neural stem cell population based on aldehyde dehydrogenase activity *Stem Cells* **24**, 975–85.
13. Reynolds, B.A., Rietze, R.L. (2005) Neural stem cells and neurospheres--re-evaluating the relationship *Nat Methods* **2**, 333–6.

Chapter 5

Magnetic Resonance Imaging of Stem Cell Migration

Eva Syková, Pavla Jendelová, and Vít Herynek

Abstract

Noninvasive cellular imaging allows the real-time tracking of grafted cells as well as the monitoring of their migration. In this review, we will focus on cell tracking using MRI, since MRI is noninvasive, clinically transferable, and displays good resolution, ranging from 50 μm in animal experiments up to 300 μm using whole body clinical scanners. In addition to information about grafted cells, MRI provides information about the surrounding tissue (i.e., lesion size, edema, inflammation), which may negatively affect graft survival or the functional recovery of the tissue. Transplanted cells are labeled with MR contrast agents in vitro prior to transplantation in order to visualize them in the host tissue. The chapter will focus on the use of superparamagnetic iron oxide nanoparticles (SPIO), because they have strong effects on T2 relaxation yet do not affect cell viability, and will provide an overview of different modifications of SPIO and their use in MR tracking in living organisms.

Key words: Stem cell, Real-time migration, Magnetic resonance imaging

1. Introduction

Stem and progenitor cells are being explored in regenerative medicine for the cell therapy of various diseases including disorders of the central nervous system. For the future success of cell transplantation in clinical use, it is crucial that the transplanted cells have the ability to migrate from the site of transplantation to the lesioned area and to survive, differentiate, and replace lost populations of cells or produce growth factors and cytokines for a prolonged period of time in order to enhance the patient's regenerative potential. The homing, migration, and engraftment factors, and the mechanisms underlying the directed migration of transplanted stem cells to the site of injury, are particularly dependent on the secretion of chemokines (chemotactic cytokines). There is a large group of cytokines whose primary function is to regulate the

Marie-Dominique Filippi and Hartmut Geiger (eds.), *Stem Cell Migration: Methods and Protocols*, Methods in Molecular Biology, vol. 750, DOI 10.1007/978-1-61779-145-1_5, © Springer Science+Business Media, LLC 2011

migration of leukocytes. Besides their well-established role in the immune system, several reports have demonstrated that these proteins also play a role in stem cell migration in the CNS. In the CNS, chemokines are constitutively expressed by microglial cells, astrocytes, and neurons, and their expression can be increased after induction with inflammatory mediators. The production of cytokines by mesenchymal stem cells and their receptor expression have been studied in relation to different types of injury. The overproduction of cytokines has been implicated in different neurological disorders such as multiple sclerosis, trauma, stroke, and Alzheimer's disease. The migration of stem cells in relation to chemokine production has been described in ischemic cerebral tissue (1), pancreatic islets (2), hypoglossal nerve injury (3), malignant gliomas (4), and ischemic myocardium (5).

Visualizing the migration of transplanted cells in vivo is essential for preclinical studies in rodents and potentially also in humans. MR tracking of cells appears to be a valuable tool for such studies, since MR imaging may serve to study cell migration to lesions, the time course of such migration and for how long the cells persist in the target region. Such information could help to identify the time window during which the transplantation of therapeutic cells can be clinically effective, the number of cells needed for functional improvement and the optimal method of their administration. For cellular MR imaging, cells need to be labeled with an MR contrast agent in order to visualize them in the host tissue. Contrast agents suitable for cell labeling must not only have high relaxation rates, but they also should not affect the cells' viability or their ability to differentiate and migrate toward the target tissue. In addition, the contrast agent must be either incorporated into the cell cytoplasm (intracellular cell labeling) or bound to the cell surface (extracellular cell labeling).

1.1. Intracellular Cell Labeling

Superparamagnetic iron oxide nanoparticles (SPIO) are currently the preferred choice mainly because of the following properties: (a) they provide the largest change in signal per unit (size) of metal, in particular on T2*-weighted images; (b) they are composed of iron, which is biocompatible and can thus be recycled by cells using normal biochemical pathways for iron metabolism; (c) their surface coating allows the chemical linkage of functional groups and ligands for specific labeling, stability, and solubility; and (d) they can be easily detected by light and electron microscopy.

SPIO require stabilization in order to prevent aggregation. Most commonly this is accomplished by a surface coating of dextran (DCSPIO – Endorem®, Feridex®, Sinerem®). Iron crystal size varies over a range of 4.3–5.6 nm; the whole particle size is 120–150 nm (6). Usually, contrast agents can be easily incorporated by endocytosis, and all the cells survive and further divide in vitro (Endorem®); alternatively, membrane permeabilization

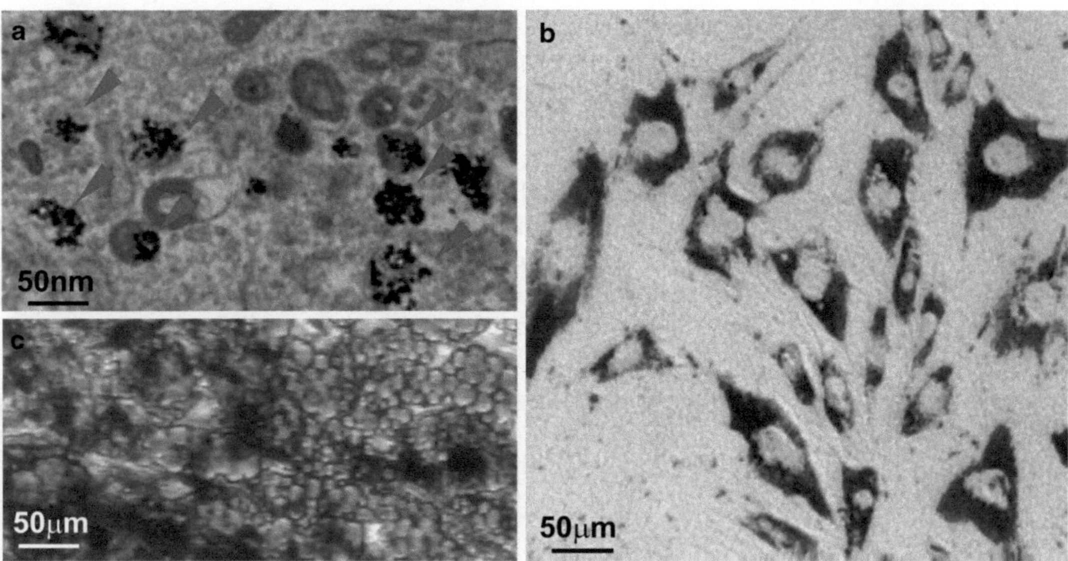

Fig. 1. (**a**) TEM microimage showing how SPIO nanoparticles are internalized in the cell cytoplasm. (**b**) Iron is detected inside rat mesenchymal stem cells by Prussian blue staining. (**c**) SPIO-labeled stem cells retain their ability to differentiate. Human MSCs differentiated into an adipogenic phenotype (Oil Red O staining).

(e.g., by lipofection) can be used to increase the efficiency of cell labeling. For the labeling of adherent cells with Endorem®, cells are incubated 2–3 days in nanoparticle-containing media (7). The nanoparticles are detected by staining for iron (Prussian blue staining) or by transmission electron microscopy (Fig. 1a, b). The efficiency of mesenchymal stem cell labeling (i.e., how many cells out of the total number of analyzed cells are labeled) is 50–70%. The labeled MSCs retain their physiological properties, e.g., they can differentiate into an osteogenic, chondrogenic, or adipogenic phenotype (Fig. 1c). Another labeling approach is based on combining a commercially available DCSPIO, such as Feridex® or Sinerem®, and a commercially available transfection agent, e.g., Superfect™, poly-L-lysine (PLL), Lipofectamin, or Fugene™. Transfection agents effectively transport nanoparticles into cells through electrostatic interactions. However, each combination of transfection agent and SPIO has to be carefully titrated and optimized for different cell cultures, since lower concentrations of the transfection agent may result in insufficient cellular uptake, whereas higher concentrations may induce the precipitation of complexes or may be toxic to the cells (8). To overcome these drawbacks, several types of maghemite nanoparticles with various coatings were prepared by the coprecipitation method (9). Surface modification of iron oxide nanoparticles with D-mannose, PLL, or polydimethylacrylamid (PDMAAm) resulted in better labeling efficiency than that seen with dextran-coated SPIO, though the mechanism of nanoparticle internalization in the cells differed depending on the type of coating (9). D-mannose-coated nanoparticles were taken up by the cells via mannose receptors on the cell surface (10). PDMAAm and PLL-coated nanoparticles were internalized in the cells due to their positive charges interacting with the negatively charged cell membrane, followed by endocytosis (11, 12). MR images of phantoms containing suspensions of those SPIO-labeled cells showed a much stronger hypointense signal than did MR images of Endorem-labeled cells (Fig. 2a). The MR detection limit in vitro was two cells in the image voxel,

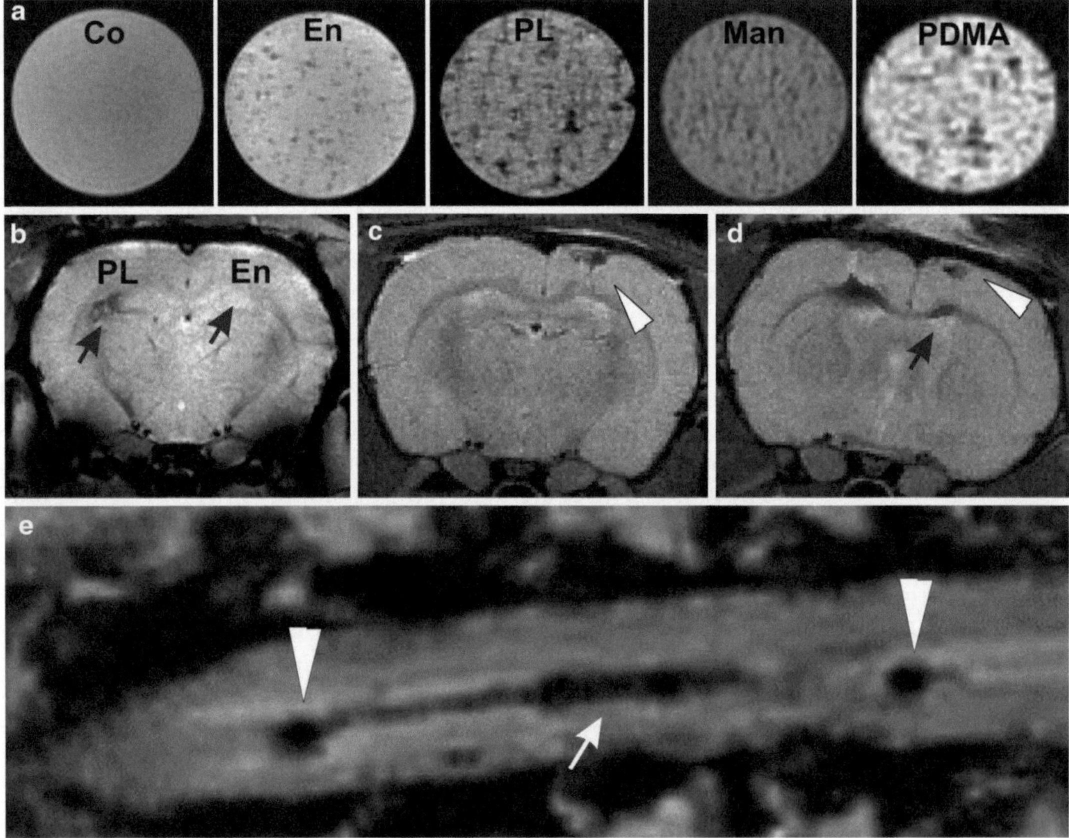

Fig. 2. (**a**) MRI images of gelatin phantoms containing unlabeled or labeled rat mesenchymal stem cells measured by a T_2-weighted turbospin echo sequence. Cells were either unlabeled (Co) or labeled with Endorem® (En), PLL-coated SPIO (PL), D-mannose-coated SPIO (Man) or PDMAAm-coated SPIO (PDMA) nanoparticles. Test tubes contained 25,000 cells in 0.5 ml, which represents two cells per image voxel on average. (**b**) A transversal MR image of a rat brain with 1,000 cells labeled with PLL-SPIO nanoparticles implanted in the left hemisphere (PL) and 1,000 Endorem-labeled cells implanted in the right hemisphere (En). The MR image was taken 3 days after implantation. Sites of implantation are marked by *arrows*. (**c**) MR image of a cortical photochemical lesion and MSCs injected into the femoral vein. A hypointensive signal (*white arrowhead*) was observed 6 weeks after intravenous injection. (**d**) MR image of a cortical photochemical lesion and mouse embryonic stem cells implanted into the contralateral hemisphere. The cell implant in the hemisphere contralateral to the lesion and the lesion itself are seen as hypointensive areas (*white arrowhead*) 15 days after grafting. Note the cell migration via the corpus callosum (*black arrow*). (**e**) A T2-weighted image of a rat spinal cord injected with nanoparticle-labeled MSCs. *Arrowheads* mark the injection sites, *arrow* indicates the lesion populated with cells.

while in vivo it was 1,000 cells injected in 5 μl of PBS (Fig. 2b). Compared to Endorem, the better internalization of PLL-SPIO particles into the cells enables their easier MRI detection and tracking in the tissue after transplantation.

Other strategies for developing MR contrast agents that are easy to detect include the utilization of viral protein cages or the use of internalizing monoclonal antibodies (13, 14).

1.2. Extracellular Cell Labeling

The disadvantage of an intracellular label is that it can affect cell metabolism and subsequently cell viability. Particles that do not internalize do not affect cell viability, and keeping cells in culture is not required for cell labeling; however, their attachment to the outer cell membrane is likely to interfere with cell surface interactions, and they may easily detach from the membrane or be transferred to other cells. Therefore, long-term follow up is so far not possible.

The possibility of labeling only for selected cell types can be very useful. Cells labeled and separated by means of immunomagnetic selection do not require in vitro culturing, since the label is attached during separation. Specific cell labeling using commercially available cell isolation kits for the magnetic separation of $CD34^+$ cells was tested (15). $CD34^+$ cells are known as hematopoietic progenitor cells. The cells were separated by means of immunomagnetic selection with anti-CD34 antibodies. For sorting, a superparamagnetic iron oxide core coated with a polysaccharide is linked to an antibody that binds to the respective cell type. The size of the label is comparable to commonly used superparamagnetic MR contrast agents, thus it can provide sufficient contrast on MR images. The average iron content per cell, determined by spectrometry, was 0.275 pg. This value is lower by two orders of magnitude than in the case of intracellular cell labels (17–38 pg/cell); nevertheless, it still provides reasonable MR contrast, though only for a limited time period.

1.3. In Vivo MR Tracking of Implanted Cells in the CNS

The use of stem cell therapy in different CNS disorders has been extensively examined. Endorem-labeled green fluorescent protein MSCs were grafted into rats in an experimental model of stroke (7). The cells were grafted either intracerebrally into the hemisphere contralateral to the lesion or intravenously into the femoral vein. Rats with grafted stem cells were examined weekly for a period of 3–7 weeks posttransplantation using a 4.7 T Bruker spectrometer. The lesion was visible on MR images as a hyperintense signal. One week after grafting, a hypointense signal was found in the lesion, which intensified during the second and third weeks regardless of the route of administration (Fig. 2c); its intensity corresponded to Prussian blue staining or GFP labeling (7). Similar experiments were done with mouse embryonic stem cells (16), (Fig. 2d). MSCs labeled with Endorem were also injected intravenously into the femoral vein 1 week after a transversal spinal cord lesion (17). MR images of longitudinal spinal cord sections from lesioned nongrafted animals showed the lesion cavity as inhomogeneous tissue with a strong hyperintensive signal. Lesions of grafted animals were seen as dark hypointense areas (Fig. 2e). Histological evaluation confirmed only a few iron-containing cells in lesioned, control animals, but strong positivity for iron in grafted animals. Compared to control rats, in grafted animals the lesion, which was populated by grafted MSCs, was considerably smaller, suggesting a positive effect of the MSCs on lesion repair. The detailed protocols provided are suitable for adherent cells, such as mesenchymal stem cells, olfactory ensheathing cells, and neural progenitor cells. It is more difficult to label undifferentiated mouse or human embryonic stem cells since they grow in compact colonies, with the result that the nanoparticles are not internalized evenly throughout the colonies, but instead are localized predominantly around only the outer edges of the colonies.

Recent evidence suggests that stem cells are useful delivery vehicles for brain tumor therapy. MR tracking was successfully used to demonstrate the incorporation of magnetically labeled bone marrow-derived precursor cells into the tumor vasculature as part of ongoing angiogenesis and neovascularization (18).

1.4. MR Tracking of Implanted Cells in Other Organs

Several successful applications of MR tracking can be found in other organs, such as the heart, liver, kidney, and pancreatic islets. The contrast agent Resovist (a DCSPIO-type nanoparticle) was used as a marker of isolated pancreatic islets for MRI. Following transplantation into the liver of a rat, the labeled pancreatic islets could be easily detected as hypointense regions in the liver on T2*-weighted MR images for up to 22 weeks following transplantation. In addition, this method can be used for monitoring islet rejection in allogenic transplantations due to a decrease in the intensity of the hypointense regions during rejection (19).

2. Materials

2.1. Cell Culturing

1. Dulbecco's Modified Eagle's Medium (DMEM) – high glucose (4.5 g/l) with L-Glutamin (PAA) supplemented with 10% fetal bovine serum (FBS, PAA) and with 0.2% primocin (Amaxa). Other antibiotics can be used instead of primocin: 100 units/ml penicillin and 0.1 mg/ml streptomycin or gentamicin (50 μl/ml).
2. Phosphate-buffered saline (PBS) pH 7.2.
3. Solution of trypsin (0.5%) and ethylenediamine tetraacetic acid (EDTA) (0.2 g/L) from Gibco/Invitrogen.
4. Solution of paraformaldehyde (4% in phosphate buffer).

5. Ficoll-Paque™ Plus (GE Healthcare Life Sciences).
6. 0.5% Trypsin (Gibco/Invitrogen).
7. Cell culture plasticware: centrifuge tubes, pipettes, Petri dishes, and tissue culture flasks.
8. Syringes (10 ml), needles (18G, Becton Dickinson).

2.2. Cell Labeling

1. Adherent cell culture, 40–50% confluent.
2. Solution of SPIO nanoparticles (e.g., Endorem, Guerbet Roissy, France). Stock solution contains 11.24 mg of iron per 1 ml of Endorem®. PLL-coated or D-mannose-coated nanoparticles are used at a concentration of 4.4 mg of iron per 1 ml of suspension.

2.3. In Vitro MR Tracking

1. Set of Eppendorf tubes 1.6 ml.
2. Cell suspension (2 million cells) labeled with SPIO nanoparticles and fixed in 4% paraformaldehyde.
3. Cell suspension (2 million cells) unlabeled and fixed in 4% paraformaldehyde (serves as control).
4. Gelatin (Sigma).

2.4. Cell Implantation and In Vivo MR Tracking

1. Suspension of cells labeled with SPIO (usually 50,000 cells/1 μl is recommended).
2. Animal with experimental model of injury or disorder. Since this protocol deals with MR tracking in living recipients, describing the induction of different models of injury is beyond the scope of this protocol.
3. MR equipment. The procedure may differ depending on the type (bore size, field strength, RF coil used, etc.) of the MRI imager. Procedures and listed sequence parameters given here are suitable for a standard horizontal bore imager 4.7 T.

2.5. Prussian Blue Staining

1. Hydrochloric acid (0.5% HCl).
2. One gram of potassium ferrocyanide ($K_4[Fe(CN)_6]$ $3H_2O$; Sigma) is diluted in 100 ml of 0.5% HCl.
3. 70% Ethyl alcohol (EtOH).
4. Aluminum sulfate (5% Al_2 $(SO_4)_3$; Sigma).
5. Nuclear fast red (Sigma) diluted in 5% aluminum sulfate.

3. Methods

MR tracking of cell migration serves for studying how certain lesions are populated with transplanted cells, at what speed the cells migrate and for how long they persist in the target organ.

High contrast effects on MR images are easily detected within an experimental time frame of about 1–2 h per animal, which is ideal for short and repetitive in vivo MRI. However, there are some disadvantages related to MR tracking in a living organism. The disadvantages include dilution of the contrast agent during cell division and a lack of information about cell viability or function. In addition, in lesioned tissue, hemorrhage products (containing iron) give rise to Prussian blue-positive deposits that are difficult to distinguish from iron-containing nanoparticles. The hemorrhage degradation products may also be partially localized in macrophages because macrophages constitute the major cellular pathway for the redistribution of iron in mammals. Furthermore, hemorrhage contributes to T2-weighted hypointensity, thus interfering with the detection of labeled cells and complicating the interpretation of MR images. Therefore, it is important to determine whether cell labels remain co-localized with cell transplants, especially under pathological conditions.

With proper attention to the limitations described above, labeling cells with superparamagnetic agents would enable us to deliver the cells and immediately verify whether the cells have populated the target organ. Moreover, it will allow us to follow the migration of such cells in clinics when transplanted into humans, establish the optimal number of transplanted cells, define therapeutic windows and monitor cell growth and possible side effects, e.g., malignancies.

The protocols described here have been prepared for the isolation, expansion, and labeling of rat or human mesenchymal stem cells and their tracking in rat brain and other organs. These protocols can be modified to label different adherent cells with SPIO nanoparticles that can be kept in culture for 48–72 h.

3.1. Cell Culturing

1. *Isolation and expansion of rat mesenchymal stem cells.* Rat bone marrow mesenchymal stem cells (rMSCs) are obtained from the tibia and femur of 4-week-old rats. The marrow is extruded with Dulbecco's Modified Eagle Medium (DMEM) by using a needle and a syringe. Human bone marrow mesenchymal stem cells (hBMSCs) can be obtained from the bone marrow of healthy donors.
2. Bone marrow aspirates are diluted in PBS and centrifuged through a density gradient (Ficoll-Paque™ Plus) for 30 min at 1,000 × *g*.
3. Nucleated cells from the interface are plated in a 75 cm^2 tissue culture flask in α-MEM for human cells and DMEM for rat cells containing 10% FBS, and 0.2% Primocin.
4. After 24 h, the nonadherent cells are removed by replacing the medium. The medium is replaced every 2–3 days as the

cells grow to confluence. The cells are lifted by incubation with 0.5% trypsin (Gibco/Invitrogen). The first passage after plating is usually taken as P0. Cells can be expanded up to passage 5 (P5) though the efficiency of cell labeling decreases with the number of passages.

3.2. Cell Labeling

1. A SPIO nanoparticle suspension (10 μl per 1 ml of culture medium corresponding to a final concentration of 112.4 μg of iron per 1 ml of culture medium) is added to a culture of rat MSCs or any adherent cells (see Note 1).
2. After 72 h the contrast agent is washed out, and the cells are harvested and used for either in vitro MR tracking or transplanted into animal models of disease or injury.

3.3. In Vitro MR Measurement

To verify what cell density can be detected, cell phantoms should be prepared and measured by MRI in vitro.

3.3.1. Preparation of Phantoms Containing Suspensions of Labeled Cells

To avoid deposition of the cells at the bottom of the test vials, the cells should be suspended in 4% gelatin. From a fixed cell suspension containing 2 million SPIO-labeled cells, several 1 ml samples can be prepared containing 100,000–1,000,000 cells/ml each. This range of concentrations ensures roughly one to ten cells in one image voxel at a standard image geometry setting. Fixed cell suspensions in PBS containing the desired number of cells are placed into an Eppendorf tube in half of the final volume (i.e., 0.5 ml). Warm 8% gelatin (0.5 ml) is poured into the tube, mixed with the cells and quickly cooled on ice (see Note 2). The same procedure should be repeated with the unlabeled cells that serve as a control.

3.3.2. Magnetic Resonance Imaging of Cell Phantoms

For in vitro measurements we use standard sequences that are used for in vivo measurements, i.e., a T2-weighted turbospin echo sequence with the following parameters: TR = 2,000 ms, effective echo time TE = 42.5 ms, turbo factor = 4, number of acquisitions AC = 16, matrix 256 × 256, slice thickness 0.5–1 mm. The field of view is chosen according to the size of the samples. For T2*-weighted images, a gradient echo sequence with the same geometry can be used (see Note 3). Parameters of the sequence: TR = 80 ms, TE = 5 ms, AC = 32.

3.4. In Vivo MRI

Superparamagnetic particles have a substantial impact on both T2 and T2* relaxation times. Therefore, both T2 (spin echo, turbospin echo) and T2*-weighted (gradient echo) sequences can be used. An implant with superparamagnetic particles will manifest itself as a hypointense area. Although T2*-weighted gradient echo is more sensitive to the presence of superparamagnetic particles, we strongly prefer a T2-weighted turbospin echo sequence, as it provides better anatomical images. The procedure may differ depending on the type (bore size, field strength, RF coil used, etc.) of MRI imager. The procedures and listed sequence parameters are suitable for a standard horizontal bore imager 4.7 T.

3.4.1. Animal Preparation

The animal should be anesthetized throughout the entire measurement. Simple MRI is relatively short, therefore intubation of the animal is not necessary and passive inhalation of the anesthetic is sufficient for maintaining anesthesia. However, either passive or active (intubated and connected to an air pump) inhalation of a suitable anesthetic can be used. The anesthetized animal is placed and fixed into a heated MR-compatible holder. An airflow of 200–300 ml per minute is maintained through a mouthpiece. The concentration of the inhalation anesthetic should be 1–2% depending on the breathing frequency of the animal. The breathing frequency is maintained at approx. one breath per second. Breathing is monitored throughout the entire experiment. A dedicated surface head coil is placed and fixed over the rat's head. A suitable surface coil is better than whole body resonators because of its higher sensitivity.

3.4.2. Magnetic Resonance Imaging

The holder with the animal is placed into the magnet bore. The RF coil must be connected and tuned. An automated procedure for shimming, setting of transmitter power, frequency and receiver gain should be sufficient for MR imaging; however, if necessary, fine tuning may be done manually. Low-resolution images are obtained in three orthogonal planes using a fast low-angle gradient echo (FLASH) sequence (flip angle 30°, repetition time TR = 100 ms, effective echo time TE = 5 ms, matrix 128 × 128, slice thickness 3 mm). The animal is repositioned if necessary. These images are used for positioning high resolution T2-weighted images. A sufficient number of slices to cover the entire volume of interest must be measured. Parameters of the sequence: TR = 2,000 ms, effective echo time TE = 42.5 ms, turbo factor = 4, number of acquisitions AC = 16, field of view FOV = 3.5 cm, matrix 256 × 256, slice thickness 0.5–1 mm, slice separation 1 mm. If the slice thickness is less than 1 mm, it is advisable to measure two sets of interleaved images to avoid saturation of neighboring slices during the measurement. This measurement can be supplemented by a T2*-weighted gradient echo sequence with the same geometry; however, the procedure will be more time consuming. Parameters of the sequence: TR = 80 ms, TE = 5 ms, AC = 32.

3.5. Staining for Iron (Prussian Blue Staining)

Fixed cells in culture wells or tissue slices are rinsed in distilled water, washed in alcohol (70%) for 2 min and twice rinsed in distilled water. Potassium ferrocyanide ($K_4[Fe(CN)_6]\cdot 3H_2O$) is applied for 30 min to produce ferric ferrocyanide (Prussian blue). Cell nuclei are counterstained with nuclear fast red, or hematoxylin can be used.

3.6. Discussion and Conclusion

The limitation of the MRI tracking of migrating iron oxide labeled cells is mainly the fact that in lesioned tissue hemorrhage contributes to T2-weighted hypointensity, thus interfering with the detection of labeled cells and complicating the interpretation of MR images. Hemorrhage products give rise to Prussian blue-positive deposits that are difficult to distinguish from iron-containing nanoparticles and may also be partially localized to macrophages. Therefore, it is important to determine whether cell labels remain co-localized with cell transplants, especially under pathological conditions.

With proper attention to the limitations described above, labeling cells with superparamagnetic agents would enable us to follow the migration of such cells when transplanted into humans, establish the optimal number of transplanted cells, define therapeutic windows and monitor cell growth and possible side effects.

4. Notes

1. *Cell labeling*: The concentration of iron can be decreased down to 15 μg of iron/ml of culture media. This positively influences cell viability; however, the labeling efficiency is lower (40–60%, depending on cell type and number of passage). Other SPIO nanoparticles, such as PLL- or D-Mannose coated, can be used in lower concentrations (15 μg of iron/ml of culture media) without adversely affecting viability or labeling efficiency. The presence of nanoparticles in the cell culture medium can be decreased to 48 h.

   We strongly recommend also using a second marker for detecting transplanted cells in histological slices. Particularly under pathological conditions, false Prussian blue positivity can result from deposits of iron in hemorrhages or in iron-containing proteins (hemosiderin).
2. *Preparation of phantoms*: Tubes with cells suspended in gelatin can be vortexed before cooling so the cells are spread evenly.
3. *MRI of phantoms*: Cell samples should be imaged at room temperature, otherwise the gelatin partially melts and the samples become inhomogeneous.

Acknowledgments

This work was supported by grants AV0Z50390703, KAN201110651, 1M0538, GACR 203/09/1242 and the EC – FP6 project DiMI: LSHB-CT-2005-512146.

References

1. Wang L, Li Y, Chen J, et al. (2002) Ischemic cerebral tissue and MCP-1 enhance rat bone marrow stromal cell migration in interface culture *Experimental hematology* **30**, 831–6.
2. Sordi V, Malosio ML, Marchesi F, et al. (2005) Bone marrow mesenchymal stem cells express a restricted set of functionally active chemokine receptors capable of promoting migration to pancreatic islets *Blood* **106**, 419–27.
3. Ji JF, He BP, Dheen ST, Tay SS. (2004) Interactions of chemokines and chemokine receptors mediate the migration of mesenchymal stem cells to the impaired site in the brain after hypoglossal nerve injury *Stem cells (Dayton, Ohio)* **22**, 415–27.
4. Birnbaum T, Roider J, Schankin CJ, et al. (2007) Malignant gliomas actively recruit bone marrow stromal cells by secreting angiogenic cytokines *Journal of neuro-oncology* **83**, 241–7.
5. Ip JE, Wu Y, Huang J, Zhang L, Pratt RE, Dzau VJ. (2007) Mesenchymal stem cells use integrin beta1 not CXC chemokine receptor 4 for myocardial migration and engraftment *Molecular biology of the cell* **18**, 2873–82.
6. Bonnemain A. (1996) Superparamagnetic and blood pool agents *Spotlight on clinical MRI. V.L.C.* Maffliers, pp. 75–88.
7. Syková E, Jendelová P. (2006) Magnetic resonance tracking of transplanted stem cells in rat brain and spinal cord *Neurodegenerative dis* **3**, 62–7.
8. Kalish H, Arbab AS, Miller BR, et al. (2003) Combination of transfection agents and magnetic resonance contrast agents for cellular imaging: relationship between relaxivities, electrostatic forces, and chemical composition *Magn Reson Med* **50**, 275–82.
9. Horák D, Babič M, Jendelová P, et al. (2009) The effect of different magnetic nanoparticle coatings on the efficiency of stem cell labeling. *J Magn Magn Mater* **321**, 1539–47.
10. Horak D, Babic M, Jendelova P, et al. (2007) D-Mannose-Modified Iron Oxide Nanoparticles for Stem Cell Labeling *Bioconjugate Chem.* **Mar 20**, E pub ahead of print
11. Babic M, Horak D, Trchova M, et al. (2008) Poly(L-lysine)-modified iron oxide nanoparticles for stem cell labeling *Bioconjug Chem* **19**, 740–50.
12. Babič M, Horák D, Jendelová P, et al. (2009) Poly(N,N-Dimethylacrylamide)-coated maghemite nanoparticles for stem cell labeling. *Bioconjugate Chem.*
13. Allen M, Bulte JW, Liepold L, et al. (2005) Paramagnetic viral nanoparticles as potential high-relaxivity magnetic resonance contrast agents *Magn Reson Med* **54**, 807–12.
14. Bulte JW, Hoekstra Y, Kamman RL, et al. (1992) Specific MR imaging of human lymphocytes by monoclonal antibody-guided dextran-magnetite particles. *Magn Reson Med* **25**, 148–57.
15. Jendelová P, Herynek V, Urdzíková L, et al. (2005) MR tracking of human CD34+ progenitor cells separated by means of immunomagnetic selection and transplanted into injured rat brain *Cell Transplantation* **14**, 173–82.
16. Jendelová P, Herynek V, Urdzikova L, et al. (2004) MR tracking of transplanted bone marrow and embryonic stem cells labeled by iron oxide nanoparticles in rat brain and spinal cord *J Neurosci Res*, 232–43.
17. Urdzikova L, Jendelova P, Glogarova K, Burian M, Hajek M, Sykova E. (2006) Transplantation of bone marrow stem cells as well as mobilization by granulocyte - colony stimulating factor promote recovery after spinal cord injury in rat *J Neurotrauma* **23**, 1379–91.
18. Anderson SA, Glod J, Arbab AS, et al. (2005) Noninvasive MR imaging of magnetically labeled stem cells to directly identify neovasculature in a glioma model *Blood* **105**, 420–5.
19. Kriz J, Jirak D, Girman P, et al. (2005) Magnetic resonance imaging of pancreatic islets in tolerance and rejection *Transplantation* **80**, 1596–603.

Chapter 6

Imaging of Schwann Cells In Vivo

Rahul Kasukurthi and Terence M. Myckatyn

Abstract

The ability to examine cells tagged with fluorescent proteins in vivo has led to exciting advances in molecular neurobiology. The integral role of Schwann cells in nerve regeneration is well characterized, but not until recently has dynamic imaging of these critical cells been possible. Unlike many static techniques, in vivo imaging of Schwann cells tagged with green fluorescent protein (GFP) allows for serial imaging and may provide important insights into their role in peripheral nerve regeneration. In this chapter, we describe a protocol for in vivo imaging of SC migration into nonfluorescent nerve grafts using transgenic *S100-GFP* mice.

Key words: In vivo imaging, Schwann cell migration, Green fluorescent protein

1. Introduction

The development of sophisticated transgenic animal models coupled with modern optics has allowed examination of neuropathology in vivo (1). The assessment of nerve regeneration has traditionally depended on time-consuming static imaging techniques such as immunohistochemistry, retrograde labeling, and electron microscopy, all of which fail to capture the dynamism of the regenerating nerve. Frequently, these static techniques are mutually exclusive, allowing the use of only one modality per specimen thus leading to increased numbers of experimental animals (2, 3). In vivo serial imaging allows monitoring of nerves over time, giving unique insight into the migration of cells in response to nerve injury.

Schwann cells (SCs) have a critical role in the peripheral nerve injury response. They quickly undergo mitosis, engulfing disintegrating myelin, releasing a variety of neurotrophic factors, and proliferating to form the scaffold that is essential for proper regeneration (4, 5).

Marie-Dominique Filippi and Hartmut Geiger (eds.), *Stem Cell Migration: Methods and Protocols*,
Methods in Molecular Biology, vol. 750, DOI 10.1007/978-1-61779-145-1_6,

Transgenic mice with fluorescently labeled SCs provide an exciting tool for exploring the behavior of these cells in response to axonal injury. *S100-GFP* mice possess SCs tagged with green fluorescent protein (GFP) driven by upstream sequences of the human *S100B* gene. Unlike some founder lines, *S100-GFP* mice are ideal for monitoring Schwann cell (SC) migration because of their fairly specific localization of GFP to SCs and relatively sparse labeling of macrophages (6). The ability to track the migration of SCs in vivo provides a novel methodology allowing visualization of these crucial cells for the first time in a living animal.

In this chapter, we describe a protocol for in vivo imaging of SC migration into nonfluorescent nerve grafts using transgenic *S100-GFP* mice.

2. Materials

2.1. Surgical Exposure and Grafting

1. Transgenic *S100-GFP* mice (Jackson Laboratory, Bar Harbor, Maine) (see Note 1) bred on a C57BL/6 background.
2. C57BL/6 mice for nonfluorescent sciatic nerve isografts.
3. Ketamine hydrochloride (Ketaset®, Fort Dodge Animal Health).
4. Medetomidine hydrochloride (Dormitor®, Orion Corporation).
5. Nair® hair removal lotion for legs (Church and Dwight Co.).
6. Sterile gauze pads.
7. Betadine pads.
8. Magnetized retractors, metal mounting plate.
9. Appropriate micro and macrosurgical instruments: Macro- and microspring scissors, Jeweler's forceps, scalpel, needle driver.
10. 11-0 sterile nylon sutures (Surgical Specialties Co).

2.2. In Vivo Imaging

1. Wide field fluorescence dissecting microscope (e.g., Olympus MVX-10; Nikon SMZ-1500, Nikon Instruments) with Z-stage (Applied Scientific Instrumentation) (Fig. 1).
2. Light source (e.g., Exfo Excite 120 Metal Halide with liquid light guide).
3. Fluorescent filter cube turret (e.g., Olympus). A shutter is a useful tool to prevent quenching of endogenous immunofluorescence. Prolonged periods of imaging will quench chromophore emission, thereby artificially reducing immunofluorescence.

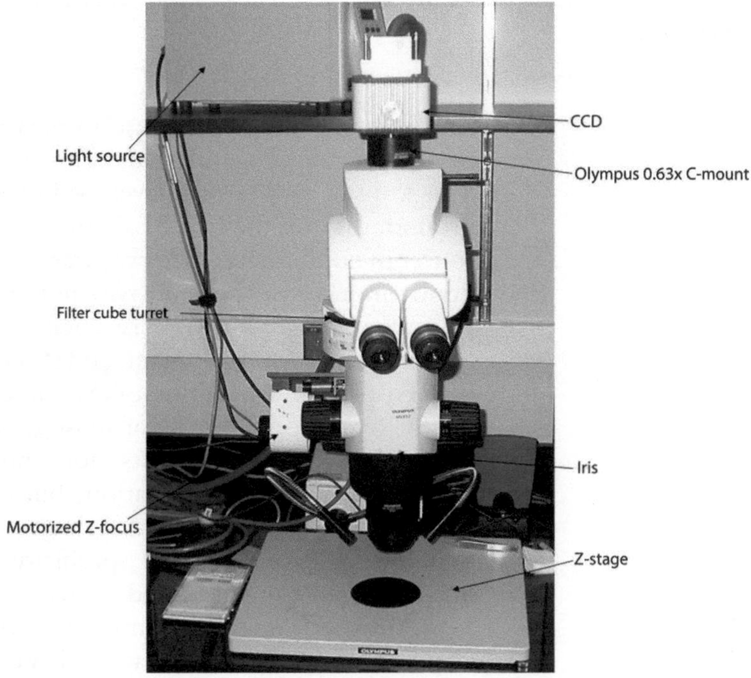

Fig. 1. A typical wide field fluorescence dissecting microscope is shown, with various components illustrated.

4. Cooled charge-coupled device (CCD) (e.g., Hamamatsu Orca-AG monochrome; CoolSNAP-ES, Photometrics; Sensicam).
5. Image acquisition and analysis software (e.g., MetaMorph®, Molecular Devices).
6. Graphics editing program (e.g., Adobe Photoshop).
7. Data analysis software (Microsoft® Excel; Statistica).
8. Evicomposite polymer resin embedded with quantum dots for image standardization (Evidots, Evident Technologies). These emit stable nanocrystalline fluorescence at a wavelength of 493 nm.

2.3. Surgical Recovery

1. Sterile normal saline.
2. 6-0 and 8-0 nylon sutures (Ethicon, NJ).
3. Atipamezole hydrochloride (Antisedan®, Orion Corporation).
4. Buprenorphine (Hospira Inc.).
5. Warming pad.

3. Methods

Implantation of a nonfluorescent nerve isograft into a *S100-GFP* mouse provides a useful model for studying SC migration in vivo. This transgenic line was developed on a C57BL/6 mouse background, making the C57BL/6 mouse an ideal nonimmunogenic source of nonfluorescent nerve grafts (see Note 2). Each C57BL/6 donor mouse can provide two nonfluorescent nerve grafts for placement into *S100-GFP* recipients. Meticulous microsurgical technique, proper equipment, and strict standardization of images help ensure the acquisition of experimentally useful data describing SC migration. The use of an isograft is ideal since rejection of a dysgeneic allograft results not only in compromised nerve regeneration and SC migration, but also elevated macrophage recruitment (2). Some macrophages express GFP in this line thereby compromising the specificity with which GFP-labeled Schwann cells can be tracked (2, 6).

Ultimately, relative quantification of fluorescence intensity of acquired images provides data which can be analyzed statistically to compare differences between groups (2). Although any good image acquisition and analysis software will suffice, our protocol employs the "linescan" command in MetaMorph® Software to quantify and analyze pixel intensity over time. Images are standardized using a 12-bit image on a grayscale ranging from 0 to 4,095. This technique, however, is still in its developmental stages and is subject to considerable interoperator variability. Therefore it is critical that any lab attempting this technique possess quality equipment and develop a strict set of protocols to reduce variability. Further, it is advisable that only one operator be used during the course of a study in order to eliminate interrater variability. In addition, we suggest in vivo imaging and quantification of SC migration not be the sole outcome measure of any experiment. Despite the promise of in vivo imaging, supplementation with static imaging techniques and functional outcome measures is ideal for confirming experimental findings.

3.1. Surgical Exposure and Grafting

1. For preoperative anesthesia induction (see Note 3) inject the C57BL/6 donor mouse with ketamine hydrochloride (50 μg/g (male) or 75 μg/g (female)) once intraperitoneally (IP) and medetomidine hydrochloride (1 μg/g) once IP.
2. Clip fur on posterolateral aspect of the femur.
3. Apply lacrilube to eyes to prevent exposure keratitis.
4. Using a cotton swab, place thin layer of Nair® and leave for 3 min before wiping off. Remove any loose or excess adherent hairs from surgical site (see Note 4).
5. Sterilize surgical area with betadine pads and allow to dry.

6. With the animal on its side, make a skin incision parallel and 2 mm posterolateral to the femur. Use a gluteal muscle-splitting approach to avoid trauma to the muscle and minimize bleeding (see Note 5). Using microsurgical instruments, develop a field of view that includes the sciatic notch proximally, and the trifurcation of the sural, peroneal, and tibial nerves distally (see Note 6).
7. Using microscissors, transect the nerve above the sciatic trifurcation distally, and again as proximally as possible to harvest a long graft. Place gauze moistened with sterile saline on top of the graft in order to prevent desiccation.
8. Anesthetize, prepare, and expose *S100-GFP* recipient mouse as described in steps 1–6. Sharply transect the sciatic nerve 5 mm above the distal trifurcation.
9. Trim the harvested graft down to the desired experimental length (we have used 1 cm in the past). Interpose the graft in a reversed orientation between the transected ends of the recipient mouse nerve using 11-0 nylon interrupted sutures. Four to six sutures may be used as needed.
10. Harvest contralateral sciatic nerve from C57BL/6 donor mouse by repeating steps 2–7. Interpose graft into second recipient mouse as in steps 1–6, and 9.
11. Sacrifice the C57BL/6 donor mouse via cervical dislocation or any other approved methodology (e.g., intracardiac pentobarbital injection).
12. The *S100-GFP* mice are now ready for in vivo imaging and analysis.

3.2. In Vivo Imaging

1. Ensure a clear surgical field of view; this can be accomplished by removing excess blood with a cotton swab, trimming away any connective tissue that may be overlying the sciatic nerve, and using retractors to pull away obscuring tissue.
2. Place the animal on the *z*-stage. Using a low power objective and bright field illumination, orient the animal so that nerve is visible both proximally and distally to the coaptation sites.
3. Use the turret on the fluorescent filter cube to change to the GFP filter (488 nm) (see Note 7), and turn off any bright field illumination. Take baseline images using low power magnification to document landmarks that may be useful later for serial imaging (see Notes 8 and 9). For the $t=0$ day images, the nonfluorescent graft should appear completely dark.
4. Switch to higher magnification as desired in order to obtain experimental images for analysis; keep the graft and adjacent nerve in view. This will reduce the depth of field, and may require taking a *z*-series of images if the desired length of

nerve is blurry (see Note 10). In MetaMorph®, use the "Acquire→Z-series" commands to obtain stacks of images. Once all desired slices have been obtained, use the "Process→Stack Arithmetic→Best Focus" commands to obtain a single image.

5. Once a suitable image has been saved, turn off the light source. The saved image will now undergo line-scanning analysis for pixel intensity. Use the "Measure→Linescan" command in MetaMorph® to accomplish this.

6. The Linescan dialog box should appear, giving a graph showing gray-pixel intensity on the *y*-axis and distance on the *x*-axis. See Fig. 2.

7. Select "open log" (lower left-hand corner), and then log measurement to a dynamic data exchange (DDE) file. A Microsoft Excel file will open. Then select "log data" to transfer values to the spreadsheet (see Note 11).

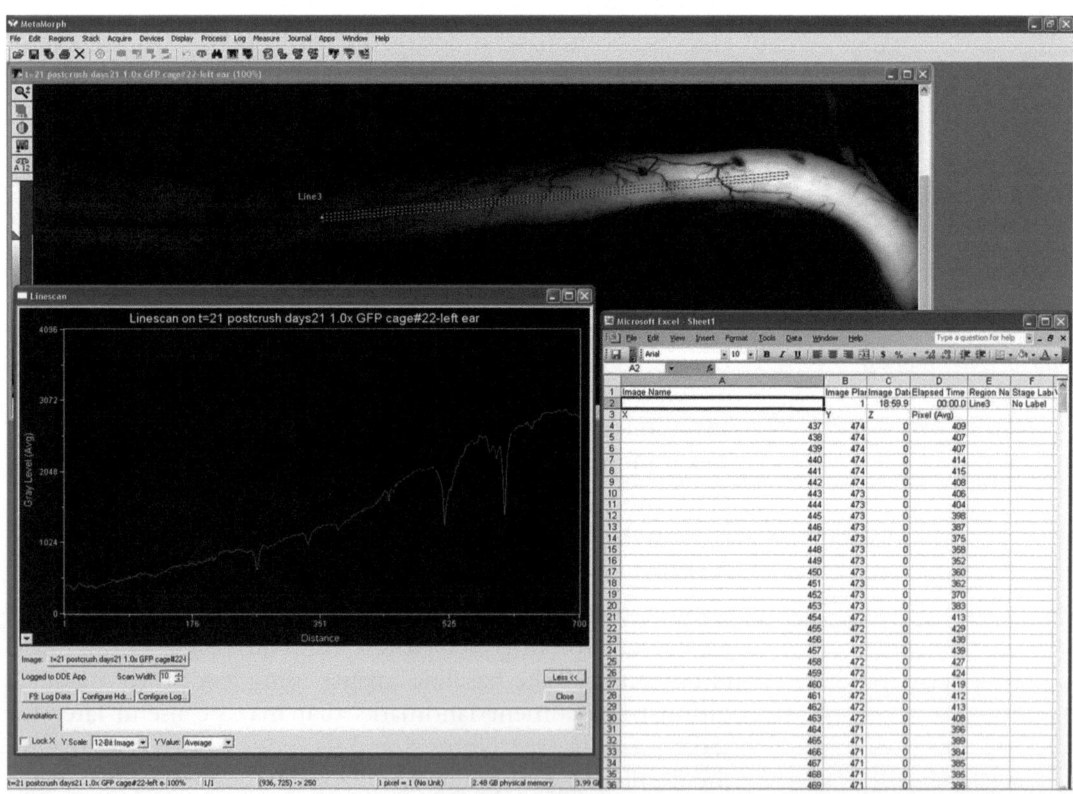

Fig. 2. *Linescan command, MetaMorph®*. A *line* is drawn across the area of interest, which results in the *gray graph* giving gray-levels in pixel intensity (0–4,096) plotted against distance on the *x*-axis. This data can be logged into an Excel file for analysis. Our lab typically breaks the graft into smaller segments, generating pixel intensity data for each individual segment which can be plotted over time.

8. Values should now appear in the spreadsheet. *X*-values represent distance along the graft, while another column gives average pixel intensity.
9. The raw data can be graphically displayed as desired by the investigator, and statistical analysis performed. See Fig. 3.

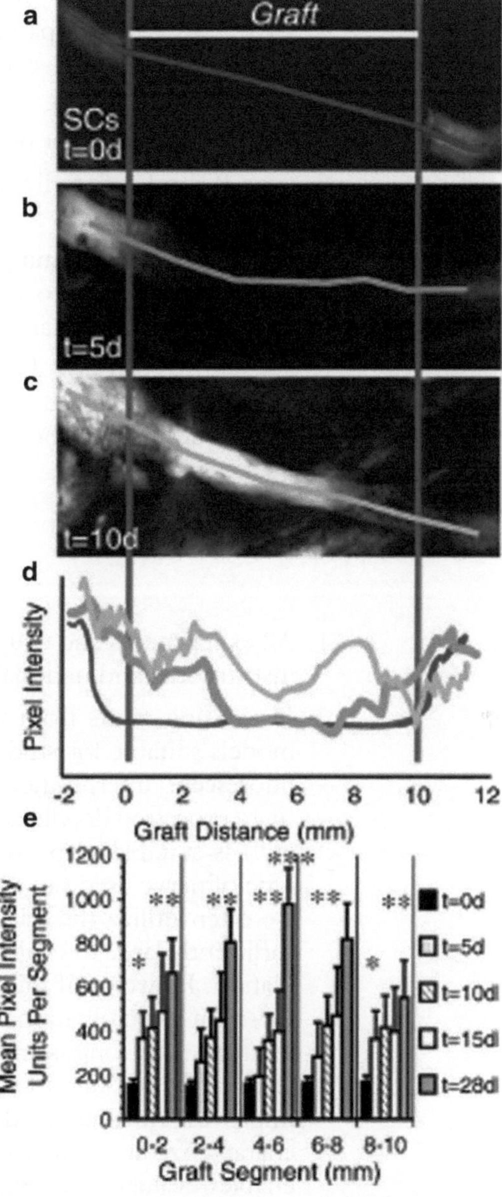

Fig. 3. (**a**) A schematic showing one possible format for presenting data. A transgenic S100-GFP mouse received a 10 mm nonfluorescent graft at $t=0$. As expected, no fluorescence is seen initially in the graft. (**b–c**) At later time points increased fluorescence is seen throughout the graft. (**d**) Fluorescent intensity is presented in graph form according to time point. (**e**) 12-Bit digitized images of the 10 mm-long graft are divided into five segments, and a mean pixel intensity (0–4,096) determined for each segment is obtained for all time points. Reprinted from Ref. (2).

3.3. Surgical Recovery

1. Remove animal from *z*-stage and return to stereomicroscope.
2. Irrigate wound with 2–3 mL of sterile normal saline. This decreases infection risk and moistens exposed nerve and graft after lengthy exposure.
3. Remove retractors from the surgical wound, and close muscle and skin in layers using 6-0 and 8-0 nylon sutures.
4. Give 1 μg/g of atipamezole hydrochloride subcutaneously once for postoperative anesthesia reversal.
5. Give 0.1 μg/g of buprenorphine subcutaneously every 8–12 h as needed for postoperative pain control.
6. Give 0.5 mL of sterile normal saline subcutaneously to prevent dehydration.
7. Place animal on warming pad. Ensure anesthesia is sufficiently reversed for animal to access food and water in its cage before transferring back to animal room.
8. The animal can now be kept alive for serial imaging sessions in the future, or can be sacrificed for various static imaging techniques to corroborate in vivo imaging data.

4. Notes

1. All experiments and drug usage should be in accordance with institutional and national guidelines.
2. Harvesting grafts from C57BL/6 mice is only one of many models suitable for studying SC migration in vivo. Any non-fluorescent interposition material with low immunogenicity (e.g., processed decellularized allograft, cold preserved allograft, etc.) is suitable, depending on the investigator's needs. The type of nerve used can also be adjusted to experimental needs; we often utilize the sciatic nerve in our laboratory since it is sufficiently large to enable microsurgical reconstructive manipulation. However, if this is not a requirement of the investigator, then the evaluation of smaller nerves, such as those in the levator aurus longus muscles near the ears, is advisable as it enables the tracking of significantly fewer SCs and axons. This improves the chances that an investigator can visualize and characterize individual cells. Evaluation of the sciatic nerve enables evaluation of SCs en masse but is far less reliable for discriminating between individual SCs.
3. Any approved anesthesia protocol is acceptable (e.g., inhalational anesthesia).
4. Diligent removal of hair from the surgical site reduces artifact and blurring during subsequent imaging steps. Nair® should

not be placed on the skin for more than 10 min, as this can cause bleeding and irritation. Also, take care to avoid the eyes when applying.

5. If excessive bleeding is encountered, apply pressure and tie off vessels as needed. Electrocautery should be avoided to prevent nerve injury that can cause fluorescent debris, which would hinder optimal imaging.
6. Repeat imaging episodes invariably lead to increased scarring that impairs evaluation by obscuring fluorescent light transmission to the nerve in question. To overcome this issue, meticulous technique, placement of nerve over muscle or fat rather than skin, and avoidance of infection can reduce, but not completely eliminate the development of scar tissue.
7. If double transgenic mice with two chromophores are being evaluated, one can use dual band filter cubes to simultaneously stimulate both chromophores, thereby reducing imaging time and quenching. Selection of chromophores with maximally disparate excitation and emission spectra is advised in order to reduce reliance on unmixing software.
8. High quality image acquisition depends on standardized light levels and orientation of the surgical site. Take care to fully darken the room, and ensure that the sciatic nerve is exposed in as horizontal a plane as possible. This will help minimize blurring, especially at high magnifications. Furthermore, the center of the CCD may be more sensitive than in the periphery; therefore, it is important to keep the image centered.
9. Experimental images should be standardized for magnification and exposure time. These values may vary depending on the transgenic line, microscope used, ambient light levels, and various other factors. The important point is to use the same magnification and length of exposure throughout your experiment. Evidots emit a stable fluorescence at a wavelength of 493 nm and can be used as a visual aid to standardize images.
10. Imaging in the lower extremity minimizes the issue of motion artifact due to respirations, but slight motion due to pulsations of the femoral and genicular arteries can still be noted at high magnifications. Many of the smaller nerves used to discriminate individual SCs are located in the head and neck. In these cases, it may be necessary to intubate and ventilate an experimental mouse. Respirations can then be arrested for a few seconds to facilitate motion-free imaging. We have found, however, that this increases morbidity. While this technique is rarely required in our laboratory, it is effectively utilized and described by others (7, 8).
11. Over time, the measured length of the nerve graft in pixels will vary slightly due to factors such as graft contraction due

to scar formation, tethering to underlying muscle by scar and neovascularization, hindlimb motion with changes in the lie of the graft over time, and intrarater variability with respect to how the line scan is drawn. This difference should not exceed 5–10% between imaging sessions, and greater variability suggests the need to improve minimally invasive surgical technique and the imaging setup. To account for this variation, we typically subdivide the nerve graft into several segments, measure the mean pixel intensity within a given subdivision, and then compare mean pixel intensity within that subdivision over time. For example, a 10 mm nerve graft can be subdivided into five equal 2 mm segments designated as subdivisions A–E. Mean pixel intensity is then calculated in "A" at the various time points and compared between time points or treatment groups.

References

1. Lichtman, J.W., and Conchello, J.A. (2005) Fluorescence microscopy *Nat Methods* **2**, 910–9.
2. Hayashi, A., Koob, J.W., Liu, D.Z., et al. (2007) A double-transgenic mouse used to track migrating Schwann cells and regenerating axons following engraftment of injured nerves *Exp Neurol* **207**, 128–38.
3. Misgeld, T., and Kerschensteiner, M. (2006) In vivo imaging of the diseased nervous system Nat Rev Neurosci **7**, 449–63.
4. Bhatheja, K., and Field, J. (2006) Schwann cells: origins and role in axonal maintenance and regeneration *Int J Biochem Cell Biol* **38**, 1995–9.
5. Bunge, R.P. (1994) The role of the Schwann cell in trophic support and regeneration J *Neurol* **242**, S19–21.
6. Zuo, Y., Lubischer, J.L., and Kang, H., et al. (2004) Fluorescent proteins expressed in mouse transgenic lines mark subsets of glia, neurons, macrophages, and dendritic cells for vital examination J *Neurosci* **24**, 10999–1009.
7. Misgeld, T., Nikic, I., and Kerschensteiner, M. (2007) In vivo imaging of single axons in the mouse spinal cord *Nat Protoc* **2**, 263–8.
8. Keller-Peck, C.R., Walsh, M.K., Gan, W.B., Feng, G., Sanes, J.R., and Lichtman, J.W. (2001) Asynchronous synapse elimination in neonatal motor units: studies using GFP transgenic mice *Neuron* **31**, 381–94.

Chapter 7

Imaging of Embryonic Stem Cell Migration In Vivo

Andrew S. Lee and Joseph C. Wu

Abstract

Conventional reporter gene technology and histological methods cannot routinely be used to track the in vivo behavior of embryonic stem (ES) cells longitudinally after cellular transplantation. Here we describe a protocol for monitoring the in vivo survival, proliferation, and migration of ES cells without necessitating animal sacrifice. Stable ES cell lines containing double fusion (DF; enhanced green fluorescent protein and firefly luciferase) or triple fusion (TF; monomeric red fluorescent protein, firefly luciferase, and herpes simplex virus thymidine kinase) reporter genes can be established within 4–6 weeks by lentiviral transduction followed by fluorescence-activated cell sorting. The cell fate and behavior of these DF or TF ES cells can subsequently be tracked noninvasively by bioluminescence and microPET imaging for a prolonged period of time.

Key words: Embryonic stem cells, Molecular imaging, Bioluminescence imaging, Positron emission tomography imaging

1. Introduction

Embryonic stem (ES) cells offer exciting promises as therapeutic donor cells for regenerative medicine. Unlike adult stem cells, ES cells can differentiate into any somatic cell of the human body and have the capacity for unlimited self renewal (1). To fully realize the therapeutic potential of ES cells, however, it is important for investigators to understand the in vivo behavior of transplanted cells following administration. At present, a number of hurdles have hindered the effective translation of ES-cell-based therapy to the clinic. These issues include teratoma formation, immune rejection, failure of cells to engraft, and cellular migration outside the area of administration. The development of sensitive and accurate methods to track the survival, proliferation, and migration of ES cells or ES cell derivatives will thus be

Marie-Dominique Filippi and Hartmut Geiger (eds.), *Stem Cell Migration: Methods and Protocols*,
Methods in Molecular Biology, vol. 750, DOI 10.1007/978-1-61779-145-1_7, © Springer Science+Business Media, LLC 2011

indispensable for future applications of ES cell therapy in human patients.

In this chapter, we describe a novel method of monitoring in vivo ES cell behavior through the use of bioluminescence imaging (BLI) and positron emission tomography (PET) reporter genes. In these types of cellular imaging, a reporter gene coding for the synthesis of a detectable protein is stably introduced into the genome of a target cell or tissue via a lentiviral vector carrying a constitutively active promoter such as ubiquitin that drives reporter gene expression. Subsequent cell-mediated synthesis of the reporter protein produces a probe that can interact with an exogenously administered substrate to generate a detectable signal (Fig. 1). In the case of BLI, the reporter gene introduced is firefly luciferase (Fluc). Interaction of Fluc with its substrate D-luciferin catalyzes production of the optically active metabolite oxyluciferin which emits low-intensity photons that can be imaged with a cool-charged couple device (CCD) camera for cell localization. In PET imaging, the reporter protein herpes simplex virus thymidine kinase (HSVtk) phosphorylates its substrate, the PET reporter probe 9-4-[^{18}F]fluoro-3-(hydroxylmethylbutyl) guanine ([^{18}F]-FHBG), to produce high-energy photons that can be captured by a PET camera. Our group has used both a double fusion (DF) construct containing enhanced green fluorescent protein (eGFP) and Fluc reporter genes, and a triple fusion (TF) construct

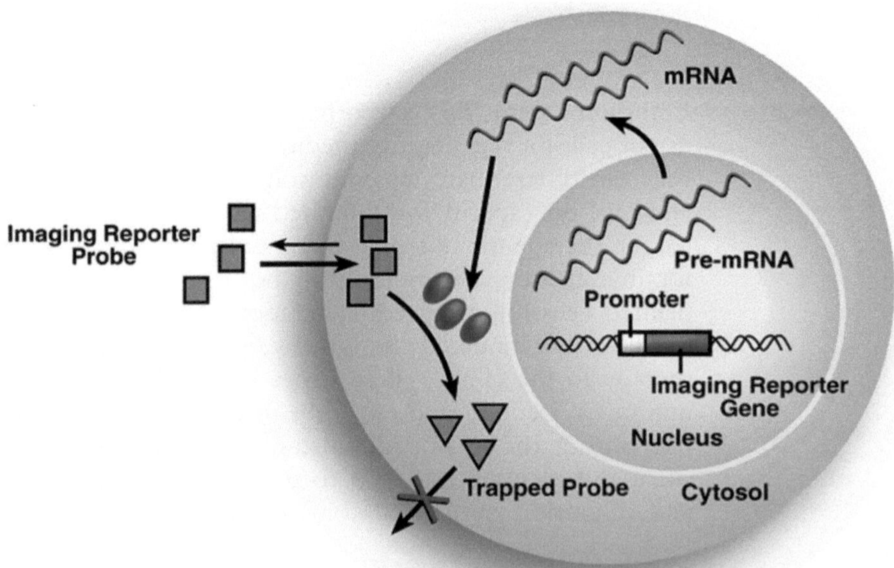

Fig. 1. Conceptual basis of reporter gene imaging. Reporter genes are stably introduced into the genomes of target cells or tissues via lentiviral transduction. Transcription of the reporter gene into mRNA and subsequent translation produces a detectable reporter protein that interacts with an administered reporter probe to generate signal.

containing monomeric red fluorescent protein (mRFP), Fluc, and HSVtk reporter genes to stably transduce ES cells for reporter gene imaging.

As compared to conventional methods of assaying cell fate such as histological analysis and staining for GFP or β-galactosidase (LacZ), reporter gene imaging allows for noninvasive and longitudinal visualization of the spatio-temporal kinetics of cell engraftment and survival in living subjects without requiring animal sacrifice. Because cellular transcription and translation must be intact for synthesis of the reporter proteins, only cells that are alive generate positive signal. In addition, genetic inheritance of the reporter gene from mother to daughter cell permits longitudinal monitoring of cellular proliferation and misbehavior (e.g., teratoma formation). Previous studies have shown that expression of the reporter genes do not significantly impact ES cell viability, proliferation, and differentiation (2–4). Our laboratory has applied reporter gene imaging to successfully monitor the in vivo survival, migration, and proliferation of transplanted ES cells (4–7) and ES cell derivatives such as cardiomyocytes (8) and endothelial cells (9, 10) over a prolonged period of time.

2. Materials

2.1. Production of DF or TF Lentiviral Transduction Vector

1. Plasmids: eGFP-Fluc double fusion (DF) construct or mRFP-Fluc-HSVtk triple fusion (TF) construct (available upon request) (Fig. 2a). The psPAX2 packaging plasmid and pMD2G envelope plasmid can be purchased from Addgene.
2. HEK 293-FT cells (Invitrogen).
3. HEK 293-FT cell medium: Dulbecco's modified Eagle's medium (DMEM) with high glucose and L-glutamine, supplemented with 10% fetal bovine serum.
4. Opti-MEM I medium.
5. Lipofectamine 2000.
6. 5× PEG-it Lentivirus Concentration Solution (System Biosciences).

2.2. Lentiviral Transduction of ESCs

1. Irradiated mouse embryonic fibroblasts (MEFs).
2. A mouse ES cell line such as D3 (ATCC) or human ES cell line such as H7 or H9 (National Stem Cell Bank, Madison, WI).
3. MEF medium: DMEM with high glucose with L-glutamine, supplemented with 10% FBS and 5% penicillin/streptomycin.
4. Mouse ES cell medium: DMEM with high glucose with L-glutamine, supplemented with 15% FBS, 2 mM L-glutamine,

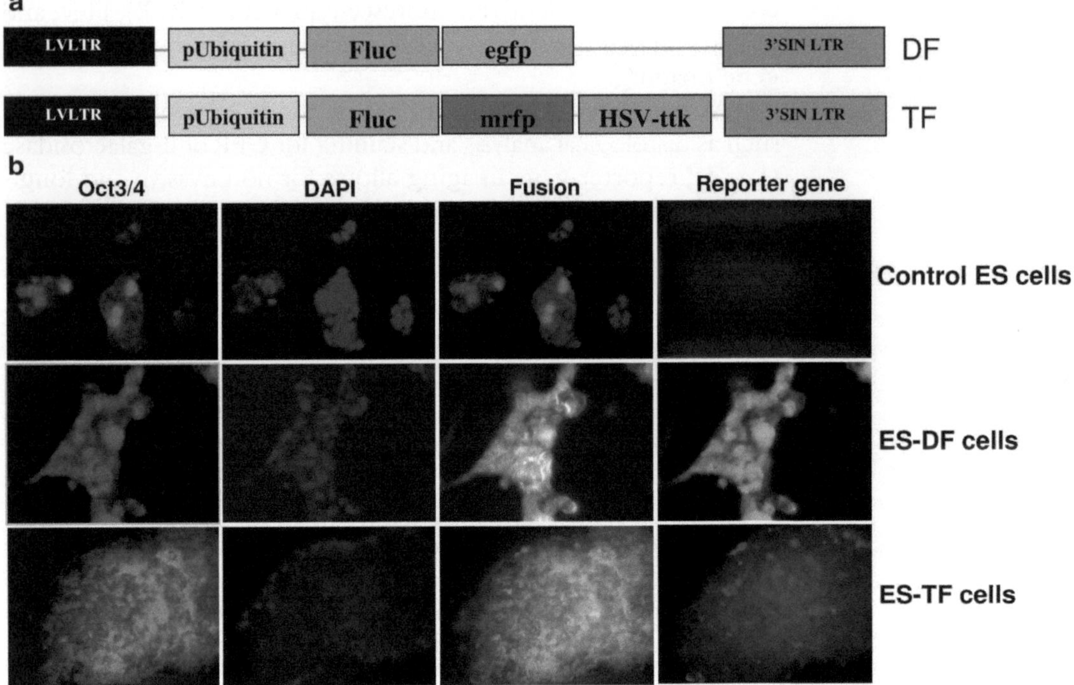

Fig. 2. Characterization of reporter constructs and stably transduced ES cells. (**a**) Schematic of the DF and TF reporter constructs cloned into a self-inactivating (SIN) lentiviral vector carrying the ubiquitin promoter. (**b**) Fluorescence microscopy reveals robust expression of mouse ES cell markers (Oct4) in control and stably transduced DF and TF ES cells. Control ES cells are negative for reporter gene expression, whereas DF and TF cells exhibit robust expression of eGFP and mRFP, respectively. (Reproduced from Ref. (7) with permission from Mary Ann Liebert, Inc).

0.1 mM β-mercaptoethanol, 0.1 mM nonessential amino acids, 1,000 IU/ml leukemia inhibitory factor (LIF).

5. Human ES cell medium: DMEM/F-12 (1:1) (Invitrogen, Gibco), supplemented with 20% Knockout Serum Replacer (Invitrogen, Gibco), 1% nonessential amino acids, 2 mM L-glutamine, 0.1 mM β-mercaptoethanol, and 4 ng/ml basic fibroblast growth factor (bFGF).
6. Gelatin: Prepare as a 0.1% solution and autoclave prior to use.

2.3. Derivation of Stable DF or TF ES Cell Lines

1. Phosphate buffer solution (PBS) without Ca^{2+} and Mg^{2+}.
2. Cell dissociation buffer.
3. 70 μm cell strainer.
4. Propidium iodide buffer.
5. Human ES cell medium + antibiotics/mycotics: add 1% penicillin/streptomycin, and 0.25 μg/ml fungizone to normal ES cell medium.
6. Mouse ES cell medium + antibiotics/mycotics: add 1% penicillin/streptomycin and 0.25 μg/ml fungizone to normal ES cell medium.

7. Collagenase IV.
8. Cell scraper.

2.4. Verification of Transduction Through In Vitro Imaging

1. D-luciferin (Biosynth): Prepare working solution at 45 mg/ml by dissolving 1 g of D-luciferin in 22 ml of PBS and separating into 1–1.5 ml aliquots. Aliquots can be frozen at –20°C for up to 6 months.
2. Xenogen In Vivo Imaging System (IVIS, Xenogen Corporation).

2.5. Transplantation of DF or TF ES Cells into Animal Models

1. Growth factor reduced, LDEV-free ES cell compatible Matrigel (BD).
2. Portable anesthesia system (Molecular Imaging Products).
3. 28.5-Gauge insulin syringe.

2.6. Longitudinal Imaging of Transplanted Cells Using BLI

1. Xenogen IVIS machine with portable anesthesia system.
2. D-luciferin prepared as above (1 g D-luciferin dissolved in 22 ml PBS).
3. 28.5-Gauge insulin syringe.
4. BLI acquisition and analysis software package such as Living Image 3.1 (Caliper Life Sciences).

2.7. Longitudinal Imaging of Transplanted Cells Using PET

1. [^{18}F]FHBG: A cyclotron is needed to produce [^{18}F]FHBG. If your institution does not have a cyclotron facility, order from a cyclotron facility with expertise in [^{18}F]FHBG synthesis.
2. Dose calibrator (Capintec).
3. Portable anesthesia machine.
4. A MicroPET scanner such as the MicroPET rodent R4 (Concorde Microsystems).
5. A MicroPET acquisition and analysis software package such as ASI Pro (Concorde Micorystems).

3. Methods

The following steps describe the derivation of murine and human ES cell lines that stably express DF (eGFP-Fluc) and TF (mRFP-Fluc-HSVtk) reporter genes, as well as imaging protocols to monitor the survival, migration, and proliferation of these cells. Because the emphasis of this chapter is on reporter gene transduction and imaging, we have not included instructions on how to make the reporter gene plasmid constructs used in Subheading 3.1 below. For a description of the PCR and standard molecular cloning techniques used to make these constructs please refer to De et al. (11), Ray et al. (12), and Cao et al. (4).

3.1. Production of DF or TF Lentiviral Transduction Vector

1. Plate approximately 5×10^6 HEK 293-FT cells onto a 100-mm culture dish using 10 ml of HEK 293-FT culture medium. Incubate at 37°C overnight.
2. After a minimum of 12 h, combine 60 µl of Lipofectamine 2000 with 1.5 ml Opti-MEM I medium in a 15 ml conical centrifuge tube. Incubate at room temperature for 5 min.
3. While the diluted Lipofectamine is incubating, mix 12 µg of the DF or TF plasmid construct, 8 µg of the psPAX2 packaging plasmid, and 4 µg of the pMD2G envelope plasmid with 1.5 ml of Opti-MEM I medium in a separate 15-ml conical centrifuge tube.
4. Combine the diluted Lipofectamine solution with the plasmid mixture and pipet gently to homogenize. Incubate the solution at room temperature for 20 min.
5. Remove the 100 mm HEK 293-FT dish from the incubator and add the plasmid:Lipofectamine 2000 mixture into the culture medium dropwise. Incubate the plate at 37°C for 6 h.
6. After 6 h, aspirate the transduction medium and add 10 ml of fresh HEK 293-FT growth medium. Incubate the cells at 37°C for 24 h.
7. The next day, the culture medium will be saturated with virus released by the HEK 293-FT cells. Collect the transduction medium and place it in a 50-ml conical centrifuge tube for storage at 4°C. Add 10 ml fresh HEK 293-FT growth medium to the cells. Incubate the cells at 37°C again for 24 h.
8. After 24 h, collect the culture medium again from the plate and combine it with the transduction medium isolated from the previous day. The total volume should be close to 20 ml. Centrifuge the combined transduction medium at $3,000 \times g$ for 15 min at 4°C.
9. Pass the supernatant through a 0.45-µm filter into a separate 50-ml conical centrifuge tube and mix with 5 ml 5× PEG-it Lentivirus Concentration Solution. Incubate at 4°C for at least 12 h. This mixture can be stored at 4°C for up to 2 weeks.
10. When ready to make the final lentivirus, centrifuge the viral particle:PEG-it mixture at $1,500 \times g$ for 30 min at 4°C. The viral particles will form a pellet. Aspirate the supernatant and centrifuge a second time at $1,500 \times g$ for 5 min at 4°C. Remove any remaining supernatant. Resuspend the pellet in 200 µl of Opti-MEM I medium. Once the lentiviral vector is created, it may be used immediately or stored at −80°C for up to 3 months.

3.2. Lentiviral Transduction of ES Cells

1. Prepare a six-well plate with a feeder layer of inactivated MEFs by coating all six wells of a six-well plate with 2 ml of autoclaved 0.1% gelatin. Incubate the plate at room temperature for at least 15 min.
2. Plate ~3×10^5 inactivated MEFs in 2 ml of MEF growth medium in each well of the six-well plate. Incubate at 37°C for 24 h to allow for the feeder cells to firmly attach to the bottom of the dish.
3. After 24 h, verify the feeder layer has firmly attached by looking under a light microscope and aspirate the MEF medium. Plate ~6×10^5 human or mouse ES cells in each well suspended in 2 ml of the appropriate ES cell medium. Incubate at 37°C for 24 h.
4. The next day, replace the culture medium with fresh ES cell medium. Look at the cells under a light microscope to measure confluence. If the cells have are close to confluence, proceed to step 5. If not, replace the existing medium with 2 ml of fresh ES cell medium and check on the cells again the next day. Mouse ES cells will typically reach confluence 2–3 days after plating. Human ES cells may take up to 4–5 days. For optimal transduction, plates should be at least at 80% confluent and colonies should be no larger than 200–400 cells per colony.
5. Once the cells have reached confluence, prepare the lentivirus transduction medium in a 15 ml conical centrifuge tube. Thaw the viral stock made in Part 3.1 above and add it at a multiplicity of infection (MOI) of 10 with 5 μL of polybrene and 2 ml of fresh ES cell medium for each well of the six-well plate to be transduced.
6. Aspirate the ES cell medium from each well of the six-well plate and replace it with 2 ml of the transduction medium. Incubate at 37°C for 24 h.
7. At 24 h, aspirate the transduction medium and replace it with 2 ml of fresh ES cell medium per well. Continue to incubate the cells at 37°C for a second day, and change the cell medium again.
8. 2 days after transduction, look at the cells under an inverted epifluorescence microscope to verify successful uptake and expression of the reporter gene. Estimate the percentage of cells that are GFP or RFP positive. Typically, at least 30–40% of total cells must be GFP or RFP positive for successful cell cytometry sorting (Fig. 2b).

3.3. Derivation of Stable DF or TF ES Cell Lines

1. The day before cell sorting, prepare a 12-well plate with a MEF feeder layer by incubating 1 ml of 0.1% autoclaved gelatin in each well at room temperature for a minimum of 15 min. Aspirate the gelatin and seed 3×10^5 MEFs per well in 1 ml of MEF medium. Incubate at 37°C for 24 h.

2. The next day, remove the transduced ES cells from the incubator and aspirate the culture medium. Wash the cells with 1 ml of PBS per well and aspirate. Add 300 μl of cell dissociation buffer to each well and incubate at 37°C for 5–10 min.
3. After 5–10 min, begin monitoring the cells under a light microscope. Once the cells begin to detach from one another, add 1 ml of ES cell medium to each well to dilute the dissociation buffer. Use a 2 ml serological pipette or cell lifter to gently break the ES cell colonies into a single cell solution and transfer the cells into a 15-ml conical centrifuge tube.
4. Pass the single cell suspension from each well through a 70-μm cell strainer and collect it in a 50-ml conical centrifuge tube. Centrifuge the cell suspension at $500 \times g$ for 2 min at 4°C. Aspirate the supernatant and add 5 ml of PBS into the centrifuge tube to wash the cells a second time. Centrifuge the cells again at $500 \times g$ for 2 min at 4°C. Aspirate the supernatant and resuspend the cells in 1 ml of PBS. Estimate the number of cells in suspension using a hemocytometer and add 10 μl of propium iodide buffer to the suspension for every 3×10^5 cells in the mixture. Transfer this solution to a fluorescence-activated cell sorting (FACS) tube and place on ice.
5. Isolate the GFP or RFP positive cells with a sterile fluorescence-activated cell sorter. Plate these cells on the 12-well plate prepared in step 1 of this section. For optimal cell survival, plate at least 50,000 cells per well. Incubate the cells at 37°C for 48 h. Use ES cell medium supplemented with 1% penicillin/streptomycin and 0.25 μg/ml fungizone for cell culture to prevent contamination. It will take close to a week to expand these cells to a sizeable number for a second sort.
6. After 48 h, aspirate the ES cell medium, replace with 1 ml of fresh medium, and incubate the cells again at 37°C. Continue to culture cells, changing medium every day and monitoring the cells once a day under a microscope. After 6–9 days, the sorted ES cells should reach a colony size of 300–500 cells per colony. When colonies of this size are reached, prepare a second 12-well plate with a MEF feeder layer as in step 1 of this section and proceed to the next step.
7. Place the 12-well dish under a fluorescence microscope and using permanent marker, clearly circle the location of all GFP/RFP positive colonies on the cover of the dish. Use a sterilized glass pipette tip or cell lifter to dislodge the GFP/RFP positive colonies and transfer them to a new MEF-coated 12-well plates using a p200 Gibson pipette. Incubate the cells at 37°C for 24 h.
8. Replace the ES cell medium every 24 h and continue to culture the cells until they are confluent (5–7 days). When the

cells near confluency, prepare another 12-well plate with a MEF feeder layer.

9. When the cells reach confluency, some of the colonies will be more differentiated than others and exhibit less of a stereotypical ES cell morphology. Examine the dish under a microscope and use a permanent marker to clearly mark all *differentiated* colonies. As in step 7, use a sterilized glass pipette tip or cell lifter to dislodge the differentiated colonies. Aspirate the culture medium and floating colonies, eliminating them from the dish and leaving only colonies with stereotypical ES cell morphology attached to the plate.
10. Replace the culture medium with 500 μl of collagenase IV per well prepared at a concentration of 1 mg/ml DMEM:F12. Incubate the cells at 37°C for 5 min. Then use a cell lifter to split these colonies onto a new 12-well plate with MEF feeder layer. Culture these plates until cells are confluent.
11. After the split cells reach confluency, prepare for a second round of cell sorting by plating another six-well culture dish with a MEF feeder layer. Harvest ES cells using the methodologies listed in steps 3–5 of this section. Sort the cells with a sterile fluorescent-activated cell sorter and plate them onto the new MEF-coated dish as outlined in step 6 of this section. These double-sorted GFP/RFP positive cells should grow stably, express the reporter genes, and can be used for in vivo transplantation. Cells can be frozen at −80°C for future use.

3.4. Verification of Transduction Through In Vitro Imaging

1. To confirm successful transduction of ES cells, it is useful to take cell plate images of bioluminescence signals and correlate to cell numbers. To image luciferase positive cells, you will need to have a working solution of the reporter probe D-luciferin already prepared. Dissolve a stock of D-luciferin at a concentration of 45 mg/ml PBS (1 g of D-luciferin in ~22 ml of PBS). Store the prepared D-luciferin in 1–1.5 ml aliquots under tin foil at −20°C for future use.
2. Aspirate the ES cell medium of 1–2 wells of a six-well plate and replace it with 1 ml of cell dissociation buffer per well. Incubate the plate at 37°C for 10 min and scrape the well with a cell scraper. Transfer the cells into a 15-ml conical centrifuge tube. Centrifuge the cells at $500 \times g$ and aspirate the supernatant. Resuspend the cells in 1 ml of PBS and transfer the cells to a 1.5 ml eppendorf tube on ice. Count the cells with a hemocytometer.
3. In a 24-well plate, seed a serially diluted series of cells (e.g., 5×10^5 cells, 1×10^6 cells, 2×10^6 cells, etc.) in a five or more wells (Fig. 3a).

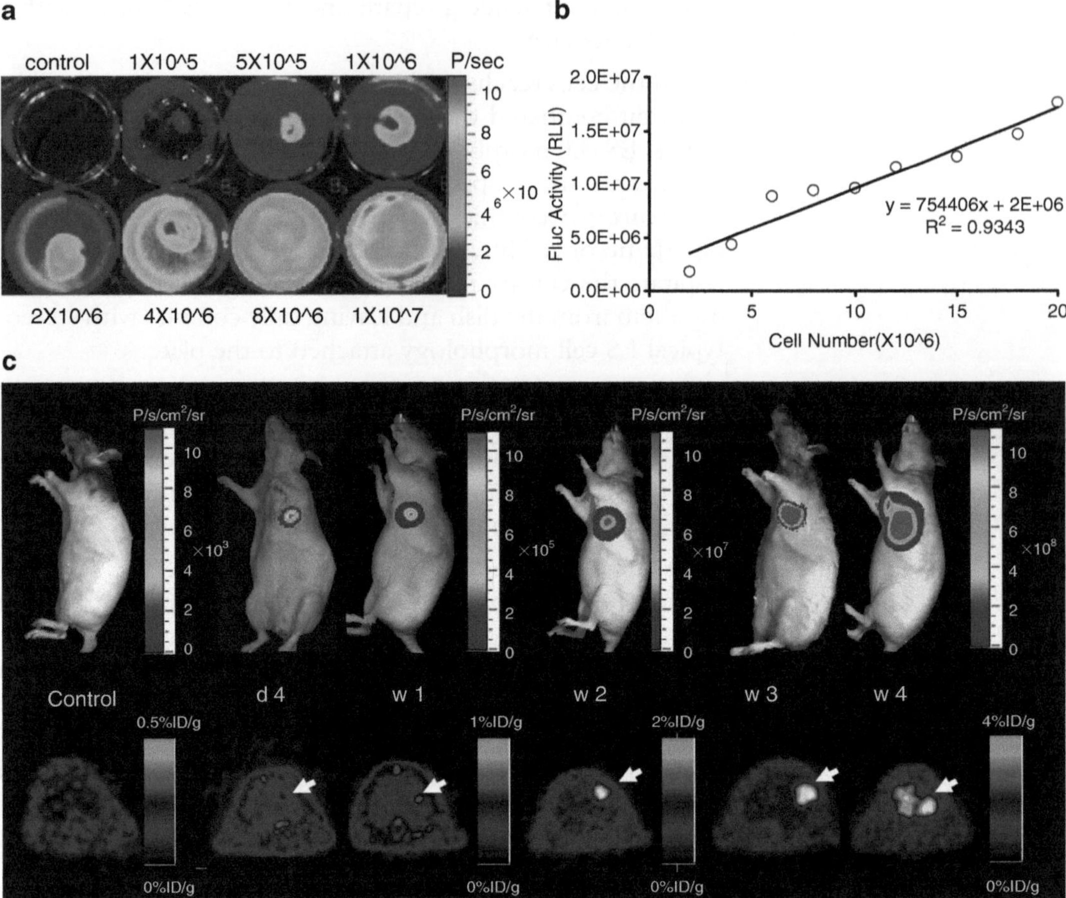

Fig. 3. Reporter gene imaging of ES cells in living animals. (**a**) Cell plate imaging of TF positive ES cells. 1×10^5, 5×10^5, 1×10^6, 2×10^6, 4×10^6, 8×10^6, and 1×10^7 mouse ES cells were plated in a six-well plate. (**b**) Linear correlation of cell number and BLI signal (photons/second/cm²/steradian) reveals a strong correlation ($R^2 = 0.93$). (**c**) Longitudinal imaging of transplanted ES cells in an immunodeficient nude rat. Transplantation of 1×10^7 ES cells into the myocardium leads to teratoma growth by week 4 as monitored by progressive increases in BLI and PET signals. For PET images, the animal was injected via tail vein with close to 100 μCi of the PET reporter probe [^{18}F]-FHBG. Images were acquired 1 h after radiotracer administration to allow for adequate biodistribution. (Reproduced from Ref. 4 with permission from Lippincott Williams & Wilkins).

4. Thaw a 1 ml aliquot of working solution of D-luciferin and dilute it in 99 ml of PBS. Add 1 ml of diluted D-luciferin to each well of the serially diluted series of cells. BLI signal should be immediately detectable.
5. Quickly image the cells using an IVIS Xenogen machine. Begin with acquisition intervals of 10 s. If the signal is saturated, reduce the acquisition interval. If signal appears to be weak, increase the acquisition interval as necessary.
6. Acquire cell plate images of bioluminescence signals repeatedly over a 5–10 min period of time.

7. Use the BLI acquisition and analysis program Living Image to draw regions of interest (ROIs) over the wells and record the BLI signal in units of maximum photons per second per centimeter square per steradian (photons/s/cm^2/sr).
8. Use a graphing software program such as Microsoft Excel to plot the bioluminescence data as maximum photons per second per centimeter square per steradian (photons/s/cm^2/sr) against cell number. Cell number should correlate with bioluminescence signal at an R^2 value of >0.90 (Fig. 3b).

3.5. Transplantation of DF or TF ES Cells into Animals

1. Expand DF/TF ES cells to a sufficient number for transplantation. For ES cell injections, our group will typically inject at least 10,000 cells. Injecting higher cell numbers (e.g., one million cells) will yield higher engraftment and faster teratoma formation.
2. For each well of a six-well plate, aspirate the ES cell medium and wash the cells with PBS. Aspirate the PBS and incubate ES cells in 1 ml of cell dissociation buffer at 37°C for 10 min. Dilute the cell dissociation buffer with 2 ml of PBS per well and use a cell scraper to dislodge the cells. Transfer the solution to a conical centrifuge tube and spin the cells down at $800 \times g$ for 2 min at room temperature. Aspirate the supernatant and resuspend the cells in as low volume as possible of PBS (start with ~100 μl and increase volume as necessary). Homogenize the solution by pipetting gently. Calculate the cell concentration using a hemocytometer.
3. Suspend the desired number of cells in a 1:1 mixture of PBS and human ES qualified Matrigel. Limit the volume of injection to less than 50 μl. Place this mixture on ice.
4. Using a portable anesthesia machine, anesthetize the animal designated for transplantation following the approved animal study protocol of your institution. In our laboratory, we have used 2% (mice) or 3% (rat) isoflurane to knock down animals. Shave the animal at the site of injection if the animal is not nude.
5. Use a 28.5 insulin syringe to administer the ES cells to the desired anatomical location. The animal can be imaged immediately after cell transplantation or alternatively hours to days later to minimize prolonged exposure to anesthesia.

3.6. Longitudinal Monitoring of Transplanted ES Cells Using BLI

1. To determine background bioluminescence levels, anesthetize a control animal that has not received cell transplantation. Image the animal with a Xenogen IVIS machine and record BLI signal in photons/s/cm^2/sr as the background signal.
2. Knock down the experimental animal with 2% isoflurane. Administer 375 mg/kg body weight of D-luciferin working

solution (45 mg/ml) by intraperitoneal injection with a 28.5 gauge insulin syringe. Wait for 10 min while keeping the animal anesthetized before imaging to allow for systemic absorption.

3. Place the animal in the imaging chamber of a Xenogen IVIS machine. Image the animal with 1 s to 5 min acquisition intervals for 30 min to catch the peak BLI signals in photons/s/cm^2/sr. Fluc signals will generally peak 20–30 min after administration (2).
4. Image the same animals at set time points over a defined period. We normally image the animals at days 0, 2, 4, 7, 10 and weekly thereafter (Fig. 3c).
5. Analyze the images using the BLI analysis package Living Image.

3.7. Longitudinal Monitoring of Transplanted ES Cells Using PET

1. For cells transduced with the HSVtk reporter gene, PET imaging can be used to produce high-intensity photons for cell localization. Prepare or order a sufficient amount of [^{18}F] FHBG to meet imaging needs. Typically you will want approximately 100 μCi [^{18}F]FHBG per animal. If your institution does not have a cyclotron facility to produce [^{18}F] FHBG, order this radiotracer from an experienced cyclotron facility.
2. Image a control animal as in step 1 of Part 3.6 to determine background signal.
3. Draw approximately 100 μCi [^{18}F]FHBG into a 28.5-gauge insulin syringe. Record the exact activity within the syringe at time of injection using a dose calibrator. Administer the entire syringe content of [^{18}F]FHBG into the animal via tail vein injection and use the dose calibrator after administration to record the remaining activity within the syringe. Record the time of measurement and time of injection. Wait for 55–60 min for the PET tracer to biodistribute before proceeding to imaging (Fig. 3c).
4. Prior to imaging, knock the animal down using 2% isoflurane. Secure the animal onto the bed of the mircoPET scanner and image the animal as per manufacturer's instructions. Record the time of imaging. Typical MicroPET scanners will have an acquisition and analysis software package such as ASI PRO.
5. Reconstruct the images with a software program provided by the MicroPET manufacturer such as ASI PRO. We have typically used filtered back projection algorithms to reconstruct the images.
6. Use a software package such as ASI PRO or A Medical Imaging Data Examiner (AMIDE) to analyze the reconstructed images.

7. Image the animal at set time points over a defined period. Because PET imaging is considerably more expensive than BLI and [^{18}F]FHBG may be hard to acquire, we usually image at a weekly or monthly intervals.

4. Notes

1. Human ES cells can be transduced on MEF feeder layers or in feeder-free conditions. While the original transductions performed in our laboratory were performed on feeder layers, feeder-free conditions using growth factor reduced, LDEV-free ES cell compatible Matrigel (BD, Franklin Lakes, NJ), and mTeSR-1 (Stem Cell Technologies, Vancouver, Canada) will maximize transduction efficiency by eliminating MEF uptake of the reporter gene. In our experience, continuous culture in feeder-free conditions leads to higher levels of ES cell differentiation as compared to culture on feeder layers. To transduce human ES cells in feeder-free conditions, use Matrigel-coated six-well plates in lieu of MEF feeder layers. mTeSR-1 should be used in place of human ES cell medium. All other steps are the same. Once DF or TF ES cell lines have been established, we recommend continuous culture on MEF feeder layers. We have found this to be more effective in keeping cells in an undifferentiated state as compared to feeder-free conditions. Mouse ES cells can be transduced directly on MEFs.
2. To calculate MOI, please refer to Tiscornia et al. (13).
3. Our laboratory uses standard WiCell public protocols for ES cell culture and maintenance. For guidance with efficacious ES cell culture please consult the WiCell Web site at: http://www.wicell.org.
4. When dissociating or splitting cells using cell dissociation buffer or collagenase IV, monitor cells under a light microscope after 5–10 min to monitor for overdigestion. Overdigestion of cells by dissociation buffers may compromise cell quality.
5. Prior to transplantation of human DF or TF ES cells, we recommend splitting to feeder-free conditions for one to two passages. This will increase purity of transplanted ES cells by eliminating presence of MEFs.
6. For intramyocardial injections, suspension of ES cells in Matrigel may lead to higher levels of mortality due to formation of clots and emboli. We recommend suspending DF or TF ES cells in PBS alone for cardiovascular injection.

7. While the maximum volume of cells we recommend to be suspended in for murine injection is 50 μl, the lower the total volume of suspended cells, the lower likelihood cells will be dispersed following transplantation. Dispersal of ES cells following transplantation leads to reduced rates of engraftment and cell survival.
8. Following cell transplantation, BLI will generally reveal acute cell death (signal decline) followed by cell proliferation (signal gain). To monitor the phenomenon of cell death followed by proliferation we typically acquire images at days 0, 2, 4, 7, 10, and 14. After day 14, cellular growth is monitored weekly.
9. When recording BLI signal, acquire images serially for a 30 min period following injection. BLI signal should peak 20–30 min after intraperitoneal administration of D-luciferin. Record and average the top three signal intensities as the peak value.

Acknowledgments

This work was supported by a Bio-X graduate student fellowship (ASL), a Howard Hughes Medical Institute research fellowship (ASL), R21 HL091453 (JCW), and R21/R33 HL089027 (JCW).

References

1. Thomson, J.A. et al. (1998) Embryonic stem cell lines derived from human blastocysts. *Science* **282**, 1145–7.
2. Wu, J.C. et al. (2006) Proteomic analysis of reporter genes for molecular imaging of transplanted embryonic stem cells *Proteomics* **6**, 6234–49.
3. Wu, J.C. et al. (2006) Transcriptional profiling of reporter genes used for molecular imaging of embryonic stem cell transplantation *Physiol Genomics* **25**, 29–38.
4. Cao, F. et al. (2006) In vivo visualization of embryonic stem cell survival, proliferation, and migration after cardiac delivery *Circulation* **113**, 1005–14.
5. Swijnenburg, R.J. et al. (2008) In vivo imaging of embryonic stem cells reveals patterns of survival and immune rejection following transplantation *Stem Cells Dev* **17**, 1023–29.
6. Swijnenburg, R.J. et al. (2008) Immunosuppressive therapy mitigates immunological rejection of human embryonic stem cell xenografts *Proc Natl Acad Sci USA* **105**, 12991–6.
7. Cao, F. et al. (2007) Molecular imaging of embryonic stem cell misbehavior and suicide gene ablation *Cloning Stem Cells* **9**, 107–17.
8. Cao, F. et al. (2008) Transcriptional and functional profiling of human embryonic stem cell-derived cardiomyocytes *PLoS ONE* **3**, e3474.
9. Li, Z. et al. (2008) Comparison of reporter gene and iron particle labeling for tracking fate of human embryonic stem cells and differentiated endothelial cells in living subjects *Stem cells (Dayton, Ohio)* **26**, 864–73.
10. Li, Z. et al. (2007) Differentiation, survival, and function of embryonic stem cell derived endothelial cells for ischemic heart disease *Circulation* **116**, I46–54.
11. De, A., Lewis, X.Z, and Gambhir, S.S. (2003) Noninvasive imaging of lentiviral-mediated reporter gene expression in living mice *Mol Ther* 7, 681–91.
12. Ray, P., De, A., Min, J.J., Tsien, R.Y, and Gambhir, S.S. (2004) Imaging tri-fusion multimodality reporter gene expression in living subjects *Cancer Res* **64**, 1323–30.
13. Tiscornia, G., Singer, O, and Verma, I.M. (2006) Production and purification of lentiviral vectors *Nat Protoc* **1**, 241–245.

Part III

Stem Cell Migration During Development

Chapter 8

In Vivo Germ Line Stem Cell Migration: A Mouse Model

Brian Dudley and Kathleen Molyneaux

Abstract

A stem cell niche is a specialized tissue environment that controls the proliferation and differentiation of its resident stem cells. The functions of these structures have been well characterized in adult organisms. In particular, the bone marrow stem cell niche in mammals has been amenable to analysis because of the ability of transplanted hematopoietic cells to home and to recolonize the bone marrow of an irradiated host. Despite clues from adult models, it remains unclear how stem cells become partitioned into appropriate niches during embryonic development. To examine the earliest steps in niche formation, we created an organ culture system to observe the development of primordial germ cells (PGCs), a migratory stem cell population that will eventually give rise to the gametes. Using this assay, we can watch PGCs as they migrate to colonize the developing gonads and can introduce growth factor agonists or antagonists to test the function of proteins that regulate this process. This provides an unprecedented opportunity to identify the cellular and molecular interactions required for the formation of the germ cell niche.

Key words: Primordial germ cells, BMP, Confocal microscopy, Time lapse microscopy, Particle tracking

1. Introduction

Primordial germ cells (PGCs) are the embryonic cells that will develop into the gametes. In the mouse, PGCs are induced to form between embryonic days 6.5 and 7.5 (E6.5–E7.5) within the posterior epiblast near the base of the allantoic bud (1). Over the next 4 days (E7.5–E11.5), PGCs proliferate and migrate through the primitive streak (2), the gut, and eventually the midline body wall (3) to colonize the genital ridges, the structures that will give rise to the gonads, kidneys, and adrenal glands. Once at the ridges, PGCs initiate a long process of differentiation dependent upon the sex of the embryo (4). In females, PGCs lose their ability to self-renew and differentiate into oocytes. In males,

Marie-Dominique Filippi and Hartmut Geiger (eds.), *Stem Cell Migration: Methods and Protocols*,
Methods in Molecular Biology, vol. 750, DOI 10.1007/978-1-61779-145-1_8, © Springer Science+Business Media, LLC 2011

PGCs retain their ability to self-renew and will eventually give rise to the spermatogonial stem cells of the testis.

Recently, BMPs have emerged as major players controlling stem cell dynamics. Members of the BMP family are important components of the stem cell niche within the fly ovary (5) and testis (6). In mammals, BMPs regulate stem cell dynamics within the hematopoietic stem cell niche (7) and within the bulge region of the hair follicle (8). BMP4, BMP2, and BMP8b are required for germ cell formation and multiple BMP family members regulate oogenesis and spermatogenesis in the adult (9, 10). To examine the role of BMPs during PGC migration, we cultured PGC containing tissue in the presence of the BMP inhibitor noggin and assayed changes in PGC number, speed, and direction of migration by using time lapse confocal microscopy (11).

2. Materials

2.1. Tissue Isolation and Culture

1. Oct4ΔPE:GFP homozygous stud males via material transfer agreement (12).
2. CD1 mouse females (Charles River). Purchased at 4 weeks of age.
3. Aerrane (Isoflurane, USP) (Baxter).
4. 70% Ethanol.
5. 10× Phosphate-buffered saline (PBS). PBS is diluted to 1× and then autoclaved and stored at room temperature.
6. Tissue culture medium: Dulbecco's Modified Eagle's Medium Nutrient Mixture F-12 (Ham) 1× (DMEM/F12) with L-glutamine and 15 mM HEPES and without Phenol Red (Gibco/Invitrogen).
7. 100× Penicillin–streptomycin solution (HyClone) (stored in 5 ml aliquots at –20°C).
8. 2% stocks of lipid-free BSA (Sigma) prepared in double-distilled water and frozen in 1 ml aliquots at –20°C.
9. 100× Glutamine (Invitrogen). Stored in 1 ml aliquots at –20°C (see Note 1).
10. Falcon Multiwell 24-well plates, 100 × 15 mm Petri Dishes, 60 × 15 mm Petri Dishes, and 35 × 10 mm Petri Dishes (non-tissue culture treated) (BD Falcon).
11. Millicell-CM Sterilized Culture Inserts (0.4 µm pore size, 12 mm diameter, PICM01250) (Fisher).
12. Mouse collagen IV (Becton Dickinson). Collagen is stored in single use aliquots (volume sufficient to coat 48 inserts)

at –80°C. The concentration of the collagen varies from lot to lot so the volume of the aliquots will also vary.

13. No. 5 Dumont super fine forceps (Fine Science Tools), curved iris scissors (Fisher), and No. 11 Feather scalpel blades (Fisher).
14. Mouse Noggin-Fc protein (R&D Systems). The Fc domain allows the protein to be efficiently expressed as a homodimer enhancing activity of the resulting fusion protein. Prepare 100 mg/ml stock solutions by dissolving the lyophilized protein in sterile PBS/0.1% BSA. Store at –20°C in manual defrost freezer in single use (10 ml) aliquots. Stock should retain activity for at least 3 months. Additional growth factor agonists and antagonists have been used in this assay (FGFs (13), Stromal Derived Factor 1(14), and Kit Ligand (15)) and in general should be prepared and used as per the manufacturer's instructions.

2.2. Time-Lapse Confocal Microscopy to Quantify PGC Velocity and Direction of Migration

1. Glass Bottom 12-well plates (14 mm culture well, no. 1.5 coverglass, uncoated) (MatTek) (see Note 2). After use, sterilize with 70% ethanol, dry, and reuse.
2. Leica TCS SP2 AOBS filter-free confocal laser scanning microscope Workstation or equivalent (equipped with an Ar/Kr laser, an inverted microscope, and a programmable stage).
3. The Cube and Box microscope stage heating system (Life Imaging Services) or similar device.
4. Velocity Software version 4.1.0 (Improvision Inc., a PerkinElmer Company).

3. Methods

3.1. Preparing Collagen-Coated Organ Culture Inserts

1. Thaw collagen IV at 4°C overnight. It must thaw slowly.
2. In a tissue culture hood, place organ culture inserts into two 24-well plates (48 inserts total).
3. Dilute collagen to a coating concentration of 55 mg/ml in sterile 0.05 M HCl (prepared in double distilled water).
4. Add 100 μl to each insert and incubate for 1 h at room temperature.
5. Aspirate the collagen and wash three times in sterile PBS. For each wash, fill both the insides of the chambers and under the chambers with PBS. Aspirate the wash.
6. Dry the chambers in the hood for 1 h. Membranes should be opaque once dry.
7. Coated inserts are stored in dry 24-well plates at 4°C.

3.2. Preparation for Culture

1. Prepare a timed mating of an Oct4ΔPE:GFP male and CD1 female. The stage of embryos is estimated by the appearance of a vaginal plug.
2. Add 5 ml of Pen/Strep to a 500-ml bottle of DMEM/F-12 media (see Note 3). Wrap the bottle with foil and store at 4°C for up to 6 months.
3. Prepare fresh tissue culture medium in a sterile hood adding 1 ml of 2% lipid-free BSA, thawed at 37°C, to 49 ml of DMEM/F12/Pen/Strep (DF12PSB). The 50 ml of medium can be stored wrapped in foil at 4°C for 1 week.
4. Fill a 100-mm Petri dish with sterile 1× PBS.
5. Fill a 60-mm Petri dish and a 35-mm Petri dish with fresh DF12PSB culture medium and warm at 37°C.
6. Add 800 μl of DF12PSB medium into two wells (one for control tissue and one for noggin-treated tissue) of a 12-well glass bottom plate (see Note 4). Fill empty wells with sterile PBS to provide humidity. Place the plate at 37°C.
7. Turn on the stage heater for the microscope. If using the Cube system, it will take at least an hour to reach 37°C.

3.3. Isolating E9.5 Embryos

1. Clean the bench, microscope, and all tools with 70% ethanol prior to starting dissections (Fig. 1).
2. In a fume hood, add one half cap full of isoflurane to the bottom of a 1-l beaker. Place the E9.5 pregnant female into the beaker and cover the top with aluminum foil. Wait 5 min, then remove the mouse and place it onto a paper towel. Euthanize the female by cervical dislocation.
3. Position the mouse on its back and soak the ventral side of the animal with 70% ethanol. Using sterile forceps, pinch and

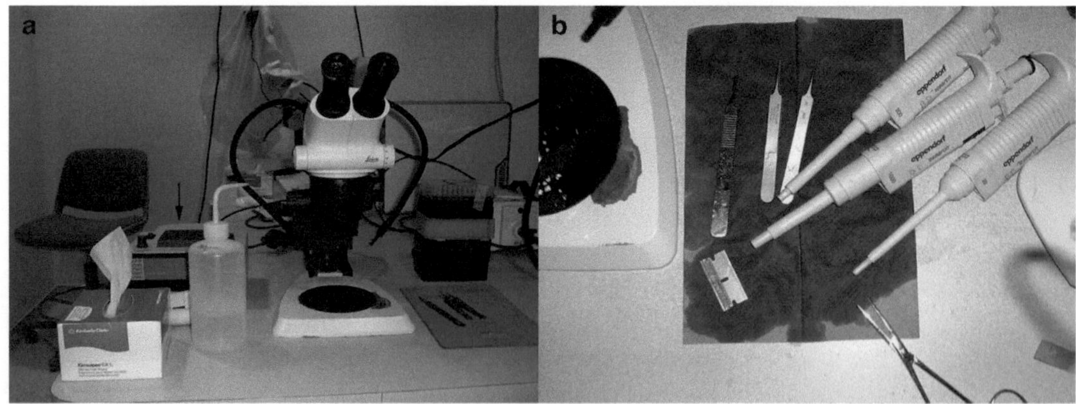

Fig. 1. Preparing the work area for tissue dissections. (**a**) The dissecting scope, 70% ethanol, warming table (*arrow*), and dissecting tools are shown. Keep tools to the right or left depending on handedness. (**b**) Sterilize the microscope stage and tools with 70% ethanol before starting dissections.

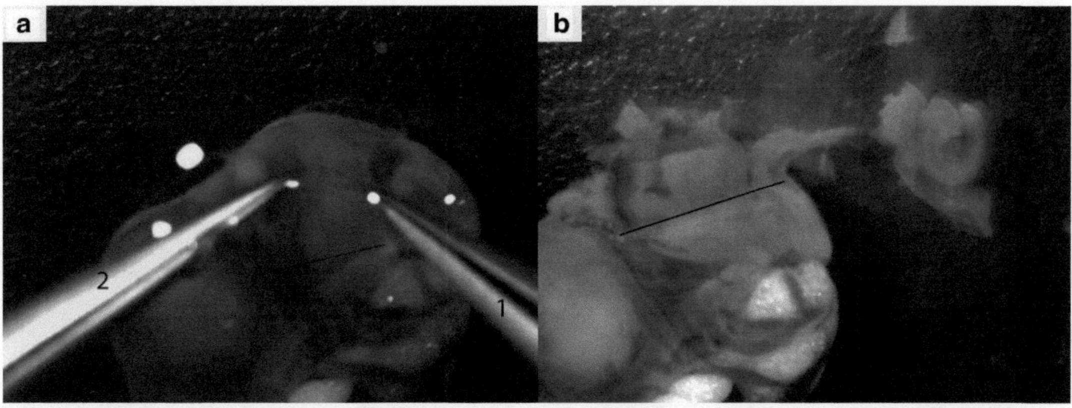

Fig. 2. Removing the embryos from the uterus. (**a**) Use forceps #1 to grasp the uterus between two implantation sites. Use these forceps to hold the tissue steady. Slide one tip of the second pair of forceps into the muscle layer near the holding forceps. Use a shallow angle to avoid tearing the amnion (see Note 5). Close the tip of forceps #2 to grasp the muscle layer. Tear the layer by gently moving forceps #2 away from the holding forceps. This should expose the embryo surrounded by the amnion. The amnion will be attached to the placenta at the proximal portion of the uterus (*near the line*). Take forceps #2 and pinch the amnion where it joins the uterus. Gently pull the embryo away. (**b**) An embryo after dissection.

pull up on the skin in the middle of the pelvis. While holding the skin, take sterile iris scissors and starting near the forceps cut up to the left arm pit and then up to the right arm pit. Be careful not to puncture the body wall. Pull the v-shaped flap of skin toward the head to expose the muscle layer. Rinse the scissors and forceps with 70% ethanol and repeat the above steps in order to cut a v-shaped flap in the body wall.

4. Use the forceps to pull out the uterus. Use the iris scissors to detach the uterus from the body cavity by making three cuts, one at the left ovary, one at the cervix, and one at the right ovary. Transfer the uterus to the 100-mm dish of sterile PBS.
5. Under a dissecting microscope remove the E9.5 embryos from the uterus (Fig. 2). Cut the tip off of a P1000 pipet tip in order to make an opening wide enough to accommodate the embryos. Transfer the embryos to the 60-mm Petri dish containing sterile DF12PSB culture medium.

3.4. Cutting Transverse Tissue Slices

1. Under a dissecting microscope position an embryo on its side. Use the scalpel to make a transverse cut immediately posterior to the forelimb buds. Discard the head. Make a transverse cut immediately anterior to the developing hindlimb buds (or if the hindlimb buds are not visible cut off the tail region where somites have yet to form). Discard the tail tissue. Keep the trunk region between the developing limb buds (see Note 6).

2. Cut the tissue into transverse sections approximately 1–2 somites thick. An ideal slice has a constant thickness and an intact gut (see Note 7).
3. Using a sterile razor blade cut the tip off of a P20 pipette tip. Transfer slices to the 35-mm Petri dish with sterile DF12PSB medium warmed to 37°C.
4. Continue to cut sections from all the remaining embryos, pooling them together in the 35-mm dish of warmed culture medium (Fig. 3).

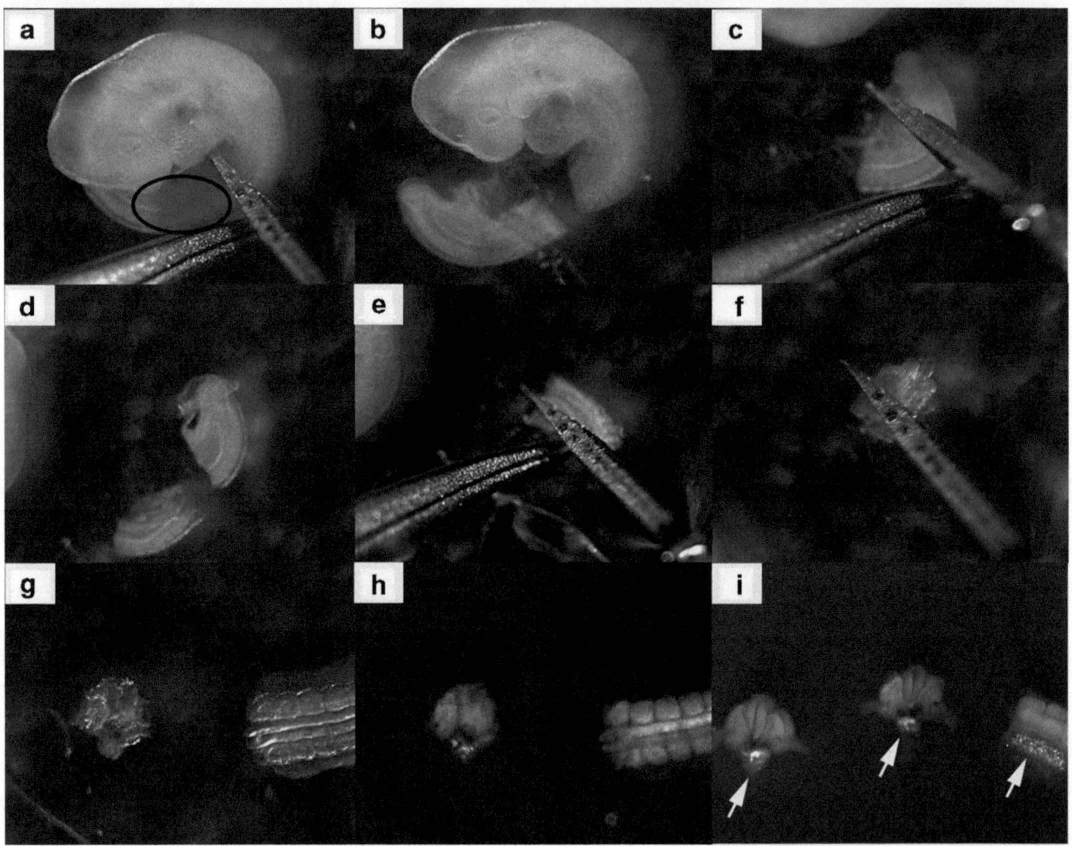

Fig. 3. Isolating PGC containing tissue pieces for organ culture. (**a–b**) At E9.5, PGCs occupy the gut and midline structures extending from the tail to the forelimb buds (*circled region*). Make the first cut just below the forelimb buds. Use the forceps in the off-hand to position the embryo while making the cut with the scalpel in the dominant hand. (**c–d**) Make the second cut just anterior to the portal vein. (**e**) Remove any ventral skin and membranes that might still be attached. This can be done by pinning the skin down on one side with the holding forceps and cutting it away with the scalpel. Rotate the piece and repeat with the skin on the other side [alternatively you can cut the skin flaps off after cutting each slice (see Note 7)]. (**f**) Cutting a transverse tissue slice. Place your scalpel about 2 somites away from the edge of the trunk piece. Hold it lightly against the tissue and use the forceps in the off hand (not shown) to position the piece before making the cut. In this way, the piece can be pivoted in order to insure a flat edge. (**g**) Bright field image of a tissue slice. (**h**) Slice in (**g**) viewed under a fluorescent dissecting scope. The position of the PGCs can be seen. (**i**) Examples of additional pieces cut from the same embryo. Typically, four to five pieces are recovered per embryo. PGCs can be seen in the slices and scattered along the midline of the trunk (*arrow*).

3.5. Culture and Time-Lapse Confocal Microscopy

1. Add 8 μl of 100 μg/ml noggin stock solution to 800 μl of media in one well of the prewarmed 12-well glass bottom plate.
2. In the tissue culture hood, place collagen-coated culture inserts into the wells of the prewarmed 12-well plate. Do not allow medium to fill the inserts. Inserts should sit on the bottom of the wells without floating.
3. Select slices that are 1–2 somites thick and that are flat (not wedge shaped). Cut the tip off of a P200 pipet tip and use it to transfer four slices to each chamber in the 12-well plate (Fig. 4). Pipet ~2 μl of control medium or noggin containing medium from the wells into the drops containing the slices. This will bring the growth factor concentration up immediately.
4. Slowly transfer the plate to an appropriate stage holder (see Note 8). On the Leica AOBS system this means that you need to remove the piezo stage holder and replace it with a tissue culture plate insert. Move slowly as the tissue pieces have not been given time to settle.
5. Leave the tissue on the stage to settle and adhere for at least 1 h.
6. Using a 10×0.4 NA objective (see Note 9) focus on a tissue piece. Adjust the image for Koehler illumination (see Note 10).
7. Find and mark the position of each tissue piece by moving from well to well. This requires a programmable stage and the controls will vary depending on the confocal system. For the Leica system, make sure that you are in the Z-wide scanning mode.

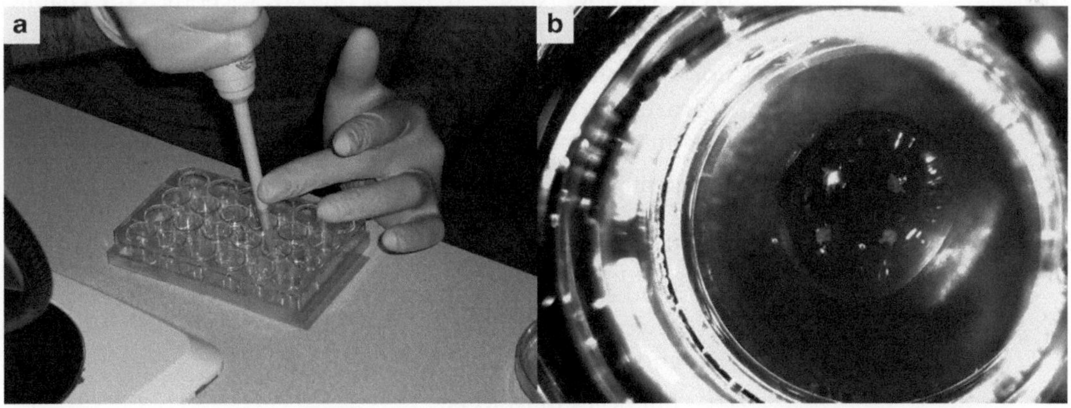

Fig. 4. Placing the slices into organ culture chambers. (**a**) Use a 200-μl pipet to transfer the tissue pieces into Millicell organ culture chambers. Cut the pipet tip to make an opening wide enough for the tissue. Pick up four tissue slices in as small a volume of media as possible. Allow gravity to pull the slices to the bottom of the pipet tip. Then pipet them directly into the center of the organ culture membrane. Following transfer, use forceps to position the pieces. They should sit flat on the membrane and should not be touching other tissue or the walls of the chamber. (**b**) Four correctly arranged tissue pieces in an organ culture chamber.

Then open the "Stage" window, select the "Mark and Find" box, and hit the "Mark" button when a slice has been centered and is in focus.

8. Once all slices have been found, move from visible to laser scanning mode. Activate the 488-nm laser (for GFP). Increase the laser output to ~80% (see Note 11). Open the pinhole to 1.8 AE. Select the appropriate PMTs to collect GFP fluorescence and transmitted light images (this varies depending on confocal microscopes). Image quality is improved by performing frame averaging so select the appropriate button to initiate averaging and perform two scans per focal plane.
9. Using the programmed stage positions, focus through each tissue piece to identify the focal plane in which the PGCs appear the brightest (saturated). Adjust gain and offset settings if necessary. Re-mark each stage position as you go. Focusing and remarking each tissue slice will insure that PGCs are in focus when you start to film.
10. After remarking all slices, perform a short time lapse series to make sure that everything has settled and there is no vibration in the room. Take one frame every 2 min for a total of 20 min (ten frames).
11. If the tissue stays in place during the short time lapse, start the longer series (see Note 12). We typically capture one frame every 7–9 min for 100 frames (12–15 h movies).

3.6. Quantifying PGC Speed and Direction of Migration

1. Import the time lapse image series into Velocity version 4.1.0 software.
2. Open one of the image series. From the menu bar select "Tools" then "Remove Noise". In the menu box, select the GFP channel and a fine filter. Hit the "Change" button to apply the changes.
3. For automated tracking (see Note 13) select the "Measurements" tab. In the "Measurements" menu bar deselect the "Automatically Update Feedback" option.
4. In the measurement window, build a tracking protocol by dragging the following tasks into the measurement box. The protocol can be named and saved for use in tracking multiple movies.
   (a) Find objects by % intensity. Select the GFP channel. The lower limit should be approximately 40 (see Note 14) and the upper limit 100.
   (b) Fill holes in objects.
   (c) Separate touching objects with a size guide of 100 μm^2.
   (d) Exclude objects based on size that are less than 40 μm^2 (too small to be a PGC).
   (e) Exclude objects based on size that are greater than 300 μm^2 (likely to be a clump of PGCs).

(f) Track objects. Once this task is dragged into the measurement window, click on the "Sun" icon in the upper right of this task box to select the following parameters. Tracking mode should be "Shortest Path" and the maximum distance between nodes should be 10 μm (since PGCs rarely move that far between frames).

5. Before tracking, check your protocol to make sure it can identify PGCs. Scroll to the first frame of the movie and select update feedback from the Measurements menu bar. Check to make sure it has found the majority of PGCs in the first frame (it will not pick them all out). Scroll to the last frame and repeat.
6. Once the protocol has been checked, select the "Measure all timepoints" option from the Measurements menu bar. Update feedback to perform tracking.
7. Data will be displayed in the panel at the bottom of the measurement window. Change the filter option to "Tracks" to display the germ cell tracking data. You can choose what data columns you want to view by going into the Measurements menu bar and selecting the "Columns" option. We display track ID, color, time span, track velocity (μm/s), displacement rate (μm/s) and meandering index (see Note 15).
8. Sort the data by using the tab at the top of the Trace time column. Arrange the data so that just the cells that were tracked for the longest time are at the top of the column.
9. Check the top 20 traces to make sure they faithfully represent PGC movement. You can do this by selecting a track and then stepping though the movie. Copy and paste the data from the 20 temporally longest traces into Excel for analysis.
10. With the top 20 traces still selected, use the Feedback Options command in the Measurements menu bar to choose to display entire traces or just the absolute displacement (as shown in Fig. 5). Traces tend to be erratic so displaying a large number of them can clutter the image. As a record take a snapshot of the traces overlying the final frame of the movie.
11. If desired, direction of the traces can be manually scored. If a line drawn through the start point and endpoint of the trace intersects the nearest genital ridge, the cell is scored as being on target.
12. PGC survival data can also be obtained from time lapse movies. You can simply count the number of PGCs in focus every five frames (or at the desired interval) and normalize that to the number of starting PGCs. Plotting PGC number versus time allows you to estimate the kinetics of PGC death and/or division (see Note 16) in response to different treatments (Fig. 5).

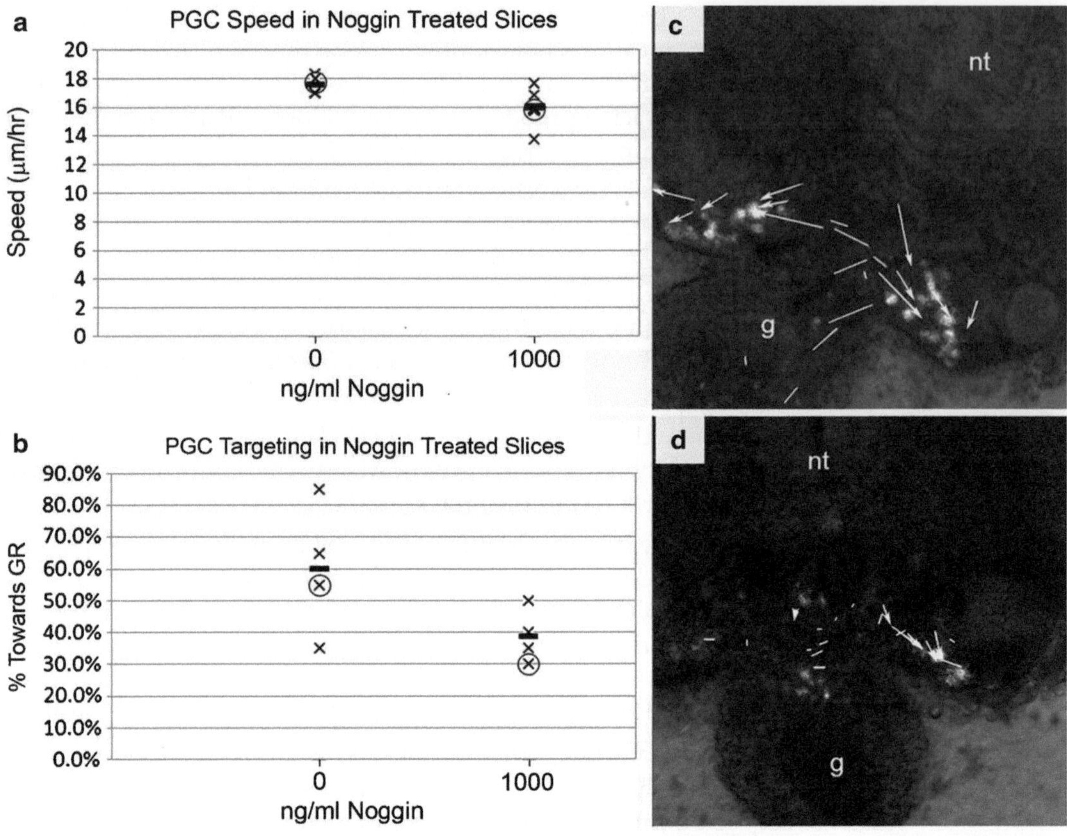

Fig. 5. Noggin treatment reduces PGC speed in culture. Untreated tissue ($n=4$) and tissue treated with 1,000 ng/ml noggin ($n=4$) were cultured and filmed for 13 h. Tracking analysis was performed using Velocity image analysis software. (**a**) Each data point represents the average migration speed for PGCs in an individual slice calculated by averaging the speeds of the 20 cells that Velocity could follow for the longest duration within each slice. (**b**) The percent of PGCs (out of the 20 tracked) that moved toward a genital ridge. Displacement of PGCs in (**c**) a control slice and (**d**) a noggin-treated slice. *White lines* indicate germ cell traces. *Lines with arrows* indicate cells that were scored as moving toward a genital ridge. The speed and direction of PGC migration of these slices are indicated by the *circled* data points in (**a**) and (**b**), respectively.

4. Notes

1. The DMEM/12 medium is supplied already supplemented with L-glutamine which is stable at 4°C for approximately 1 month (~65% remaining after 24 days in storage (Sigma)). We supplement the medium with additional L-glutamine if the stock has been in storage for >1 month.
2. You must get glass bottom plates with the 14-mm microwell insert. This is to accommodate the size of the organ culture chambers. Smaller microwells will not allow the chambers to be level.

3. Penicillin–streptomycin is only suitable for short-term assays (~24 h). PGC survival is better in long-term culture (2–4 days) in the absence of antibiotics.
4. Additional wells can be used to test additional concentrations or different compounds. We have used all 12 wells in a single experiment. You should make sure that your stage controller can move between wells quickly enough to accommodate your desired frame rate. On our system, we had to increase the frame rate to 9 min between exposures to accommodate all 48 samples.
5. If you puncture the amnion while dissecting the embryo from the uterus, the embryo will squeeze either completely or partially out of the resulting hole. You can still use the embryo, but you need to be careful not to tear the trunk. Continue trying to dissect away the muscle layer and avoid grabbing anywhere near the tail of the embryo.
6. Often the resulting trunk piece is crescent shaped. Using the scalpel cut a thin wedge-shaped slice from each end of the trunk, leaving behind a straight edge. It will now be easier to obtain flat slices.
7. Once a slice has been cut there also may be flaps of skin hanging down from the ventral sides of the slice extending toward the gut. These should be cut off.
8. The tissue slices do not adhere very strongly to the collagen. Move the plate slowly in order to avoid dislodging the tissue. You want the slices to stay separated.
9. The tissue culture chambers have feet ~1 mm high so you will need a lens with at least a 1-mm focal length.
10. Having the scope adjusted for Koehler illumination will yield good quality bright field images. You do not need to adjust for Koehler illumination if you just plan to take pictures using the GFP channel.
11. Confocal systems often have default parameters (laser output, pinhole settings, etc.) for capturing good quality images on various fluorescent channels. However, these default settings are not ideal for imaging in microwell chambers. We recommend increasing laser power and opening the pinhole slightly in order to compensate for the loss of brightness incurred by trying to image thick tissue suspended 1 mm above the coverslip. The tissue is fairly resistant to damage from the 488-nm laser.
12. If the tissue moves during the short time lapse, adjust the plate or chambers and try another short movie. Issues that can cause the tissue to drift include bubbles trapped under an organ culture chamber, too much media in the well, or a chamber that is not level (e.g., the chamber is not centered

and is sitting at an angle with one foot on the plastic rim of the microwell).

13. Previously, we performed tracking by hand using NIH image (11). However, this is labor-intensive and subject to individual bias. Automated tracking eliminates these drawbacks but it requires that the images be impeccable. Take time to get good image quality on the confocal. Velocity also has a manual tracking feature should your images be too dim or your cells too crowded for the automated tracking algorithm.
14. To find more PGCs, you can set the threshold slightly lower, but we do not recommend setting the lower threshold at less than 25. You will start picking up features that are not PGCs (as the somatic tissue retains a low level of GFP signal). This background signal is caused by the transgene driving GFP expression in the epiblast prior to gastrulation.
15. The meandering index (MI) is equal to the displacement rate divided by the velocity of a cell. It can be thought of as the percent of time a cell spends actually moving toward its destination. A cell with an MI of 1 moves in a straight line. PGCs migrate inefficiently and have an average MI of 0.36 ± 0.07 ($n = 1{,}020$ cells tracked in 51 slices). This parameter can be used as a measure of how efficiently a population of cells migrates. However, we have found that MI is inversely related to the log of the trace time. This makes it difficult to compare MIs using automated tracking as Velocity will trace cells for different lengths of time (e.g., until the program looses the cell).
16. In vivo, PGCs divide slowly (once every 16 h) so you are unlikely to detect changes in this parameter in a short-term movie. Also, PGCs do not move much in the *Z*-axis during filming, however, the tissue may compress a bit in culture bringing new cells into focus. This may affect the accuracy of your counts.

References

1. Lawson, K.A., Dunn, N.R., Roelen, B.A., et al. (1999) Bmp4 is required for the generation of primordial germ cells in the mouse embryo *Genes Dev* **13**, 424–36.
2. Anderson, R., Copeland, T.K., Scholer, H., Heasman, J., and Wylie, C. (2000) The onset of germ cell migration in the mouse embryo *Mech Dev* **91**, 61–8.
3. Molyneaux, K.A., Stallock, J., Schaible, K., and Wylie, C. (2001) Time-lapse analysis of living mouse germ cell migration *Dev Biol* **240**, 488–98.
4. Brennan, J., and Capel, B. (2004) One tissue, two fates: molecular genetic events that underlie testis versus ovary development. *Nat Rev Genet* **5**, 509–21.
5. Song, X., Wong, M.D., Kawase, E., et al. (2004) Bmp signals from niche cells directly repress transcription of a differentiation-promoting gene, bag of marbles, in germline stem cells in the Drosophila ovary *Development* **131**, 1353–64.
6. Kawase, E., Wong, M.D., Ding, B.C., and Xie, T. (2004) Gbb/Bmp signaling is essential for

maintaining germline stem cells and for repressing bam transcription in the Drosophila testis *Development* **131**, 1365–75.

7. Zhang, J., Niu, C., Ye, L., et al. (2003) Identification of the haematopoietic stem cell niche and control of the niche size *Nature* **425**, 836–41.
8. Zhang, J., He, X.C., Tong, W.G., et al. (2006) Bone morphogenetic protein signaling inhibits hair follicle anagen induction by restricting epithelial stem/progenitor cell activation and expansion *Stem Cells* **24**, 2826–39.
9. Zhao, G.Q. (2003) Consequences of knocking out BMP signaling in the mouse *Genesis* **35**, 43–56.
10. Shimasaki, S., Moore, R.K., Otsuka, F., and Erickson, G.F. (2004) The bone morphogenetic protein system in mammalian reproduction *Endocr Rev* **25**, 72–101.
11. Dudley, B.M., Runyan, C., Takeuchi, Y., Schaible, K., and Molyneaux, K. (2007) BMP signaling regulates PGC numbers and motility in organ culture *Mech Dev* **124**, 68–77.
12. Anderson, R., Fassler, R., Georges-Labouesse, E., et al. (1999) Mouse primordial germ cells lacking beta1 integrins enter the germline but fail to migrate normally to the gonads *Development* **126**, 1655–64.
13. Takeuchi, Y., Molyneaux, K., Runyan, C., Schaible, K., and Wylie, C. (2005) The roles of FGF signaling in germ cell migration in the mouse *Development* **132**, 5399–409.
14. Molyneaux, K.A., Zinszner, H., Kunwar, P.S., et al. (2003) The chemokine SDF1/CXCL12 and its receptor CXCR4 regulate mouse germ cell migration and survival *Development* **130**, 4279–86.
15. Runyan, C., Schaible, K., Molyneaux, K., Wang, Z., Levin, L., and Wylie, C. (2006) Steel factor controls midline cell death of primordial germ cells and is essential for their normal proliferation and migration *Development* **133**, 4861–9.

Chapter 9

Live Microscopy of Neural Stem Cell Migration in Brain Slices

Jin-Wu Tsai and Richard B. Vallee

Abstract

In the developing central nervous system (CNS), neural stem cells undergo a complex series of morphogenetic and motile events. Errors in neural stem cell proliferation or migration cause serious brain developmental disorders. However, the relative importance of each step in neurogenesis and migration and the identity of genes affecting these processes has only begun to be explored. Using live imaging in brain slices, neural stem cells and their progeny labeled by in utero gene transfer can be monitored at high spatial and temporal resolution for as long as several days. Cell cycle progression, mitosis, morphogenesis, and migratory behavior can each be documented in detail. Furthermore, the behavior of subcellular structures, including nuclei, centrosomes, and microtubules, can also be observed using fluorescent marker proteins. This chapter describes the application of these approaches in combination with RNA interference to investigate normal developing brain and the role of genes involved in brain developmental disorders, such as lissencephaly.

Key words: Neural stem cell, Cell migration, In utero electroporation, Brain slice culture, In vivo imaging, Neocortex, Brain development, Radial glia, Lissencephaly

1. Introduction

During development, neurons are generated within the germinal layers of the nervous system by proliferation of progenitor cells. In the developing cerebrum (neocortex), pyramidal neurons are generated from radial glial progenitor cells (1–3). These cells exhibit an extraordinary form of "interkinetic" nuclear oscillations that are coordinated with cell cycle progression (4). During interphase, the nucleus ascends toward the basal end of the cell and then descends to the apical end, located at the ventricular surface, where cell division occurs (Fig. 1). Symmetric divisions produce two radial glial progenitor cells, which repeat the nuclear

Marie-Dominique Filippi and Hartmut Geiger (eds.), *Stem Cell Migration: Methods and Protocols*,
Methods in Molecular Biology, vol. 750, DOI 10.1007/978-1-61779-145-1_9, © Springer Science+Business Media, LLC 2011

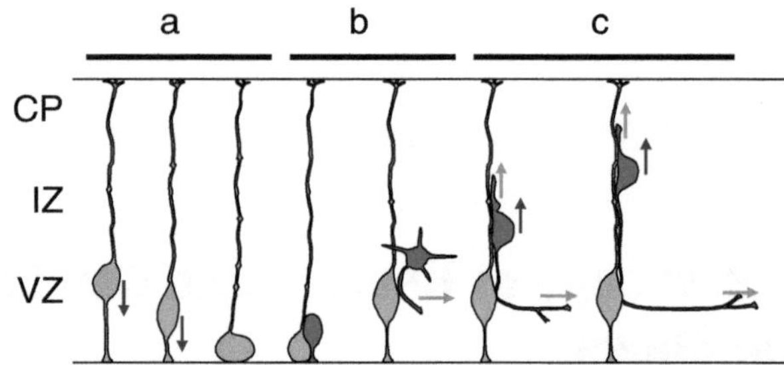

Fig. 1. Pathway for neural progenitor cell morphogenesis and migration in the fetal neocortex. Proliferating radial glial cells expand through symmetric divisions, and generate neurons through asymmetric divisions (**a**), which ascend to the subventricular zone, where they reside as multipolar cells (**b**). The latter transform into bipolar cells, which locomote along radial glial fibers toward the pial surface of the developing cortex (**c**). *VZ* ventricular zone, *SVZ* subventricular zone, *IZ* intermediate zone, *CP* cortical plate. Modified from (10).

oscillatory cycle. Asymmetric divisions produce postmitotic neurons, which migrate to the subventricular zone (SVZ), where they convert to a multipolar nonmigratory phase (5–7). After about a day they convert to a bipolar state and undergo glial-directed radial migration. Because the overall behavior of these cells is complex and has not been recapitulated in vitro, the molecular mechanism of this complex progression in cortical development and how defects in this pathway contribute to developmental diseases such as lissencephaly have not been extensively explored until recently.

Neural stem cells and their subcellular structures can be fluorescently labeled in embryonic mouse (8, 9) or rat brain (10) using in utero intraventricular injection of cDNAs followed by electroporation. Brain slices can then be prepared, cultured, and imaged using epifluorescence or confocal microscopy. Neuronal precursor cells continue to progress through the neurogenesis and migration pathway under these in vitro conditions, allowing detailed examination of cell behavior. Precursors can also be allowed to migrate out from slices (11, 12) for higher resolution imaging, or into juxtaposed slices (13) to test the effects of altered genetic background on migratory behavior.

2. Materials

2.1. cDNA Constructs and siRNA

1. cDNAs are prepared using Endonuclease Free MaxiPrep kit (QIAGEN) and dissolved in H_2O or Tris–EDTA (TE) buffer at 1–5 μg/μl.
2. For RNA interference (RNAi) experiments, shRNA encoding sequences are introduced into the pRNAT-U6.1/Neo vector

(GenScript), which expresses a GFP marker along with a short hairpin RNA.

3. Fluorescently labeled synthetic siRNA oligonucleotides (Dharmacon) can also be used.

2.2. In Utero Electroporation

1. Animal strain: Sprague Dawley rat (Taconic).
2. Anesthetic: 75–95 mg/kg ketamine + 5 mg/kg xylazine of animal body weight.
3. DNA injection: PCR mircopipets 1–10 μl (Drummond) pulled by a needle puller and beveled by a beveller (World Precision Instruments).
4. Electroporator: BTX EMC 830 Electro Square Porator with 7 mm TweezerTrode (Harvard Apparatus).
5. Antibiotics: Antibiotic–antimycotic (100×) contains 10,000 units of penicillin, 10,000 μg of streptomycin, and 25 μg of amphotericin B/ml utilizing penicillin G, streptomycin sulfate, and amphotericin B in 0.85% saline.

2.3. Brain Sectioning and Slice Culture

1. Artificial cerebral spinal fluid (ACSF): 125 mM NaCl, 2.5 mM KCl, 1 mM $MgCl_2$, 2 mM $CaCl_2$, 1.25 mM NaH_2PO_4, 25 mM $NaHCO_3$, 25 mM glucose. The solution should be approximately 310 mOsm, pH 7.4 when bubbled with 5% CO_2/95% O_2. The solution should be freshly made and filtered with 0.25 μm filter.
2. Embedding gel: 4% low melt agarose (gelling temperature ≤28°C) in ACSF. Kept at 37°C before use.
3. Brain slice culture medium: 25% Hank's balanced salt solution, 47% Basal modified Eagle's medium, 25% normal horse serum, 1× Pen/Strep/glutamine, and 0.66% glucose.

2.4. Matrigel and Immunofluorescence

1. Matrigel (BD Biosciences): keep frozen, thaw on ice before use. It will solidify at room temperature to 37°C.
2. Fixatives: 4% paraformaldehyde (PFA; EMS) and 0.1% Triton X-100 in 0.1 M PBS, pH 7.4.
3. Blocking solution: 10% goat serum, 0.1% Triton X-100, and 0.2% gelatin in PBS.

3. Methods

We describe here the basic method for in utero electroporation in rat brain tissue. We include our own modifications of this method, and its adaptation for RNAi and live imaging of cellular and subcellular markers (10, 11).

RNAi has proven to be a powerful approach for investigating molecular mechanisms involved in neural stem cell morphogenesis and migration. RNAi permits relatively acute and often very severe inhibition of gene expression, allowing for primary effects of altered gene expression to be monitored with minimal complications from changes in expression of other genes. Also, because in utero electroporation results in transfection of a subset of brain cells, it is possible to explore the complete behavior of individual cells in a nontransfected and wild-type background. When genes are introduced into the ventricles of the embryonic brain, electroporation results exclusively in the transfection of the radial glial progenitor cells, the endfeet of which line the ventricular surface. Thus, these cells can be tagged and gene expression can be manipulated from the earliest stage in the proliferation/migration pathway. Here, we use the lissencephaly gene, LIS1, a cytoplasmic dynein regulator, as an example to demonstrate the flexibility of this approach.

Following in utero electroporation the brains are removed at a series of time points for fixed cell analysis or cultured for live cell imaging. Time-lapse imaging of live cells can be accomplished by automated epifluorescence, confocal, or multiphoton microscopy. It is essential to keep the slices in an environment controlled for temperature, CO_2 level, and moisture.

3.1. In Utero Electroporation

1. This method for gene transfer into rodent neocortex has been described in detail (8, 9, 14).
2. In brief, pregnant Sprague Dawley rats at E16 (see Note 1) are deeply anesthetized by intraperitoneal injection of ketamine/xylazine (see Note 2).
3. 1–2 μl of cDNAs (1–5 μg/μl) or siRNA (1 μg/μl) mixed with Fast Green (Sigma) are injected through the uterine wall of the embryonic rat brain into the lateral ventricles.
4. A pair of copper alloy oval plates attached to the electroporation generator ECM 830 is used to transmit 5 × 50 V electric pulses of 50 ms duration at 1 s intervals through the uterine wall (see Note 3). Contact by the electrodes with the placenta and other parts of the embryos should be avoided. Note that the anode should be placed over the site of injection.
5. The abdominal wall and the skin are then surgically sutured, and the pregnant rat is allowed to recover from anesthetics on a heating pad kept at 37°C. It usually takes 1–1.5 h for the animals to recover.
6. It is important to monitor postsurgical animal welfare before dissecting the brains from the embryos (see Note 4).
7. A typical time course of the distribution of GFP-labeled cells after in utero electroporation of the empty form of the pRNAT RNAi vector is shown in Fig. 2.

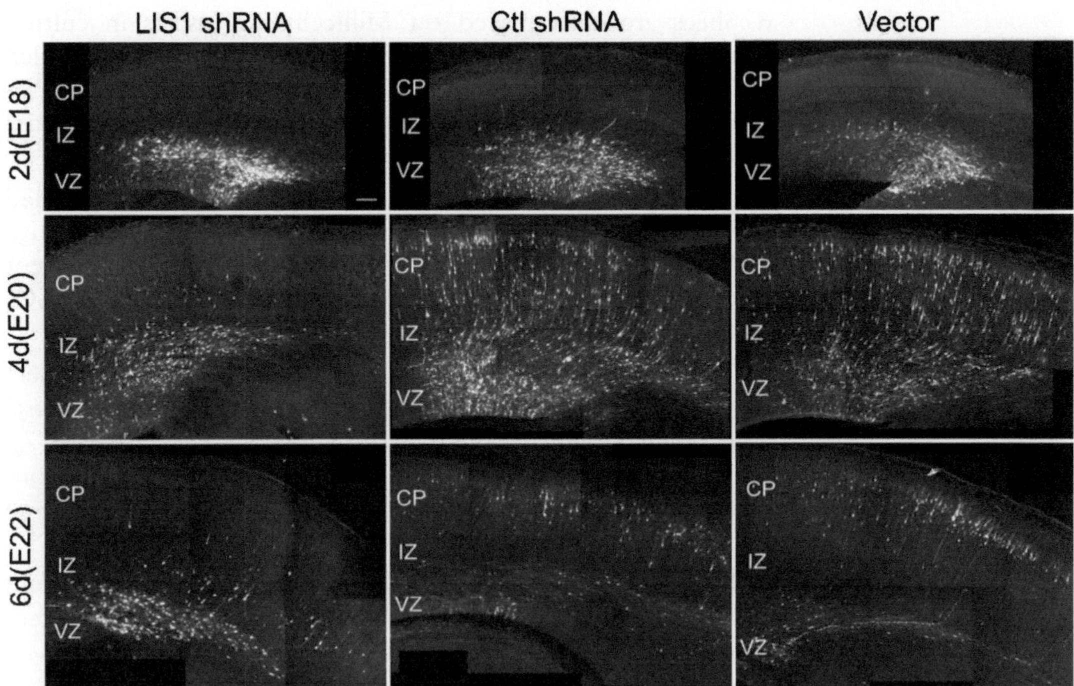

Fig. 2. Redistribution of GFP-labeled neural progenitor cells after in utero electroporation. Coronal sections of rat brain 2, 4, and 6 days after electroporation at E16 with LIS1 shRNA, control shRNA, or empty vector. Cells transfected with control shRNA or empty vector migrated radially from the VZ to the CP with increasing time (*middle* and *right panels*). In contrast, cells expressing LIS1 shRNA were largely restricted to the VZ/SVZ, though some appeared within the lower IZ by days 4 and 6 (*left panels*). Note the additional lateral spread of VZ/SVZ cells in the control. Bar: 100 μm. Modified from (10).

3.2. Live Imaging in Brain Slices

1. Rat embryos subjected to in utero electroporation at E16 are subsequently harvested at E17–22. The embryos are surgically exposed within the uterus and washed in cold ACSF. The brain is then removed carefully without damaging its gross structure.
2. Using a spatula, the brain is then embedded into 4% low-melting agarose kept at 42°C. It is important to remove most of the ACSF before putting the brain into the agarose. Swirl the agarose around the brain a few times to remove air bubbles and ensure that the agarose is in close contact with the brain.
3. The brains embedded in the agarose in Petri dish are then put on ice for about 5 min to let the agarose solidify.
4. A cubic block is cut out from the agarose and trimmed to the boundaries of the brain. The brain is then sliced into 300–400-μm thick sections using a Vibratome. In our case, coronal sections are made.
5. For live imaging the slices are collected with a brush or Pasteur pipette and placed in ACSF at room temperature bubbled with 95% oxygen/5% CO_2 gas mixture (see Note 5).

6. Slices are then placed on Millicell-CM inserts in culture medium in coverglass-bottom Petri dishes (MatTek), which have been preincubated at 37°C in 5% CO_2 in an incubator for at least 1 h. Note that the level of medium inside the Millicell insert should be sufficient to just cover the slices.
7. The slices with transfected cells are then placed on an inverted microscope with a long-working distance 40× objective (NA = 0.55) in a home-made on-stage constant-environment enclosure (see Note 6). The slice is kept at 37°C in 5% CO_2 and humid conditions.
8. Time-lapse images are captured by Coolsnap HQ camera (Roper Scientific) using MetaMorph software (Universal Imaging) at intervals of 10 or 15 min for 10–18 h (see Note 7). Epifluorescence images from several focal planes are deconvolved using AutoDeblur software (AutoQuant Imaging) to produce sharp images (see Note 8).
9. Examples of interkinetic nuclear oscillations in neural stem cells and radial migration of neurons in live brain slices are shown in Fig. 3.
10. The neural stem cells can also be electroporated with one or more cDNAs to label different cell structures. We have had good success imaging centrosomes with dsRed-centrin II, chromatin using CFP-histone H1, and microtubules using GFP-EB3. The latter reagent has been particularly useful in neuronal precursors because it labels growing microtubule plus ends, which can be resolved because of their staggered distribution along the processes of the cell. Motility events of a neural precursor cell labeled with the nucleus, centrosome, and cell body are revealed by this approach (Fig. 4).

3.3. In Vitro Neural Culture in Matrigel

1. For in vitro neural culture, coronal sections are prepared from rat embryos as described above.
2. The cortical plate is then surgically removed. The remainder of the slice is embedded in a thin layer of Matrigel of comparable thickness. A thin layer of brain slice culture medium should be added to keep the tissue moist.

Fig. 3. Live cell imaging of neural stem cell behavior within the neocortex. Rat brains were electroporated with LIS1 (*lower panel*) or control shRNA (*upper panel*) constructs at E16 and the brains were sectioned and cultured 2–3 days later. (**a**) Cell body of a control progenitor cell at the radial glial stage migrates away from and then toward the ventricular surface (*dotted line*), where it divides by the last time point (*upper panels*). Cell body of LIS1 shRNA-transfected cell is relatively immobile over a 14-h time period (*lower panels*). (**b**) Images from bipolar cells within the IZ were taken every 10 min. Control cells extended a leading process toward the CP and the cell body followed, resulting in forward locomotion with a process of relatively constant length (*upper panels*). When transfected with LIS1 shRNA, the leading process of the cells continued to grow, but the cell body remained immobile. The leading process also extended many short projections along its length (*lower panels*). Time in hh:mm. Bar: 5 μm. Modified from (10).

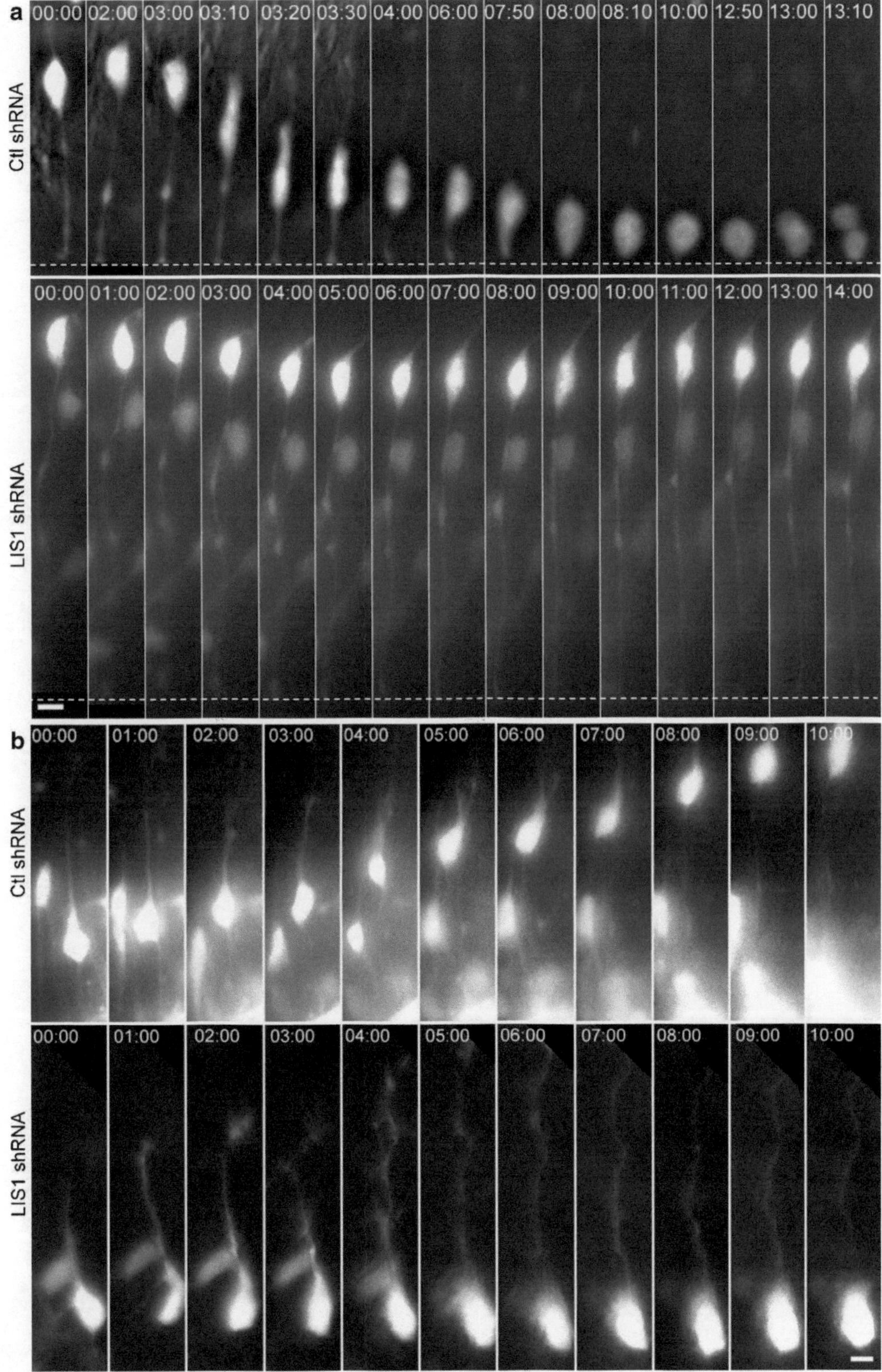
a
Ctl shRNA
00:00
02:00
03:00
03:10
03:20
03:30
04:00
06:00
07:50
08:00
08:10
10:00
12:50
13:00
13:10
LIS1 shRNA
00:00
01:00
02:00
03:00
04:00
05:00
06:00
07:00
08:00
09:00
10:00
11:00
12:00
13:00
14:00
b
Ctl shRNA
LIS1 shRNA
00:00
01:00
02:00
03:00
04:00
05:00
06:00
07:00
08:00
09:00
10:00

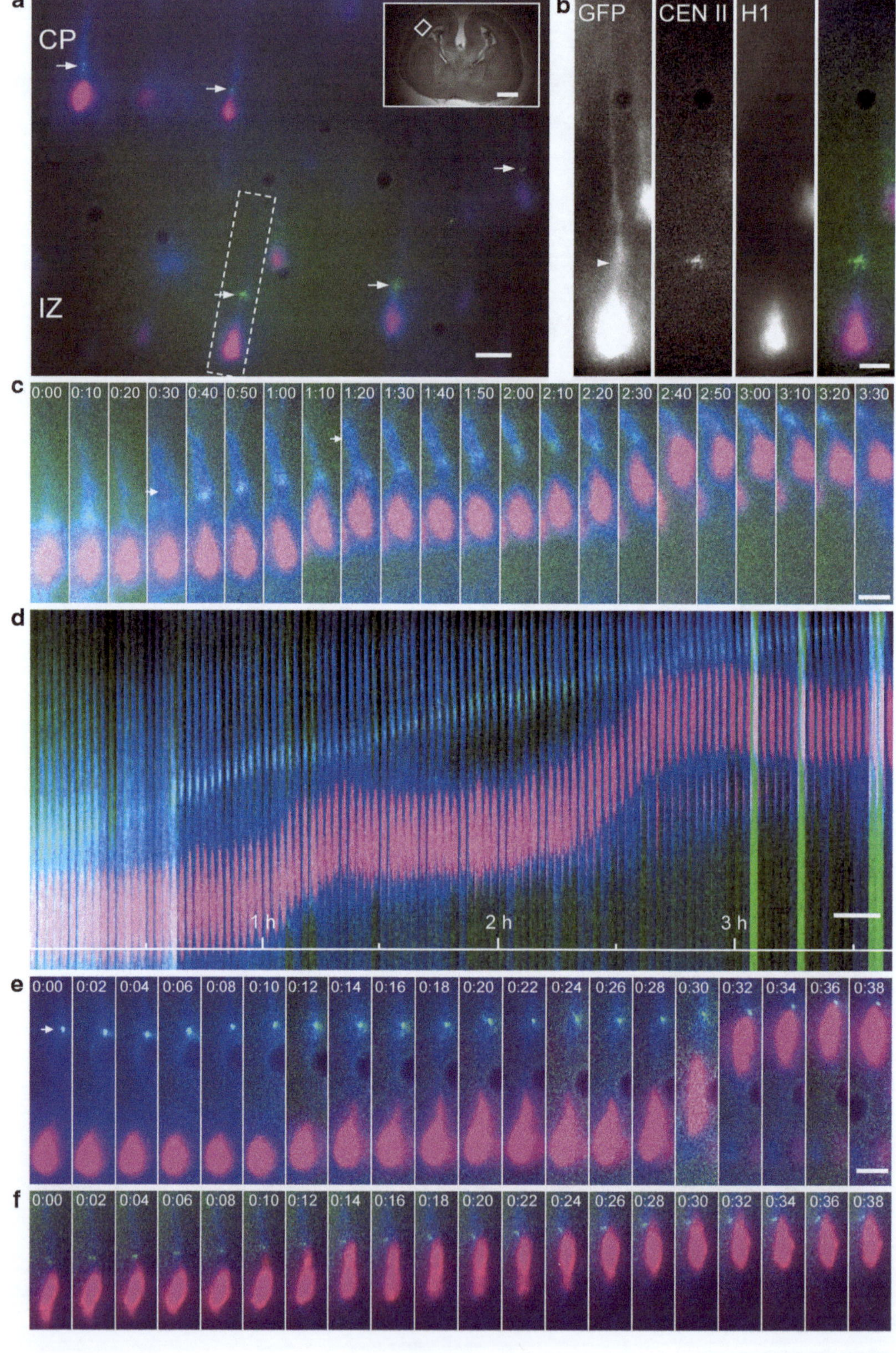
a
CP
IZ
b
GFP
CEN II
H1
c
0:00
0:10
0:20
0:30
0:40
0:50
1:00
1:10
1:20
1:30
1:40
1:50
2:00
2:10
2:20
2:30
2:40
2:50
3:00
3:10
3:20
3:30
d
1 h
2 h
3 h
e
0:00
0:02
0:04
0:06
0:08
0:10
0:12
0:14
0:16
0:18
0:20
0:22
0:24
0:26
0:28
0:30
0:32
0:34
0:36
0:38
f
0:00
0:02
0:04
0:06
0:08
0:10
0:12
0:14
0:16
0:18
0:20
0:22
0:24
0:26
0:28
0:30
0:32
0:34
0:36
0:38

3. Radially migrating cells emerging from the intermediate zone into the surrounding Matrigel 1–2 days later can be monitored by time-lapse phase contrast microscopy (Leica) using a 10× objective (NA = 0.4).
4. An example of neural precursor migration in Matrigel is shown in Fig. 5.

3.4. Immunocytochemistry and Confocal Microscopy

1. For immunocytochemistry in brain sections, rat embryos are perfused transcardially with ice-chilled saline followed by 4% PFA in 0.1 M PBS, pH 7.4. Brains are postfixed in PFA overnight (see Note 9) and sectioned at 100 μm on a Vibratome.
2. To fix the cells migrating in the Matrigel, the cells are fixed by 4% PFA in 0.1 M PBS, pH 7.4 overnight (see Note 9).
3. The brain sections or cells are then blocked at room temperature for 1 h with 10% goat serum, 0.1% Triton X-100, and 0.2% gelatin in PBS.
4. Primary antibodies are applied overnight at the following concentrations: mouse anti-dynein IC74.1 1:100 (Chemicon), rabbit anti-pericentrin 1:150 (Covance).
5. Cells are then washed with PBS and incubated in secondary antibodies and counter stained with the general protein stain dichlorotriazinyl aminofluorescein (DTAF, Sigma–Aldrich).
6. Fixed cells in brain slices or in dissociated culture are imaged using Zeiss LSM 510 META or LSM 510 NLO multiphoton laser-scanning confocal microscope with 40× water immersion objective (NA = 0.8).
7. Excitation/emission wavelengths are 488/515 nm (GFP, DTAF), 543/580 nm (DsRed, Cy3), 633/690 (Cy5), 458/490 nm (CFP), and 800/460 nm (DAPI). Z-series images were collected at 2–3 μm steps.

Fig. 4. Time-lapse fluorescence microscopy of triple labeled neural precursor cells in live brain slices. (**a**) Neural precursor cells expressing GFP (*blue*), DsRed-centrin II (*green*), and CFP-histone H1 (*red*) 3 days after electroporation in utero. One or two centrosomal spots (*arrows*) can be readily seen. In each transfected cell, substantial distance between centrosome and nucleus can be observed. Bar: 10 μm. *Inset*: phase contrast image of the coronal section of the brain. The *solid box* shows the region where the cells were imaged. Bar: 1 mm. (**b**) High magnification view of a triply labeled cell (*dashed box* in (**a**)). The cell body, the processes, and the swelling located within the leading process (*arrowhead*) are visible. In this case the centrosome has reached the swelling (see text). (**c–f**) Time-course of centrosome and nucleus movement in neural precursor cells. (**c**) At the beginning of the sequence a swelling had formed within the proximal part of the leading "migratory" process (*arrow*). The centrosome moved continuously into the swelling. The nucleus then followed the centrosome in a saltatory manner. Time in hh:mm. (**d**) Kymograph of same cell produced from images of a narrow strip of the cell as it migrated. The centrosome moved at a relatively constant rate, whereas nuclear translocation was saltatory. (**e, f**) Distinct modes of nuclear movement. (**e**) The centrosome had separated by as much as 18 μm from the nucleus and had reached the swelling in the migratory process (*arrow*). The nucleus showed dramatic distortion prior to advancing into the migratory process. (**f**) Nuclear movement was much more continuous in ~10% of cells imaged, as shown here, with small or barely detectable steps. Bar: 5 μm in (**b–f**). Modified from (11).

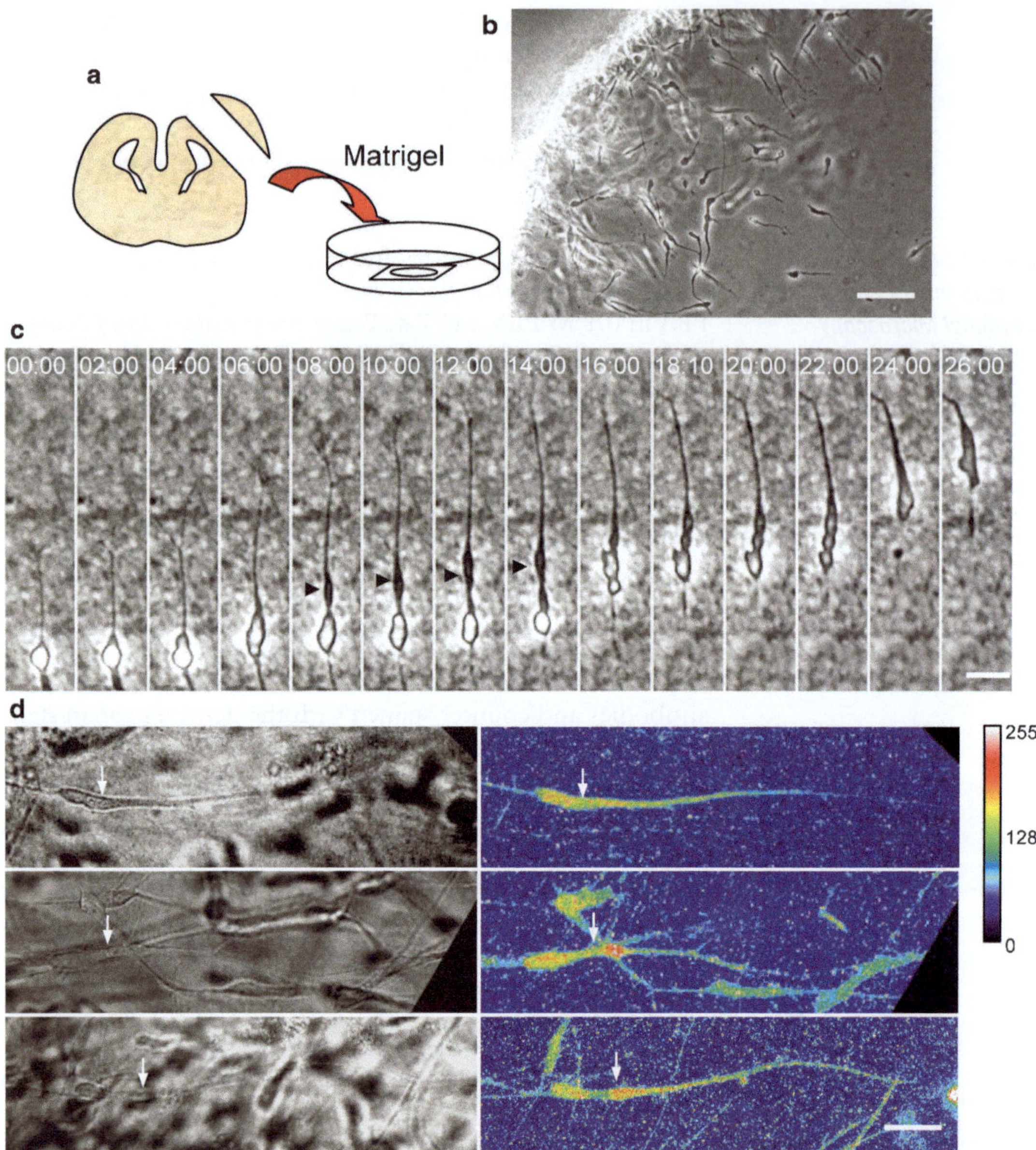

Fig. 5. Neural precursors migrating in a three-dimensional matrix for high-resolution imaging. (**a**) Brain slices with cortical plate surgically removed were cultured in Matrigel in a coverglass-bottomed culture dish to obtain dispersed neural precursor cells. (**b**) Low magnification phase contrast image showing exposed intermediate zone (IZ) and outward spread into Matrigel by bipolar neural precursor cells. Bar: 50 μm (**c**) Time-lapse images of migrating bipolar cell with prominent leading migratory process with proximal swelling (*arrowhead*) and a finer trailing axon. The cell soma advanced into the swelling in a saltatory manner. Bar: 5 μm. (**d**) Immunofluorescence image showing dynein distribution. Cells in Matrigel were fixed and immunostained with anti-pericentrin (*Left panels, green*), anti-dynein, and dichlorotriazinyl aminofluorescein (DTAF). The ratio signal of dynein versus DTAF was normalized into an 8-bit grayscale (0–255) and pseudocolored (*right panels*). In cells exhibiting a juxtanuclear centrosome and lacking a defined swelling, dynein was distributed diffusely throughout the cell (*top panels*). In cells where the swelling had formed, dynein was specifically concentrated within the swelling (*middle panels*). In cells with the characteristic elongated cell body, staining of the soma as well as the swelling was increased (*bottom panels*). *Arrows* indicates the location of the centrosome. Diagram on the *right* depicts relative localization of dynein (*red*), microtubules (*blue*), and centrosome (*magenta*) at intermediate stage in migration cycle. Bar: 10 μm. Modified from (11).

8. To gain insight into the local concentration of the antigen of interest relative to cell thickness or total cell protein, the ratio of the immunocytochemical signal versus DTAF in each image is calculated by MetaMorph.
9. An example for high-resolution immunostaining of cells migrating in Matrigel is shown in Fig. 5.

4. Notes

1. The neuronal progenitors labeled at E16 generate neurons that are primarily destined to comprise layer 2/3 neurons. Labeling of neurons for other layers can be accomplished by performing electroporation at earlier or later different ages. We find, however, that in utero electroporation before E14 and after E17 is difficult because of the size of the embryos and thickness of the skull.
2. Other anesthetics, e.g., isoflurane can also be used with pregnant rats. Consult with local veterinarian in your institution.
3. The electrodes used for in utero electroporation should be just large enough to cover the head of the embryos. Voltage settings may need to be adjusted for younger or older embryos (14).
4. Consult your institutional IACUC for guidelines.
5. Bubble the ACSF with 95% O_2/5% CO_2 for at least 1 h before use. The pH should reach ~7.4.
6. In this setup, the microscope must image through the coverglass, the layer of medium covering the brain slice, and the filter membrane into the thick brain slice. Thus, very long working distance objectives (≥1–3 mm depending on the setup) are needed. Upright microscopes can also be used, in which case water immersion lenses are required for imaging.
7. The brain slices usually exhibit some degree of lateral and focal drift during prolonged imaging periods. Adjust the stage position and focus accordingly. The drifts often decrease after a few hours of imaging.
8. To achieve even better image quality, confocal or multiphoton microscopes can be used. However, laser power should be kept minimal to reduce phototoxicity.
9. Do not fix the brains for longer than 24 h, which cause fluorescence from paraformaldehyde to become prominent in brain slices.

Acknowledgments

The authors thank Dr. Wei-Nan Lian and Shahrnaz Kemal for critical reading of this manuscript. This work was supported by NIH grants HD40182 and GM47434, the March of Dimes Birth Defects Foundation to RBV, and the New York State Spinal Cord Injury Research Board to JWT.

Note: Another application of the methods described in this article has recently been published by our lab (15).

References

1. Malatesta, P., Hartfuss, E., and Gotz, M. (2000) Isolation of radial glial cells by fluorescent-activated cell sorting reveals a neuronal lineage *Development* **127**, 5253–63
2. Miyata, T., Kawaguchi, A., Okano, H., and Ogawa, M (2001) Asymmetric inheritance of radial glial fibers by cortical neurons. *Neuron* **31**, 727–41
3. Noctor, S. C., Flint, A. C., Weissman, T. A., Dammerman, R. S., and Kriegstein, A. R. (2001) Neurons derived from radial glial cells establish radial units in neocortex *Nature* **409**, 714–20
4. Sauer, F. C. (1935) Mitosis in the neural tube *J Comp Neurol* **62**, 377–405
5. Noctor, S. C., Martinez-Cerdeno, V., Ivic, L., and Kriegstein, A. R. (2004) Cortical neurons arise in symmetric and asymmetric division zones and migrate through specific phases *Nat Neurosci* 7, 136–44
6. Rakic, P., Stensas, L. J., Sayre, E., and Sidman, R. L. (1974) Computer-aided three-dimensional reconstruction and quantitative analysis of cells from serial electron microscopic montages of foetal monkey brain *Nature* **250**, 31–4
7. Tabata, H., and Nakajima, K. (2003) Multipolar migration: the third mode of radial neuronal migration in the developing cerebral cortex *J Neurosci* **23**, 9996–10001
8. Saito, T., and Nakatsuji, N. (2001) Efficient gene transfer into the embryonic mouse brain using in vivo electroporation *Dev Biol* **240**, 237–46
9. Tabata, H., and Nakajima, K. (2001) Efficient in utero gene transfer system to the developing mouse brain using electroporation: visualization of neuronal migration in the developing cortex *Neuroscience* **103**, 865–72
10. Tsai, J. W., Chen, Y., Kriegstein, A. R., and Vallee, R. B. (2005) LIS1 RNA interference blocks neural stem cell division, morphogenesis, and motility at multiple stages *J Cell Biol* **170**, 935–45
11. Tsai, J. W., Bremner, K. H., and Vallee, R. B. (2007) Dual subcellular roles for LIS1 and dynein in radial neuronal migration in live brain tissue *Nat Neurosci* **10**, 970–9
12. Schaar, B. T., and McConnell, S. K. (2005) Cytoskeletal coordination during neuronal migration *Proc Natl Acad Sci USA* **102**, 13652–7
13. McManus, M. F., Nasrallah, I. M., Pancoast, M. M., Wynshaw-Boris, A., and Golden, J. A. (2004) Lis1 is necessary for normal non-radial migration of inhibitory interneurons *Am J Pathol* **165**, 775–84
14. Saito, T. (2006) In vivo electroporation in the embryonic mouse central nervous system *Nat Protoc* **1**, 1552–8
15. Tsai, J. W., Lian, W. N., Kemal, S., Kriegstein, A. R., and Vallee, R. B. (2010) Kinesin 3 and cytoplasmic dynein mediate interkinetic nuclear migration in neural stem cells *Nat Neurosci* **13**, 1463–71

Chapter 10

Whole Embryo Imaging of Hematopoietic Cell Emergence and Migration

Michael J. Ferkowicz and Mervin C. Yoder

Abstract

The use of transgenic mice in which tissue or lineage-specific, cell-restricted promoters drive fluorescent reporters has recently been reported as a means to follow the in vivo migration of various hematopoietic cells during murine development. At present there is limited ability of these approaches to image the emergence of the first hematopoietic cell subsets due to lack of unique markers that define those hematopoietic cells. We have utilized whole embryo analysis via immunostaining and confocal laser-scanning microscopic (CLSM) imaging to define the emergence of the first hematopoietic elements in the yolk sac of the developing conceptus. The methods employed to examine yolk sac hematopoiesis may be applied to hematopoietic cell emergence in the embryo proper or fetal liver in the generation of a complete map of hematopoietic ontogeny.

Key words: Developmental hematopoiesis, Yolk sac, Aorta–gonad–mesonephros, Blood island

1. Introduction

A prerequisite to studying stem cell migration during development is to discover the spatial and temporal origins of these cells. In general, detection of a specific hematopoietic stem or progenitor may be restricted to the availability of an in vivo assay that permits detection of a specific activity of the test cells. For example, if one proposes that the definition of a hematopoietic stem cell is defined by whether or not a test cell engrafts in a lethally irradiated adult mouse, then the in vivo assay system may not readout the emergence of the first hematopoietic stem cells during ontogeny (stem cells that seed the fetal liver may not seed the marrow).

In the last century, studies focusing on microscopic examination of live chick embryos revealed that the vertebrate yolk sac is the first site of hematopoietic cell emergence (1). Similar results

Marie-Dominique Filippi and Hartmut Geiger (eds.), *Stem Cell Migration: Methods and Protocols*,
Methods in Molecular Biology, vol. 750, DOI 10.1007/978-1-61779-145-1_10, © Springer Science+Business Media, LLC 2011

were reported for the murine and human systems (2, 3). More recent evidence has suggested that emergence of cells giving rise to lymphoid and myeloid lineages may occur simultaneously within the murine yolk sac, embryo proper, and placenta (4).

The earliest hematopoietic and endothelial stem/progenitor cells arise from extraembryonic mesodermal precursors that seed the murine yolk sac. At embryonic day 7.25 (E7.25) primitive erythroid (Ery^P) and mixed lineage progenitors are detectable (via progenitor assays) that produce mainly primitive erythroid and some macrophage cells within the yolk sac. The first definitive myelo-erythroid progenitors arise around E8.25 beginning in the yolk sac and later in the embryo proper. It remains unclear whether the precursors of these myeloid progenitors also give rise to the lymphoid lineages and how these precursors relate to the hemogenic endothelium-producing stem cells that persist to adulthood (5, 6). Furthermore, the interactions among these cells, the migration of these cells to later sites of hematopoiesis, and the long-term regulation of their potentials within each stem cell niche during ontogeny are not fully understood. The emergence of adult-repopulating HSCs in the aorta–gonad–mesonephros (AGM) region of the E10.5 (and later) embryos has been demonstrated by the use of Sca-1/GFP transgenic embryos (7). However, the HSCs are a small subset of the Sca-1/GFP^+ cells of the embryo precluding the precise identification of the HSC. Furthermore, the Sca-1/GFP construct fails to detect the earlier (<E10.5) yolk sac hematopoietic population with lymphomyeloid potential (6).

A global, whole embryo approach to detection of the first hematopoietic cells would be very informative when studying the complex temporal process of cell migration and seeding of hematopoietic sites. Historical approaches to imaging HSC emergence, development, and migration relied heavily on histological stains and immunostaining of tissue sections. This was necessary because whole mouse embryos were not sufficiently optically clear to be directly imaged with the existing technologies. Recent advances now allow live confocal laser-scanning microscopic (CLSM) imaging of cultured mouse embryos (8). Live imaging of transgenic reporter strain(s) embryos allows direct observation of processes, such as vascular remodeling and cardiovascular development, initiation of blood flow, and changes in shear stress during vascular remodeling (9–12). Currently, the greatest limitation of this technology is that the usual CLSM imaging depth of 200–300 μm is not achievable due to light scattering and absorption by the live embryo and thus only structures and processes occurring near the outside surface of the intact embryo/yolk sac are visible. Innovations in submicron multicolor imaging technologies and the generation of more varied reporter lines will greatly improve this promising approach.

Deeper penetrating technologies such as X-ray computer tomography (CT) scanning and magnetic resonance imaging

(MRI) have begun to be applied to the imaging of early vertebrate embryos (13), however, these modalities still lack cellular resolution necessary to identify all the composite cells of tissues and organs of the early embryo. Optical projection tomography (OPT) is an exciting technology that can image large embryos but the resolution in the order of tens of microns and thus, can discern the cellular staining patterns of tissues and organs using most visible dyes (14). Recent advances in OPT have increased the resolution achievable to about 1 μm (15), however, cellular resolution using different excitation wavelengths of light (for the detection of multiple fluorochromes) is not yet possible.

CLSM is an optical technique with subcellular (0.25 μm) resolution, but which has not been developed to image whole embryos three dimensionally. The two main reasons for this are the lack of methods for clearing and imaging whole embryos via CLSM and the lack of software and hardware capable of handling such massive amounts of raw data. Here, we describe the methods necessary to stain, clear, mount, image, and assemble the data from 3D imaged whole embryos.

The approach described focuses on E7.0–E9.5 embryos that are isolated from accurately timed pregnancies and embryo dissection (Fig. 1). The limitation of this approach is that static time points are used, making it necessary to image multiple litters of carefully staged embryos to reconstruct the ontogeny of stem cell migration. Whole embryo CLSM is performed on the fixed, cleared, and stained embryos via the collection of multiple volume stacks of imaged data that are subsequently "stitched" together slice by slice via custom software plugins for ImageJ, to generate massive volume stacks with near-perfect registration and subcellular resolution. Use of this method with slight modifications allows these techniques to be applied to other stem/progenitor niches of the embryo such as the developing fetal liver and AGM region of older embryos and even adult tissues.

2. Materials

Most of the materials necessary for the collection, fixation, antibody staining and clearing of embryos are listed with a detailed manufacturer's record. These materials are only suggestions and may be modified due to availability or personal preferences.

2.1. Embryo Collection and Fixation

1. Time-mated pregnant mice (see Note 1).
2. Iscove's Modified Dulbecco's Medium (IMDM), supplemented with 10% fetal bovine serum (FBS) (HyClone, Ogden, UT).
3. Halogen lamp with two flexible swan-neck transmission tubes (Intralux, 150H, Volpi AG, Urdorf-Zurich, Switzerland).

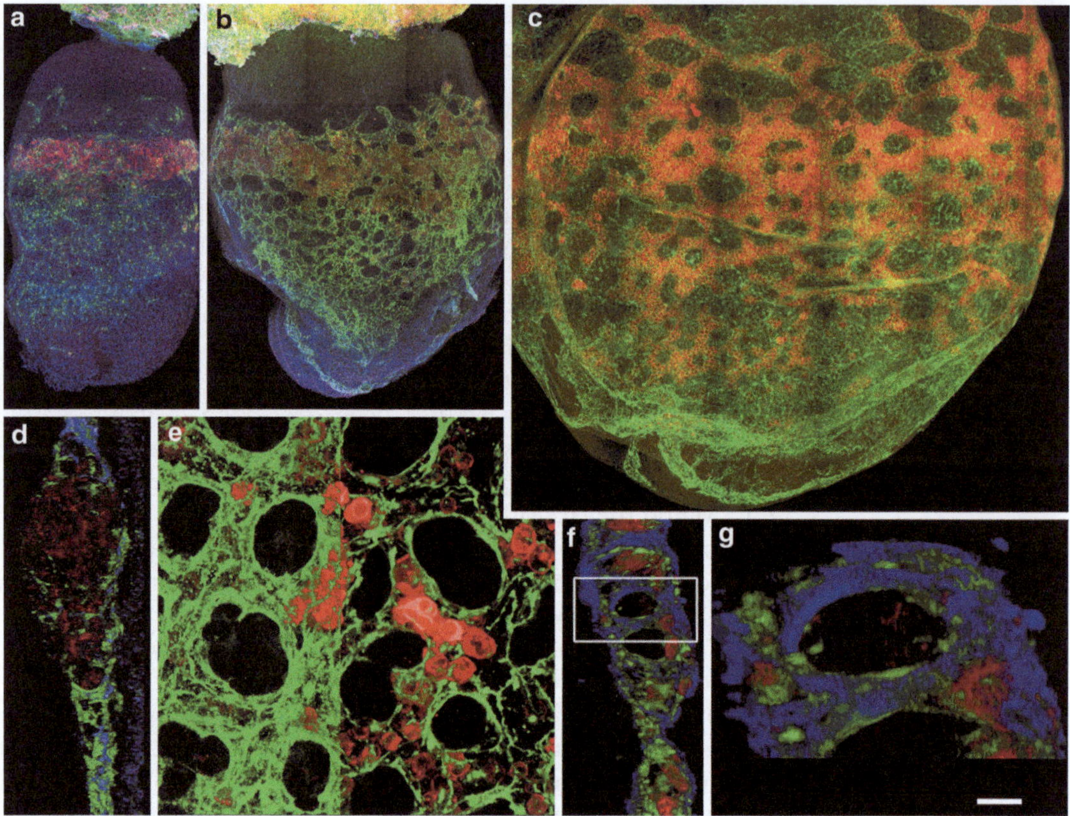

Fig. 1. Confocal images of intact whole embryos. (**a–c**) Left lateral global views of composite maximum projections (MetaMorph) of stitched stacks from the embryonic midline to the left surface with embryonic anterior to the left. *Red*: CD41$^+$ primitive progenitors and erythrocytes. *Green*: VE-Cadherin/Flk-1$^+$ endothelial cells and progenitors. *Blue*: fibronectin expression. (**a**) Precirculation E7.75 embryo with blood cell band (*red color*) in the blood island region of the yolk sac. (**b**) E8.25 2-somite pair embryo near the onset of circulation. Note the remodeling of the blood band. One of the paired dorsal aortae is visible at the distal apex of the conceptus. (**c**) E8.5 8-somite pair embryo a few hours after the onset of circulation with most of the primitive erythrocytes still resident in the yolk sac. Note the paired dorsal aortae at the distal apex (*bottom*) of the embryo (Fibronectin expression (*blue*) not shown). (**d–g**) Alpha projections (Voxx) of the blood island region of the yolk sac. (**d**) E7.75 blood island with incomplete endothelial covering of the blood band within the inner layer (only mesothelial covering). (**e**) Complete endothelial covering of E8.5 blood island region with surface cut away on the right side to reveal the red blood cells inside. (**f**) Orthogonal view of the remodeled blood island where primitive erythrocytes have been removed to reveal the vascular lumens. (**g**) 3.5× zoom of a region of (**f**) to reveal the detail achievable. Endodermal side is to the *right* in (**d**, **f**, and **g**). Scale bar = 100 μm in **a–c**, 20 μm in **d–f** and 5.7 μm in **g.**

4. Dissecting microscope (Leica MZ75, X6.3-50; Leica Microsystems AG, Heerburg, Switzerland).
5. Fine tipped forceps and scissors (Fine Science Tools).
6. Plastic petri dishes (diameter = 100 × 20 mm).
7. Disposable polyethylene graduated transfer pipettes with ends cut and fired to appropriate size so as not to damage the aspirated embryos (Fisherbrand (13-711-9AM)) precoated with 10% serum to prevent the embryos from sticking to the inner walls of the pipette.

8. Phosphate-buffered saline (PBS): Prepare a 10× stock solution by dissolving 2 g of KCl, 80 g of NaCl, 17.8 g of $Na_2HPO_4 \cdot 2H_2O$, and 2.4 g of KH_2PO_4 in 800 ml of distilled water. Make up volume to 1 l with distilled water. Sterilize by autoclaving. Store at room temperature. Prepare working solution by diluting one part with nine parts of water.
9. 24-Well polypropylene round bottom plate (UNIPLATE, 7701-5102, Whatman, Piscataway, NJ) (see Note 2).
10. Fixative: –20°C 100% acetone (ACS grade, Sigma–Aldrich) (see Note 3).

2.2. Whole Embryo Immunolabeling

1. Affinity purified antibodies: Endothelial markers: VEGF R2 (Flk1) (clone AVAS12) (BD Biosciences) goat anti-mouse VE-Cadherin polyclonal antibody (R&D Systems). Early hematopoietic marker: rat anti-mouse CD41 monoclonal antibody (clone MW Reg 30) (BD Biosciences). Mesodermal marker: goat anti-mouse Brachyury polyclonal antibody (T) (Santa Cruz Biotechnology). ECM marker: rabbit anti-mouse Fibronectin polyclonal (FN) (Sigma–Aldrich).
2. Alexa Fluor 488, 546, and 647 antibody labeling kits (Invitrogen). One for each wavelength to be visualized via CLSM (see Note 4).
3. Triton X-100. Dilute to 10% in H_2O (v/v).
4. Blocking serum (bovine or of the same species of secondary antibody if secondary antibodies are used (Jackson Immuno-Research Laboratories)).
5. PBS for washing (see Subheading 2.1, step 8).

2.3. Embryo Clearing and Mounting

1. Anhydrous glycerol (Sigma).
2. Adjustable flow airline fitted with 18-gauge needle.
3. Glycerol dehydration series 5, 10, 20, and 30% v/v with PBS, 40, 50, 60, 70, and 80% v/v with H_2O.
4. 1,4-Diazabicyclo(2.2.2)octane (DABCO) antifade agent (Sigma).
5. Murray's Clear mounting medium (1:1 benzyl alcohol:benzyl benzoate with 1% (w/v) DABCO).
6. Disposable 5 ml polystyrene tube.
7. #1.5 microscope cover slips (18×18×0.17 mm, Corning Glass Works).
8. WillCo glass bottom dishes 50×0.17 mm (70674-02, Electron Microscopy Sciences).
9. Modeling clay.

10. Disposable polyethylene graduated transfer pipettes with ends cut to appropriate size not to damage embryos and precoated with 10% serum and then with 80% glycerol.
11. Rubber cement (Elmers) freshly loaded into a 3-ml plastic syringe then fitted with #18-gauge needle.

2.4. Confocal Immunofluoresence and Image Acquisition

1. Olympus FV1000-MPE confocal microscope mounted on an Olympus IX81 inverted microscope equipped with a krypton–argon (488, 568, and 647 nm excitations) laser, a Prior ProScan II motorized stage (Prior Scientific, Rockland, MA) and a UPLSAPO 40× (oil) 0.90 NA objective lens.
2. Immersion oil.
3. Cotton swabs and Sparkle glass cleaner (Sparkle).
4. ImageJ software (http://rsb.info.nih.gov/ij) with custom stitching macros (available upon request: mferkow@iupui.edu).
5. Windows computer running XP with at least 500 Mb video card.

3. Methods

The choice of antibody combinations used depends on the specific cellular lineage of inquiry. Markers for early hematopoietic stem/progenitor cells are stage-dependent. Embryonic human, mouse, and chicken hematopoietic stem/progenitors express CD41 (αIIb integrin) (16–22). CD41 is expressed by the first murine hematopoietic progenitors at about E7.25 as detected by primitive progenitor assays (17). Thus, $CD41^{dim}$ expression serves as a marker for the onset of primitive megakaryopoiesis, macrophage CFC, and Ery^{P} progenitor cell emergence in the murine yolk sac. Definitive hematopoietic progenitor cells are first detectable in the E8.25 yolk sac (23). These definitive progenitors are $CD41^{bright}$ cells that first arise in the blood island region of the yolk sac (17, 19) possibly from a hemogenic endothelial intermediate (24). Thus, murine CD41 is a unique marker of the earliest hematopoietic cells that emerge from the mesodermal ($brachyury^{+}$) compartment and contribute first to primitive hematopoiesis and also definitive hematopoiesis in the blood island region of the murine yolk sac. The yolk sac endothelium also emerges from the $brachyury^{+}$ extraembryonic mesoderm and uniquely expresses VE-cadherin and VEGF R2 (Flk-1) proteins during development. Fibronectin (FN) is the predominant extracellular matrix (ECM) protein of the early yolk sac and is required for normal hematovascular formation.

Intact embryos with surrounding yolk sac are fragile and subject to distortion and damage when using tissue handling techniques that would be adequate for solid tissues such as the fetal liver AGM region or even the whole embryos. It is of utmost importance that the embryos remain completely submerged in the specific mediums at ALL times. Embryo trapping in the surface tension of the liquid interface will damage all stages of embryos and distort or tear tissues. All changes of solutions of different density or osmolarity must be done gradually. It is better to keep embryos in one well and change solutions than to move embryos from solution to solution.

3.1. Embryo Collection and Fixation

1. Set up timed matings in the early evening and check the next morning for the presence of a vaginal plug (0.5 days postcoitus (E0.5)).
2. Euthanize pregnant dams with carbon dioxide and subsequent cervical dislocation on days E7.0–E9.5 of embryonic development.
3. Carefully dissect the embryos free from the decidua and Reichert's membrane with the yolk sac and ectoplacental cone intact (25). Use 100 mm petri dishes containing sufficient IMDM w/10% FBS to completely submerge specimens.
4. Gather embryos with a coated transfer pipette into one well of the round bottom 24-well plate. Embryos must remain submerged at all times to prevent damage from trapping in the surface tension.
5. Remove excess IMDM/FBS so that embryos are just slightly submerged. Wash three times with 2 ml of PBS for 5 min each removing as much of the previous wash as possible without damaging the embryos.
6. Fix embryos 10 min at -20°C with 2–3 ml of -20°C 100% acetone. The solution should turn cloudy as salts in the residual PBS precipitate out.
7. Carefully rinse three more times with 2 ml of PBS as above. The first wash should be added gradually with mixing (5–10 min) as the embryos will be very buoyant and need to be gradually equilibrated to prevent surface tension damage.
8. Replace PBS with 200–1,000 μl of blocking solution (see Note 5). Fixed embryos may be stored for a few days at 4°C in blocking solution without noticeable degradation of morphology and image quality.

3.2. Whole Embryo Immunolabeling

1. Directly conjugate the primary antibodies to fluorochromes using Alexa Fluor 546 (fibronectin or brachyury), Alexa Fluor 488 (VECad and Flk-1), and Alexa Fluor 647 (CD41)

monoclonal antibody labeling kits according to the manufacturer's protocol (see Note 6).

2. Distribute embryos in blocking solution with a transfer pipette into separate wells of the round bottom plate for each antibody concentration or combination to be tested. One to ten embryos are typically incubated in each well. See Table 1.
3. Carefully adjust the volume of each well by using a reference well similarly wetted and containing blocking solution.
4. Incubate embryos in blocking buffer 2–3 h at room temperature with gentle rocking block to prevent nonspecific antibody binding. See Table 1.
5. Incubate embryos with directly conjugated primary antibodies at optimal concentrations (see Note 7) for 12–18 h at 4°C with gentle rocking. Most directly conjugated antibodies are used at a final concentration of 5 μg/ml, while primaries detected by secondary antibodies are diluted to 2.5 μg/μl.
6. Perform 9 × 20 min washes with gentle rocking removing as much of the previous wash as possible.
7. Reblock in 200 μl of blocking solution for 1 h.
8. Incubate with secondary antibody at a final concentration of 5 μg/ml with gentle rocking for 1 h.
9. Repeat washes from step 6.

3.3. Embryo Clearing and Mounting

1. Gradually equilibrate embryos into 5% glycerol/PBS by incremental addition, prolonged mixing, and gradual replacement of the medium (see Note 8).
2. Prepare a glycerol step gradient containing 0.5 ml steps of 80, 70, 60, 50, 40, 30, 20, 10, and 5% glycerol in a 4-ml disposable tube. Gently layer lower concentration steps onto previous layers with a P-1000 pipette.
3. Load embryos onto the freshly prepared glycerol step gradient with a coated transfer pipette.
4. Seal lid and leave upright at 4°C overnight. If embryos have not settled on the bottom by morning, warm to room temperature and gently tap and spin tube(s) intermittently to facilitate settling.
5. Remove most of the overlying glycerol gradient and recover embryos from the round bottom of the tube.
6. Transfer embryos to a fresh clean well of the staining plate. Perform two changes with 80% glycerol 1 ml each. Use a gentle stream of air passed though an 18-gauge needle to spin/mix the embryos for this and subsequent reagent exchanges.
7. Superior tissue clearing may be achieved by gradually equilibrating the embryos/tissues into Murray's Clear (MC)

mounting medium using at least three intermediates (1:3, 1:1, and 3:1 MC:80% glycerol) (see Note 9).

8. Additional nuclear or cytoplasmic or membrane dyes may be added to the penultimate clearing step (e.g., DAPI, Sytox Green).
9. Prepare glass bottom dish for mounting embryo by placing 3–4 clay feet slightly larger than the specimen around the center of the dish within the perimeter of the coverslip (see Note 10).
10. Deposit embryo and sufficient mounting medium into center of dish. Manipulate embryo into desired position with clean forceps (usually a lateral view) (see Note 11).
11. Gently coverslip. Slide the coverslip to attain the optimal mounting angle (usually a lateral or frontal view is desirable). Carefully compress the clay feet with even pressure at alternate corners of the coverslip so that the embryo is just slightly compressed.
12. Add/remove excess mounting medium and completely seal the edge with rubber cement. Allow cement to dry clear (15–30 min).
13. Store embryos prior to imaging at 4°C in dark until for up to 2 weeks. Image quality slowly degrades after 4–5 days.

3.4. Confocal Image Acquisition and Stitching

1. Equilibrate mounted embryos to ambient temperature before positioning under oil-immersion objective.
2. The imaging depth and thus the size of the embryo that can be imaged completely is twice the working distance of the chosen objective with a two-sided imaging chamber. Oil immersion objectives between 20× and 40× offer the best compromise between optical resolution and long working distance required to collect larger, older embryos in their entirety.
3. Collect data in a cubic format (e.g., if the XY resolution of the 40× objective is 0.62 μm, collect z-slices at 0.62 μm intervals). The XY stage controller should be set to acquire adjacent fields with no overlapping of pixels.
4. Map the XYZ coordinates of all fields to be collected in the confocal acquisition software. Our stitching macros currently require left to right and then top to bottom collection of the adjacent fields beginning in the upper left most quadrant of the specimen.
5. Imaging depth is determined by the working distance of the objective. For whole embryo reconstructions, it is best if the total height of embryo is less than twice the working distance.

6. Remove the imaging chamber and gently clean immersion oil from imaged side with cotton swabs and Sparkle. Invert the imaging chamber to image the second side.
7. Embryos may be recovered postimaging, washed with PBS, and further processed (e.g., PCR genotyping).

4. Notes

1. The size and total cell number of embryos from inbred mouse strains such as C57BL/6 are about one half the size of outbred CD-1 embryos and thus easier to image completely (see Table 1).
2. A round bottom vessel is essential for efficient exchange of fixation, blocking, staining, and wash solutions. Polypropylene vessels are desirable if using acetone as the fixative. Reducing the plate height by about half provides better access for washes.
3. Alternative fixatives such as 4% paraformaldehyde may be necessary for different antibody combinations but acetone-fixed embryos are the clearest and produce the best whole embryo images especially in the E8.0 and larger embryos.
4. The exact fluorophores used are determined by the laser lines on the confocal microscope of choice. Antibodies with the weakest signals (determined empirically) are matched to the most

Table 1
Blocking and staining guidelines for various stages of embryos

Stage	Approximate number of cells/embryo[a]	Volume of 10% Triton X-100/1 ml (μl)	Maximum number of embryos/400 μl incubation	Minimum blocking time (h)
<E7.5	NA	12	50	0.5
E7.5	8×10^3	12	20	1
E8.5	56×10^3	24	10	2
E9.5	480×10^3	48	5	3
E10.5	$3{,}365 \times 10^3$	96	3	3
>E10.5	NA	96	2	3

[a] Cell numbers were derived from outbred CD-1 embryos (23) and applied to inbred strains (e.g., C57 Bl6 embryos) by dividing by 2. Primary antibody incubations are carried out overnight for all stages

sensitive photomultiplier tube (PMT) detectors. Alternatively, secondary antibodies may be used to amplify the primary antibody signal, but this requires more incubations and washings and control images to rule out nonspecific changes.

5. Blocking/staining solution: 10% serum in 1× PBS with varying amounts of 10% Triton X-100 depending on the age of the embryos. For E7.5 and younger embryos, use 12 μl of 10% Triton X-100 per 10 ml of serum/PBS, 24 μl for E8.5 embryos, 48 μl for E9.5 embryos, and 96 μl for E10.5 and older embryos.
6. It is important that the antibodies are free of amine-containing compounds, such as Tris-containing buffers and protein stabilizers such as BSA or gelatin. Stabilizers can be added after the labeling reaction. Store the eluted labeled antibody at 4°C for several months or freeze aliquots at –20°C or –80°C for long-term storage.
7. Each lot of antibody is titrated individually for the concentration giving the optimal signal-to-noise ratio. Typically, final concentrations approximating 5 μg/ml yield the best results. Similarly labeled isotype control antibodies should produce no specific staining at these concentrations but must be validated for each stage.
8. It is crucial that embryos be *gradually* equilibrated into glycerol-containing solutions to prevent crushing and distortion of the fragile intact embryos. Mix with a gentle stream of air blown through an 18-gauge needle obliquely positioned to the surface to promote mixing but not embryo damage. Mix each incremental step at least until the swirling refractive mismatch mixing lines disappear between steps. Mounting the needle on one of the swan necks with a bulldog clamp allows easy positioning and reduces error and fatigue.
9. This step requires patience and practice and may take 6–8 h.
10. Defined height spacers may be cut as 1 mm strips from layered (~20 μm/layer) mylar tape.
11. Yolk sacs from older larger embryos may be mounted flat on slides if the exact spatial location is not critical. This is preferred for high-resolution imaging (60× objective or higher). Arrange the more opaque outer visceral endoderm side down and coverslip and image from the inner mesothelial side.

Acknowledgment

Jason Byars at the Indiana Center for Biological Microscopy wrote the Image J stitching macros.

References

1. Sabin, F. R. (1920) Studies on the origin of blood-vessels and of red blood-corpuscles as seen in the living blastoderm of chicks during the second day of incubation *Contr Embryol* **9**, 215–262.
2. Luckett, W. P. (1978) Origin and differentiation of the yolk sac and extraembryonic mesoderm in presomite human and rhesus monkey embryos *Am J Anat* **152**, 59–97.
3. Moore, M. A., and Metcalf, D. (1970) Ontogeny of the haemopoietic system: yolk sac origin of in vivo and in vitro colony forming cells in the developing mouse embryo *Br J Haematol* **18**, 279–96.
4. Rhodes, K. E., Gekas, C., Wang, Y., Lux, C. T., Francis, C. S., Chan, D. N., Conway, S., Orkin, S. H., Yoder, M. C., and Mikkola, H. K. (2008) The emergence of hematopoietic stem cells is initiated in the placental vasculature in the absence of circulation *Cell Stem Cell* **2**, 252–63.
5. North, T. E., de Bruijn, M. F., Stacy, T., Talebian, L., Lind, E., Robin, C., Binder, M., Dzierzak, E., and Speck, N. A. (2002) Runx1 expression marks long-term repopulating hematopoietic stem cells in the midgestation mouse embryo *Immunity* **16**, 661–72.
6. Samokhvalov, I. M., Samokhvalova, N. I., and Nishikawa, S. (2007) Cell tracing shows the contribution of the yolk sac to adult haematopoiesis *Nature* **446**, 1056–61.
7. de Bruijn, M. F., Ma, X., Robin, C., Ottersbach, K., Sanchez, M. J., and Dzierzak, E. (2002) Hematopoietic stem cells localize to the endothelial cell layer in the midgestation mouse aorta *Immunity* **16**, 673–83.
8. Jones, E. A., Crotty, D., Kulesa, P. M., Waters, C. W., Baron, M. H., Fraser, S. E., and Dickinson, M. E. (2002) Dynamic in vivo imaging of postimplantation mammalian embryos using whole embryo culture *Genesis* **34**, 228–35.
9. Fraser, S. T., Hadjantonakis, A. K., Sahr, K. E., Willey, S., Kelly, O. G., Jones, E. A., Dickinson, M. E., and Baron, M. H. (2005) Using a histone yellow fluorescent protein fusion for tagging and tracking endothelial cells in ES cells and mice *Genesis* **42**, 162–71.
10. Jones, E. A., Baron, M. H., Fraser, S. E., and Dickinson, M. E. (2004) Measuring hemodynamic changes during mammalian development *Am J Physiol Heart Circ Physiol* **287**, H1561–9.
11. Jones, E. A., Baron, M. H., Fraser, S. E., and Dickinson, M. E. (2005) Dynamic in vivo imaging of mammalian hematovascular development using whole embryo culture *Methods Mol Med* **105**, 381–94.
12. Larina, I. V., Shen, W., Kelly, O. G., Hadjantonakis, A. K., Baron, M. H., and Dickinson, M. E. (2009) A membrane associated mCherry fluorescent reporter line for studying vascular remodeling and cardiac function during murine embryonic development *Anat Rec (Hoboken)* **292**, 333–41.
13. Jacobs, R. E., and Fraser, S. E. (1994) Magnetic resonance microscopy of embryonic cell lineages and movements *Science* **263**, 681–4.
14. Sharpe, J., Ahlgren, U., Perry, P., Hill, B., Ross, A., Hecksher-Sorensen, J., Baldock, R., and Davidson, D. (2002) Optical projection tomography as a tool for 3D microscopy and gene expression studies *Science* **296**, 541–5.
15. Walls, J. R., Sled, J. G., Sharpe, J., and Henkelman, R. M. (2007) Resolution improvement in emission optical projection tomography *Phys Med Biol* **52**, 2775–90.
16. Corbel, C., and Salaun, J. (2002) AlphaIIb integrin expression during development of the murine hemopoietic system *Dev Biol* **243**, 301–11.
17. Ferkowicz, M. J., Starr, M., Xie, X., Li, W., Johnson, S. A., Shelley, W. C., Morrison, P. R., and Yoder, M. C. (2003) CD41 expression defines the onset of primitive and definitive hematopoiesis in the murine embryo *Development* **130**, 4393–403.
18. Hashimoto, K., Fujimoto, T., Shimoda, Y., Huang, X., Sakamoto, H., and Ogawa, M. (2007) Distinct hemogenic potential of endothelial cells and CD41+ cells in mouse embryos *Dev Growth Differ* **49**, 287–300.
19. Mikkola, H. K., Fujiwara, Y., Schlaeger, T. M., Traver, D., and Orkin, S. H. (2003) Expression of CD41 marks the initiation of definitive hematopoiesis in the mouse embryo *Blood* **101**, 508–16.
20. Mitjavila-Garcia, M. T., Cailleret, M., Godin, I., Nogueira, M. M., Cohen-Solal, K., Schiavon, V., Lecluse, Y., Le Pesteur, F., Lagrue, A. H., and Vainchenker, W. (2002) Expression of CD41 on hematopoietic progenitors derived from embryonic hematopoietic cells *Development* **129**, 2003–13.
21. Ody, C., Vaigot, P., Quere, P., Imhof, B. A., and Corbel, C. (1999) Glycoprotein IIb-IIIa is expressed on avian multilineage hematopoietic progenitor cells *Blood* **93**, 2898–906.
22. Tavian, M., and Peault, B. (2005) Embryonic development of the human hematopoietic system *Int J Dev Biol* **49**, 243–50.

23. Palis, J., Robertson, S., Kennedy, M., Wall, C., and Keller, G. (1999) Development of erythroid and myeloid progenitors in the yolk sac and embryo proper of the mouse *Development* **126**, 5073–84.
24. Yokomizo, T., Takahashi, S., Mochizuki, N., Kuroha, T., Ema, M., Wakamatsu, A., Shimizu, R., Ohneda, O., Osato, M., Okada, H. et al. (2007) Characterization of GATA-1(+) hemangioblastic cells in the mouse embryo *Embo J* **26**, 184–96.
25. Hogan, B. L., Beddington, R. S., Costantini, F., and Lacy, E. (1994) Manipulating the mouse embryo. Plainview: *Cold Spring Harbor Laboratory Press.*

Chapter 11

Stem Cell Migration: A Zebrafish Model

Pulin Li and Leonard I. Zon

Abstract

Compared with other vertebrate animal models, zebrafish (*Danio rerio*) has its superior advantages for studying stem cell migration. Zebrafish have similar tissues and organs as mammals, where tissue-specific stem cells reside in. Zebrafish eggs are externally fertilized and remain transparent until most of the organs are fully developed. This allows imaging stem cells in vivo very easily. Recently, a zebrafish double pigmentation mutant, *casper*, became a new popular imaging model in the zebrafish field due to its completely transparent bodies in adulthood. It has been used as an excellent model to study adult hematopoietic stem cell (HSC) in the transplantation setting. The unparalleled imaging power of zebrafish provides great opportunities of tracing stem cells in vivo in the developmental and regenerative context. In this chapter, we use HSC as an example and combine the powerful imaging techniques in zebrafish, to provide protocols for in vivo imaging fluorescence-labeled stem cell migration, stem cell fate tracing in zebrafish embryos, HSC transplantation, and in vivo imaging in both zebrafish embryos and adults. These techniques can also be applied to other types of stem cells in zebrafish embryos and adults.

Key words: Zebrafish, Hematopoietic stem cell, Fluorescence proteins, Confocal fluorescence microscopy, Cell tracing, Casper

1. Introduction

In the past two decades, zebrafish has been proved to provide unprecedented opportunity for stem cell research, because of its conserved stem cell regulatory pathways as mammals and more extraordinary regenerative capability than mammals. Multiple types of stem/progenitor cells have been studied in zebrafish embryos and/or adults, such as HSCs (1, 2), neural crest stem cells (3, 4), and neural stem cells (5, 6). During the fast embryonic development, these stem cells migrate to the appropriate microenvironment, and undergo series of self-renewal and differentiation to form functional tissues and organs. In adult zebrafish, the homeostasis of most tissues and organs are still

Marie-Dominique Filippi and Hartmut Geiger (eds.), *Stem Cell Migration: Methods and Protocols*,
Methods in Molecular Biology, vol. 750, DOI 10.1007/978-1-61779-145-1_11, © Springer Science+Business Media, LLC 2011

maintained by stem cells, which allows the fast regeneration of damaged organs, such as the caudal tail fin, which includes multiple different tissue and stem cell types, such as vascular progenitors, mesenchymal stem cells, and melanocyte stem cells (7).

One of the many advantages of studying stem cells in zebrafish is the unparalleled imaging power, which allows directly tracing stem cells in vivo in the developmental and regenerative context. Zebrafish eggs are externally fertilized and remain transparent until most of the organs are fully developed. Imaging in adult zebrafish had been challenging because the pigmented cells reduced the visibility of internal organs. Recently, White et al. developed a transparent double pigment mutant zebrafish, called *casper*, in which organs such as the heart and blood vessels can be seen using standard stereomicroscopy. Using *casper* as the stem cell transplantation recipient allows for in vivo assessment of stem cell migration and engraftment (8).

Among the different stem cell types, HSC is one of the best studied stem cells. Zebrafish also provides great opportunities for studying HSC migration. Zebrafish embryonic hematopoiesis greatly resembles mammalian embryonic hematopoiesis, including primitive and definitive waves (1, 9). The primitive wave only gives rise to lineage-committed erythroid and myeloid progenitors, while the definitive wave generates functional HSCs and all lineages found in the adult. The definitive HSCs can be identified with markers, such as *runx1* (10, 11), *cmyb* (12, 13), and CD41 (14). The adult hematopoietic tissue, kidney marrow, the equivalent of mammalian bone marrow, is highly spatially localized in adult zebrafish. Transplanting adult zebrafish whole kidney marrow (WKM), which contains all the hematopoietic stem and progenitor cells, into lethally γ-irradiated fish can rescue the recipients with all the blood lineages fully repopulated (15).

Here, we use HSCs as an example and provide protocols for (1) live-imaging of fluorescence protein-labeled hematopoietic stem cell migration in zebrafish embryos; (2) in vivo stem cell tracing in zebrafish embryos; (3) and (4) hematopoietic stem cell transplantation and in vivo imaging in both zebrafish embryos and adults. These techniques can be adapted to study stem cells of other tissue/organ origins.

2. Material

2.1. Zebrafish Husbandry

Zebrafish are maintained and bred according to the protocols in *The Zebrafish Book* (16). Adult fish are maintained in aquaria at 28.5°C with Fish water circulating. Newborn embryos are kept in still E3 water in petri dish for 5–7 days without food, and then transferred to the aquaria. Adult zebrafish as transplantation recipients are kept off flow for 5 days in the special ICU water, and then moved to stand-alone circulating system designated only for postirradiation care, to reduce potential pathogen infection.

1. Fish water: Purified water is supplemented with instant ocean salts to 300–600 μS, and buffered with sodium bicarbonate and crushed coral to pH of 6.8–7.1.
2. E3 water: 0.2922 g of NaCl, 0.0127 g of KCl, 0.0366 g of $CaCl_2$, and 0.0390 g of $MgSO_4$ should be dissolved in 1 L of distilled water.
3. ICU water: regular Fish water for adults supplemented with 0.25 mL/L Stress Coat (fish and tap water conditioner, Aquarium Pharmaceuticals), 0.125 mL/L Melafix (Antibacterial fish remedy, Aquarium Pharmaceuticals), 0.125 mL/L Primafix (Antifungal fish remedy, Aquarium Pharmaceuticals).

2.2. Special Zebrafish Lines

1. *Tg*(*cmyb:GFP*) transgenic reporter line was created by homologous recombination of a 3.7-kb EGFP construct downstream of the 5′ untranslated region and precisely before the start site of a five phage artificial chromosome (PAC) clone containing *cmyb*[2].
2. *Tg*(*CD41:GFP*) transgenic reporter line was created by cloning and placing the CD41 promoter upstream of the GFP cDNA in the expression vector pEGFP-1 (Clontech). A 6-kb DNA sequence containing putative CD41 promoter elements was identified and cloned from a zebrafish PAC clones containing the CD41 gene (14).
3. *Tg*(*gata1:DsRed*) transgenic reporter line was created by cloning a 7-kb zebrafish gata1 promoter fragment into the multiple cloning site of the pDsRed2-1 vector (Clontech) (17).
4. *Tg*(*lmo2:DsRed*) transgenic reporter line was created by subcloning a 2.5 kb *lmo2* promoter fragment into *pDsRed2-1* vector (Clontech) upstream of the DsRed fluorescent reporter gene (18).
5. *Tg*(*β-actin:GFP*) ubiquitous transgenic line was created by cloning a 10 kb fragment containing the first exon and upstream sequence of zebrafish β-actin into pBluescript upstream of a 1.5-kb EGFP (19).
6. *Casper* is a double homozygous pigmentation mutant created by crossing *nacre* (*mitf*$^{-/-}$) and *roy* (mutation unknown) (8).
7. Red GloFish® is a DsRed2 transgenic fish driven by ubiquitous promoters. They were commercially purchased through 5D-Tropical.

2.3. Other Reagents

1. Pronase: Type XXV protease, 30 mg/mL in dH_2O.
2. Tricaine-S (3-amino benzoic acid ethyl ester): Sigma–Aldrich makes fresh stock at 4 mg/mL in fish water. Long-term storage should be at –20°C, and short-term stock can be kept at 4°C.
3. Low-melting point agarose: dissolved in E3 water by heating up.
4. Caged rhodamine–dextran 10,000 (Molecular Probes).
5. Liberase Blendzyme II (Sigma–Aldrich).
6. PBS: phosphate-buffered saline 1×, without calcium or magnesium.
7. FBS: fetal bovine serum, heat-inactivated.
8. Trypan blue: 0.4% trypan blue stock can be diluted with dH_2O at 1:4 ratio and the diluted solution can be added to cells at 1:1 ratio (final 1:10 dilution).
9. Penicillin and streptomycin: 10,000 units/mL penicillin and 10 mg/mL streptomycin.

3. Methods

3.1. Live-Imaging of Fluorescence Protein-Labeled Hematopoietic Stem Cell Migration in Zebrafish Embryos

1. The procedures of making lineage-specific fluorescence protein-labeled transgenic zebrafish have been thoroughly described before (20). During the definitive hematopoiesis in zebrafish embryos, the hematopoietic stem cells express *cmyb* and CD41, which can be labeled as the GFP^+ cells in the *Tg*(*cmyb:GFP*) and *Tg*(*CD41:GFP*) transgenic reporter lines (2, 14).
2. Adult *Tg*(*cmyb:GFP*) or *Tg*(*CD41:GFP*) fish are outcrossed with wild-type zebrafish. Embryos are collected the next morning and staged for the desired developmental time points to look at the specific hematopoietic tissues, such as the aorta-gonad-mesonephron (AGM) region between 36 and 40 hpf and the caudal hematopoietic tissue (CHT) between 3 hpf and 5 dpf (Fig. 1a).
3. Embryos are dechorionated by pronase treatment in E3 water and anesthetized in E3 water containing 0.04 mg/mL Tricaine-S first and then embedded in 1% low-melting point agarose (36–48 hpf) with the same concentration of Tricaine-S on 35-mm glass-bottomed dishes (Iwaki, Chiba, Japan). The embryos should be transferred to the agarose when the agarose is cooled down and flattened with fine forceps before it is solidified. A coverslip can be added on the top (see Note 1).
4. To follow the dynamic process of single cell migration, time-lapse confocal imaging is performed on a Zeiss LSM510 confocal microscope (Carl Zeiss, Le Pecq, France). For longer

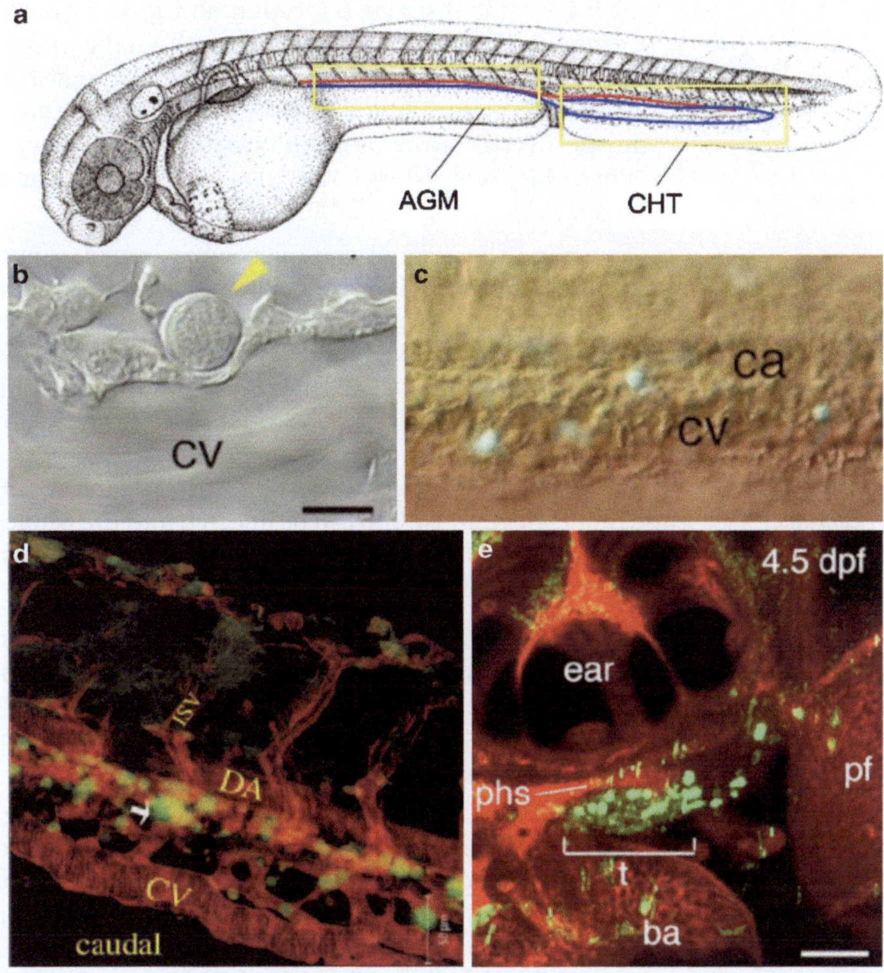

Fig. 1. (**a**) Illustration of zebrafish embryonic hematopoietic tissues; (**b**) DIC image of a HSC-like cell (*arrowhead*) rolling along the vessel in the CHT region; (**c**) fluorescence and DIC image overlay of the CHT in the 35-hpf old *Tg*(*CD41:GFP*) embryo. The CD41⁺ cells start to colonize the CHT region from 35 hpf. (This research was originally published in *Blood.* Kissa, K. et al. Live imaging of emerging hematopoietic stem cells and early thymus colonization. *Blood.* 2008;**111**:1147–56. © the American Society of Hematology) (22); (**d**) fluorescence image of the caudal tissue of a 48-h old *Tg* (*CD41:GFP; lmo2:DsRed*) embryo. The CD41:GFP⁺ cells (*the arrow points* to a HSC-like cell) are in close contact with the blood vessels (*red*) (This research was originally published in *Blood.* Lin, H.F. et al. Analysis of the thrombocyte development in CD41-GFP transgenic zebrafish. *Blood.* 2005;**106**:3083–10. © the American Society of Hematology) (14); (**e**) CD41:GFP⁺ cells (*green*) colonize the left thymus at 4.5 dpf. The surrounding tissues are nonspecifically labeled by BODIPY TR (*red*) (This research was originally published in *Blood.* Kissa, K. et al. Live imaging of emerging hematopoietic stem cells and early thymus colonization. *Blood.* 2008;**111**:1147–56. © the American Society of Hematology) (22).

time period of imaging, the imaging stage should be covered in a PeCon open chamber (PeCon, Erbach, Germany) and the temperature should be set at 28°C. A CCD color video camera is used for capturing the images over time.

5. Besides using the fluorescent reporter embryos, the migration of the blood cells can be directly visualized and recorded with differential interference contrast (DIC) video because

they have the distinctive location and movement (Fig. 1b). Although DIC video can image the cells in the microenvironment surrounding the HSC, it cannot distinguish the stem cells from mature blood cells. So the DIC and fluorescence image overlay gives more information than either by itself alone. Figure 1c shows the CD41:GFP^{+} cells (fluorescence image) colonizing CHT (DIC image) at 35 hpf.

6. To image the cells in the microenvironment surrounding the HSC, other fluorescent reporters can be used together with the *Tg*(*CD41:GFP*) and *Tg*(*cmyb:GFP*). For example, *Tg*(*lmo2:DsRed*) have blood vessels marked in red. By recording both GFP and DsRed, the relative localization of HSCs to the blood vessels can be monitored (Fig. 1d) (see Note 2).

3.2. In Vivo Stem Cell Tracing in Zebrafish Embryos

To follow the fate of a specific HSC or progenitor cell, caged rhodamine–dextran 10,000 can be injected into the whole embryo and uncaged at single cell level later to mark a specific stem cell. The procedure is described as follow:

1. Using the regular microinjection technique as described by Rosen et al. (21), 0.5 ng of caged rhodamine–dextran 10,000 can be injected into the 1- to 4-cell stage *Tg*(*CD41:GFP*) embryos. The embryos are allowed to develop in the dark to reduce the illumination and bleaching.
2. The embryos are uncaged under the inverted microscope with the power of single cell resolution. A 365-nm Micropoint pulsed nitrogen laser system (Photonic Instruments, St Charles, IL) can be oriented through the epifluorescence port to target the embryos. A laser pulse of 10–20 s each is recommended for labeling cells in the AGM region around 36 hpf.
3. After uncaging, the embryos are allowed to develop in the dark until the stage is ready for imaging. Uncaged rhodamine–dextran cells can be observed by confocal microscopy as described in Subheading 3.1. Figure 2 is an example of 10 CD41:GFP^{low} HSCs in the AGM region uncaged at 40 hpf, which later seed the thymus by 4.5 dpf (see Notes 3 and 4).

3.3. Transplantation and In Vivo Imaging of Hematopoietic Stem Cell Migration in Zebrafish Embryos

Hematopoietic stem cells can migrate to appropriate microenvironment after transplantation. In addition, transplanting stem cells from a donor into a recipient can help distinguish autonomous versus nonautonomous effects of certain signaling pathways or genes. The transplantation of zebrafish hematopoietic stem cells can be performed in three different manners: embryo to embryo, adult to embryo, and adult to adult, while transplanting embryonic HSCs into adults is still facing some technical challenges. Because of the transparency, zebrafish embryos are great recipients for visualizing transplanted cells in vivo. The transplantation technique into embryos is introduced in this subheading,

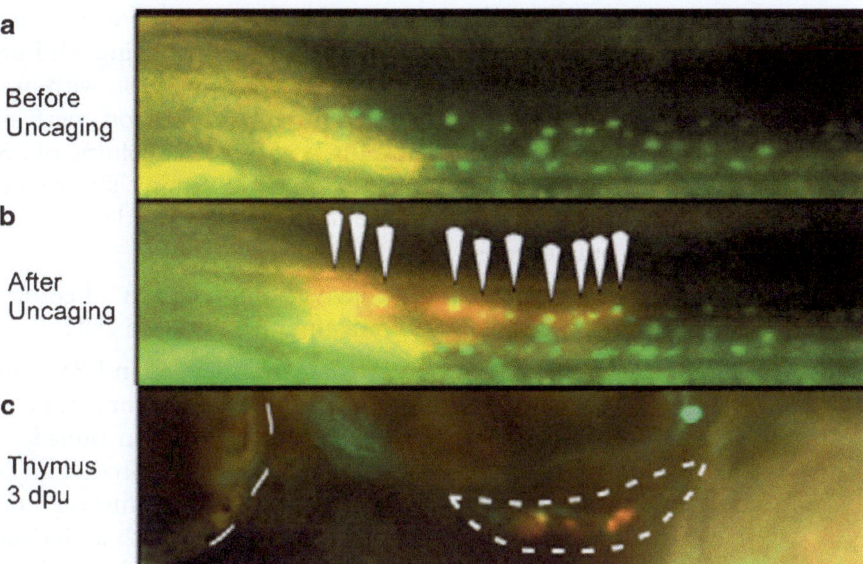

Fig. 2. Uncaging rhodamine–dextran in the CD41+ cells in the AGM region and observing the uncaged cells seeding the thymus. (**a**) CD41+ cells in the AGM before uncaging at 40 hpf; (**b**) uncaged cells in the AGM at 40 hpf right after uncaging (*arrowhead, orange*); (**c**) the rhodamine positive cells seed the thymus 3 days after the uncaging. (Reproduced from Bertrand, J.Y. et al., 2007 with permission from The Company of Biologists) (23).

while transplanting from adult to adult fish is described in Subheading 3.4.

1. Primary donor cells can be harvested from either embryonic or adult fluorescence reporter fish.
2. From embryos: Single cell suspension from younger embryos can be prepared by mechanical dissociation. For older embryos, the combination of mechanical and protease dissociation is proved to be effective. Three-day-old *Tg*(*CD41:GFP;gata1:DsRed*) embryos are rinsed with sterile PBS and then mince up the embryos with a clean razor. Then Liberase Blendzyme II is added into the PBS to 0.14 units/mL. Incubate the embryos with the enzyme at 33°C until the embryos are dissociated into single cells (see Note 5). To stop the digestion, FBS is added to 5%. To get rid of the undigested tissues, the suspension is filtered through 40-μm nylon mesh, washed once with PBS and spinned down at 1,500 rpm (400 g) for 5 min. The pellets are resuspended in PBS and undergo fluorescence-activated cell sorting. The hematopoietic stem/precursor cells fall into the $CD41^{+}/gata1^{-}$ gate. The cells are collected into PBS supplemented with 5% FBS.
3. From adults: Adult *Tg*(*β-actin:GFP; gata1:DsRed*) zebrafish are sacrificed with overdosed Tricaine-S. To dissect WKM, a ventral midline incision is made on the donor fish and the pigmented kidney is located right behind the dorsal side of

the swim bladders. Whole kidneys are dissected out and placed into 1 mL of ice-cold PBS containing 5% FBS. Single-cell suspensions are generated by aspiration with pipettes followed by filtering through a 40-μm nylon mesh filter. The flow-though part is diluted with a final volume of 25 mL and spinned down at 1,500 rpm for 8 min. The supernatant is discarded and the pellet cells are resuspended in 1 mL of PBS containing 5% FBS.

4. The number of cells alive is counted with a hemocytometer, using Trypan blue to distinguish the dead cells.
5. Cells are spinned down and resuspended in PBS, and injected into the sinus venous of wild-type or mutant embryos through borosilicate glass capillary needles (1 mm outside diameter, no filament) made with a Flaming/Brown micropipette puller. Cell suspensions are back-loaded into each needle and injected into circulation by forced air with a Narishige injection station and a Narishige micromanipulator (17).
6. Recipient embryos are maintained in E3 medium containing 5% penicillin and streptomycin during and for several hours after transplantation to prevent infection.
7. Transplanted embryos are visualized at the desired time point after transplant with a fluorescent microscope to monitor donor cell migration as described in Subheading 3.1. Figure 3a is the wild-type recipient embryos transplanted with GFP^{+};$DsRed^{-}$ cells from *Tg(CD41:GFP;gata1:DsRed)* donor embryos at 3 dpf. One day after transplantation, the GFP^{+} cells already colonize the thymus and CHT. Figure 3b shows a *bloodless* recipient with donor-derived blood cells circulating at 8 week after being transplanted with *Tg(β-actin:GFP; gata1:DsRed)* whole kidney marrow cells.

3.4. Transplantation and In Vivo Imaging of Stem Cell Migration in Adult Zebrafish

1. Ten- to fourteen-week-old *casper* recipients are sublethally irradiated with 30 Gy γ-irradiation, split by two doses, 15 Gy each at both 2 days and 1 day before transplantation.
2. Whole kidney marrow from adult *Tg(β-actin:GFP)* or other fluorescence reporter zebrafish, e.g., Red GloFish®, are harvested as described in Subheading 3.3 (see Note 5).
3. Adult *casper* is used as peripheral blood donors. *Casper* is anesthetized with 0.2% Tricaine-S. Peripheral blood is obtained by cardiac puncture with micropipette tips coated with heparin and collected 0.9× PBS containing 5% FBS. Single-cell suspensions of peripheral blood cells are generated by filtering through a 40-μm nylon mesh filter into a 50-mL conical tube. The number of cells is counted with a hemocytometer.
4. The peripheral blood and kidney marrow cells are spinned down at 1,500 rpm for 5 min and resuspended in 0.9× PBS

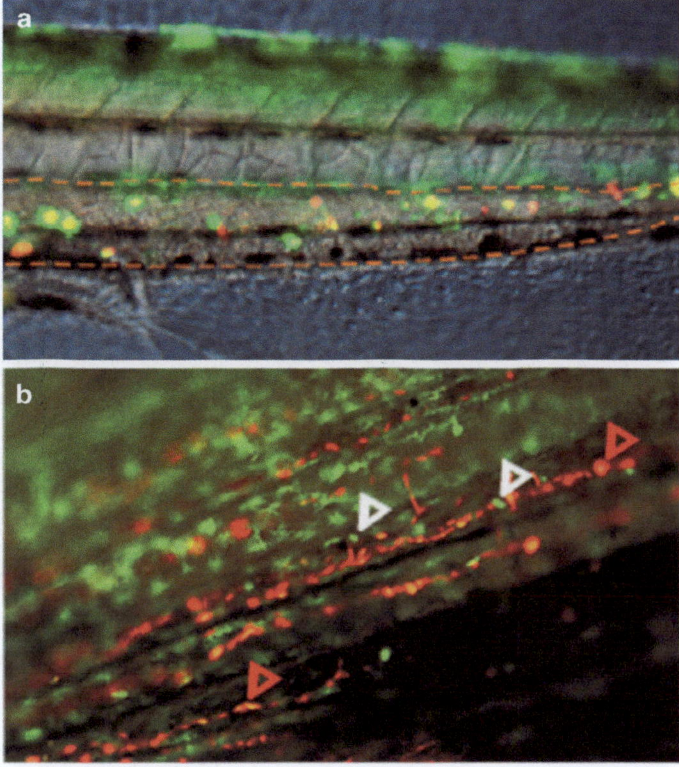

Fig. 3. (**a**) Transplanted CD41:GFP+ cells robustly colonize the caudal hematopoietic tissues 1 day after transplantation. Transplanted CD41+ cells also carried a gata1:DsRed transgene to visualize erythroid differentiation. (Reproduced from Bertrand, J.Y. et al., 2008 with permission from The Company of Biologists) (12); (**b**) transplanting adult WKM from *Tg(β-actin:GFP;gata1:DsRed)* donors into bloodless mutant embryos. The recipient embryos sustain the donor-derived hematopoiesis up to 8 week old. The figure represents GFP+ leukocytes are (*white arrowhead*) and DsRed+ erythrocytes (*red arrowhead*) circulating in the dermal capillaries. (Reproduced from Traver, D. et al., 2003 with permission from Nature Publishing Group) (17).

containing 5% FBS at the desirable final concentrations, and mixed together for transplant.

5. *Casper* recipients are anesthetized in Tricaine-S. 3–5 μL of the cell suspension mixture above is injected into the circulation retro-orbitally through a Hamilton syringe (26 s gauge, 10 μL volume), as demonstrated in Fig. 4a. Usually 50,000–200,000 marrow cells plus 100,000–200,000 peripheral blood cells should be able to rescue the recipient.
6. Recipients are kept in still ICU water to reduce fungal and bacterial infection for 1 week with minimal feeding, and then placed into system with circulating water and reduced amount of food.
7. Transplanted recipients can be anesthetized in Tricaine-S and placed on a plastic or agarose plate. Donor cells can be

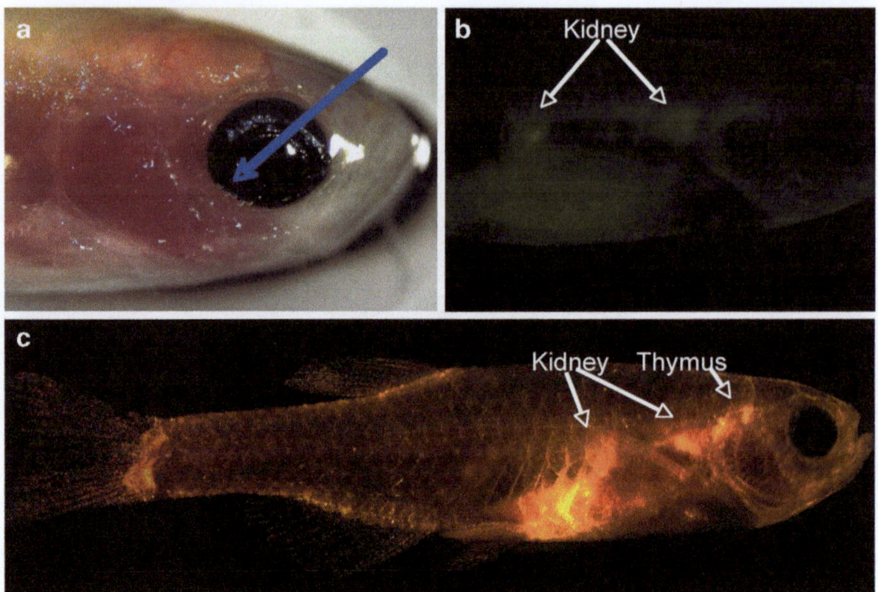

Fig. 4. Adult WKM transplant in *casper*. (**a**) Illustration of retro-orbital injection into *casper*, (**b**) kidney engraftment at 4 week posttransplant in casper transplanted with 100,000 *Tg(β-actin:GFP)* WKM cells; (**c**) multiple organ engraftment at 6 week posttransplant in *casper* transplanted with 200,000 Red GloFish® WKM cells.

visualized multiple times over time without sacrificing the animals. For fluorescent and magnification views, a Zeiss Discovery V8 stereomicroscope with a 1.2× PlanApo lens and GFP/DsRed2 filters is used. Images are captured using AxioVision software (Fig. 4b, c).

8. To achieve single cell resolution, transplant recipient zebrafish are embedded in 1% low-melting point agarose containing 0.04 mg/mL Tricaine-S in glass-bottom culture dishes. Confocal microscopy is performed for GFP positive cells using a Zeiss LSM Meta confocal microscope.

4. Notes

1. For the confocal imaging step, the older embryos (3–5 dpf) can be mounted with 3% methylcellulose.

2. Another choice to light up the surrounding cells is to use nonspecific dyes, such as CellTrace BODIPY TR (Molecular Probes). Embryos are soaked in E3 water with 0.1 mM BODIPY TR for 1 h and washed with plain E3 water three times. Filter sets suitable for Texas Red can be used for imaging BODIPY TR labeled tissue. Figure 1e is an example of the $CD41^+$ HSCs that colonize the thymus at 4.5 dpf. The thymus and other surrounding tissues are shown as red.

3. To follow up the uncaged cells, extensive imaging/light is required for the uncaged embryos, which might cause unwanted uncaging of other cells. To reduce this noise, additional long-pass filters can be installed in both the epifluorescence and transmitted light paths.
4. Unfortunately, the production of rhodamine–dextran 10,000 has been discontinued. Instead, caged fluorescein–dextran can be used in a similar way. The uncaging procedure is the same as described before. The detection of uncaged fluorescein requires fixation and immunohistochemistry, which has been described by Murayama et al. in details (24). Another alternative is photoconvertible fluorescent protein as described by (25).
5. The dissociation time depends on the enzyme concentration and the stage of the embryos. The dissociation can be monitored with a microscope. Triturating the suspension with pipette can help accelerate the dissociation.
6. Due to the lack of sophisticated stem cell markers, it has been difficult to identify hematopoietic stem cells in adult zebrafish. Potentially, CD41^{low} population or side population can be considered as candidates. Transplanting these more purified populations might lead to better observation of stem cell migration.

Acknowledgments

We thank Dr. Owen Tamplin for reading the manuscript, Dr. Richard M. White for developing the zebrafish retro-orbital injection technique, and the rest of the Zon lab for the constant advice and help.

References

1. de Jong, J.L., and Zon, L.I. (2005) Use of the zebrafish system to study primitive and definitive hematopoiesis *Annu Rev Genet* **39**, 481–501.
2. North, T.E., Goessling, W., Walkley, C.R., et al. (2007) Prostaglandin E2 regulates vertebrate haematopoietic stem cell homeostasis *Nature* **447**, 1007–11.
3. Dorsky, R.I., Moon, R.T., and Raible, D.W. (1998) Control of neural crest cell fate by the Wnt signalling pathway *Nature* **396**, 370–3.
4. White, R.M., and Zon, L.I. (2008) Melanocytes in development, regeneration, and cancer *Cell Stem Cell* **3**, 242–52.
5. Chapouton, P., Adolf, B., Leucht, C., et al. (2006) her5 expression reveals a pool of neural stem cells in the adult zebrafish midbrain *Development* **133**, 4293–303.
6. Stigloher, C., Chapouto, P., Adolf, B., and Bally-Cuif, L. (2008) Identification of neural progenitor pools by E(Spl) factors in the embryonic and adult brain *Brain Res Bull* **75**, 266–73.
7. Johnson, S.L., and Bennett, P. (1999) Growth control in the ontogenetic and regenerating zebrafish fin *Methods Cell Biol* **59**, 301–11.
8. White, R.M., Sessa, A., and Burke, C., et al. (2008) Transparent adult zebrafish as a tool for in vivo transplantation analysis *Cell Stem Cell* **2**, 183–9.

9. Galloway, J.L., and Zon, L.I. (2003) Ontogeny of hematopoiesis: examining the emergence of hematopoietic cells in the vertebrate embryo *Curr Top Dev Biol* **53**, 139–58.
10. Kalev-Zylinska, M.L., Horsfield, J.A., Flores, M.V., et al. (2002) Runx1 is required for zebrafish blood and vessel development and expression of a human RUNX1-CBF2T1 transgene advances a model for studies of leukemogenesis *Development* **129**, 2015–30.
11. Lam, E.Y., Chau, J.Y., Kalev-Zylinska, M.L., et al. (2008) Zebrafish runx1 promoter-EGFP transgenics mark discrete sites of definitive blood progenitors *Blood* **113**, 1241–1249.
12. Bertrand, J.Y., Kim, A.D., Teng, S., and Traver, D. (2008) CD41+ cmyb+precursors colonize the zebrafish pronephros by a novel migration route to initiate adult hematopoiesis *Development* **135**, 1853–62.
13. Mucenski, M.L., McLain, K., Kier, A.B., et al. (1991) A functional c-myb gene is required for normal murine fetal hepatic hematopoiesis *Cell* **65**, 677–89.
14. Lin, H.F., Traver, D., Zhu, H., et al. (2005) Analysis of thrombocyte development in CD41-GFP transgenic zebrafish *Blood* **106**, 3803–10.
15. Traver, D., Winzeler, A., Stern, H.M., et al. (2004) Effects of lethal irradiation in zebrafish and rescue by hematopoietic cell transplantation *Blood* **104**, 1298–305.
16. Monte Westerfield IoN, University of Oregon. 2000 *The Zebrafish Book. 4 ed*: University of Oregon Press.
17. Traver, D., Paw, B.H., Poss, K.D., Penberthy, W.T., Lin, S., and Zon, L.I. (2003) Transplantation and in vivo imaging of multilineage engraftment in zebrafish bloodless mutants *Nat Immunol* **4**, 1238–46.
18. Zhu, H., Traver D., Davidson, A.J., et al. (2005) Regulation of the lmo2 promoter during hematopoietic and vascular development in zebrafish *Dev Biol* **281**, 256–69.
19. Gillette-Ferguson, I., Ferguson, D.G., Poss, K.D., and Moorman, S.J. (2003) Changes in gravitational force induce alterations in gene expression that can be monitored in the live, developing zebrafish heart *Adv Space Res* **32**, 1641–6.
20. Lin, S. (2000) Transgenic zebrafish *Methods Mol Biol* **136**, 375–83.
21. Rosen, J.N., Sweeney, M.F., and Mably, J.D. (2009) Microinjection of zebrafish embryos to analyze gene function *J Vis Exp.*
22. Kissa, K., Murayama, E., Zapata, A., et al. (2008) Live imaging of emerging hematopoietic stem cells and early thymus colonization *Blood* **111**, 1147–56.
23. Bertrand, J.Y., Kim, A.D., Violette, E.P., Stachura, D.L., Cisson, J.L., and Traver, D. (2007) Definitive hematopoiesis initiates through a committed erythromyeloid progenitor in the zebrafish embryo *Development* **134**, 4147–56.
24. Murayama, E., Kissa, K., Zapata, A., et al. (2006) Tracing hematopoietic precursor migration to successive hematopoietic organs during zebrafish development *Immunity* **25**, 963–75.
25. Hatta, K., Tsujii, H., and Omura, T. (2006) Cell tracking using a photoconvertible fluorescent protein *Nat Protoc* **1**, 960–7.

Chapter 12

Imaging Pluripotent Cell Migration in *Drosophila*

Michael J. Murray and Robert Saint

Abstract

Drosophila melanogaster offers a powerful system for the analysis of cell migration. In the embryo, pluripotent cells of the mesodermal and endodermal primordia undergo epithelial–mesenchymal transitions and cell migration, while primordial germ cells migrate through an endodermal barrier to form the gonads. Visualisation of these migrations has traditionally been achieved by staining fixed embryos at different developmental stages or through live imaging of cells using tissue-specific expression of marker fluorescent proteins. More recently, photoactivatable fluorescence proteins have allowed the labelling of small groups of cells or single cells so that their migratory patterns and fate can be followed. By fusing the photoactivatable fluorescent protein to proteins that mark different subcellular components, it is now possible to visualise different aspects of the cells as they migrate. Here, we review previous studies of the migration of pluripotent embryonic cells and describe, in detail, methods for visualising these cells.

Key words: *Drosophila*, Mesoderm, Photoactivatable GFP, Cell migration, Cell fate mapping

1. Introduction

After more than a century of research, *Drosophila melanogaster* continues to be a key model system for the study of cell and developmental biology. *Drosophila* provides a wide range of elegant genetic tools and resources (1, 2), and a well-developed body of knowledge concerning the cellular and genetic basis of development (3). In addition, since the earliest days following the discovery of green fluorescent protein (GFP) (4, 5), flies have been at the forefront of model systems utilising new fluorescent protein (FP) technology (6).

In this chapter, we begin with an overview of the use of FPs for live cell imaging in *Drosophila*. We then focus on three pluripotent cell types in the early embryo – primordial germ cells (PGCs), mesodermal cells, and endodermal cells – and explain how live cell imaging has enhanced our understanding of their

Marie-Dominique Filippi and Hartmut Geiger (eds.), *Stem Cell Migration: Methods and Protocols*, Methods in Molecular Biology, vol. 750, DOI 10.1007/978-1-61779-145-1_12, © Springer Science+Business Media, LLC 2011

migration. Finally, we describe in detail a relatively new approach that uses photoactivatable GFP (PAGFP) to optically highlight cells in the early embryo via spatially restricted exposure to short wavelength light. Since detailed protocols for studying cell migration in *Drosophila* using standard immunohistochemical (7) and live imaging (6, 8) techniques have been published previously, we focus here on the particular advantages and challenges associated with photoactivatable FPs.

1.1. Using Fluorescent Proteins to Visualise Cells in Drosophila

Live imaging of cells expressing FPs has become an integral component of *Drosophila* research. Live imaging is possible at several stages of the life cycle of flies but is easiest during embryogenesis. Following the removal of a waxy outer covering called the chorion, the embryo is conveniently encased in a transparent shell called the vitelline membrane. Dechorionated embryos are easily visualised by light microscopy and develop normally, provided they are protected from desiccation by a covering of water or gas-permeable oil (see Subheading 2). Cell movement can also be imaged in early pupae through the semitransparent pupal case, and in later pupae by dissecting away a portion of the increasingly opaque pupal case (9), and in adult tissues such as the ovaries by dissecting the organs and culturing them (10).

FP expression is most simply achieved by placing FP transgenes directly under the control of regulatory sequences of DNA that confer a desired pattern of expression. For example, studies on epidermal cell behaviour have utilised promoters from ubiquitously expressed genes (e.g. ubiquitin) fused to FP fusion proteins that localize to the cell cortex (11). Alternatively, one can highlight particular subsets of cells by using promoters from genes that are specifically transcribed in those cells (e.g. *nanos* sequences to label PGCs – see Subheading 1.2).

The binary GAL4/UAS transcriptional activation system provides an alternative and more flexible approach to FP expression (12, 13). Here, the yeast transcriptional activator GAL4, under the control of various regulatory sequences, drives transcription of a reporter line, such as GFP, which itself is placed downstream of the Upstream Activation Sites (UAS) to which GAL4 binds. When the driver and reporter are brought together in the same fly, the GAL4 protein binds to the UAS and the reporter is expressed in the GAL4 pattern. For example, a GAL4 line using the promoter of the *slow border cell* was used to drive expression of a UAS-GFP-Moesin line in border cells, a small cluster of somatic cells that detach from the follicular epithelium, allowing visualisation of their migration through the cellular milieu of the egg chamber (10). Several more complicated schemes for genetic manipulation are also available (2, 13).

1.2. Live Imaging of Pluripotent Cell Migration in Drosophila

In *Drosophila*, stem cells have been identified in the central nervous system (14), at the tips of ovaries and testes (15), in the larval lymph gland (16), and in the epithelia of the posterior adult midgut (17, 18), hindgut (19), and malpighian tubules (20). These cells are not migratory, so the use of live imaging has been confined to visualising cell divisions and not movement (21, 22). There are, however, three migratory pluripotent cell types in the early embryo: PGCs, mesodermal cells and endodermal cells. Although not stem cells at this stage, these undifferentiated cells all give rise to multiple differentiated cells types and/or stem cells later in development.

PGCs begin migration as a tight cluster of cells within a pouch formed by the posterior midgut (PMG) epithelium. PGCs migrate through this epithelium, away from the midline and towards the overlying mesoderm where they eventually coalesce with somatic gonadal precursor cells (SGPs) to form the gonads. Although PGC migration has received extensive genetic analysis over the years (23), our understanding of the cellular mechanisms involved has been significantly advanced recently by the use of live FP reporters. For example, it has been known for some time that transepithelial migration requires the G-protein coupled receptor, *Tre1* (24). Recently, however, expression of an eGFP-Moesin actin binding domain reporter using regulatory regions from the *nanos* gene, has revealed that *tre1* mutant PGCs fail to dissociate and polarise towards the midgut epithelium (25). This same reporter also showed that the lipid phosphatases *wunen* and *wunen2*, which were known to be required for lateral migration, are essential in preventing PGCs from aberrantly crossing the midline (26). PGC migration, therefore, is both genetically tractable, and also accessible for live imaging via genetically controlled FP reporters.

Mesodermal cells arise as a band of cells in the ventral epidermis that furrows inwards at gastrulation to form an internalised epithelial tube. The cells then undergo an epithelial to mesenchymal transition (EMT), dissociate, and migrate out over the inner surface of the ectoderm (27). Genetic analysis of mesoderm migration is also well advanced, with two signalling pathways having been identified: the Heartless FGF receptor pathway (28–31) and the Pebble Rho GEF pathway (32–34). In contrast to PGCs, however, expression of FPs using mesoderm-specific promoters, has not been useful, due to the relatively short period of time between mesoderm specification and migration. To circumvent this problem, we turned to PAGFP which allowed us to visualise mesoderm migration live, as well as track the fate of individual cells (see below). With this approach we found that mesodermal cells spread out using a combination of directed group migration over the ectoderm, "cell-hopping," in which internal cells move past outer cells, and intercalation (35). More recently, multiphoton imaging of embryos ubiquitously expressing an FP nuclear marker, combined with automated tracking software, has provided an alternative way of viewing these cell rearrangements (36).

Finally, endodermal cells derive from the PMG and anterior midgut (AMG) primordia, which are located at the opposite ends of the blastoderm. Cells of each primordium are internalised during gastrulation, undergo an EMT, and migrate towards each other along the visceral mesoderm (Fig. 1) (37). The genetic regulation of endoderm migration is less well understood than that of the PGCs and mesoderm, but is known to involve integrins and Rho GTPase family members (37–39). The behaviour of

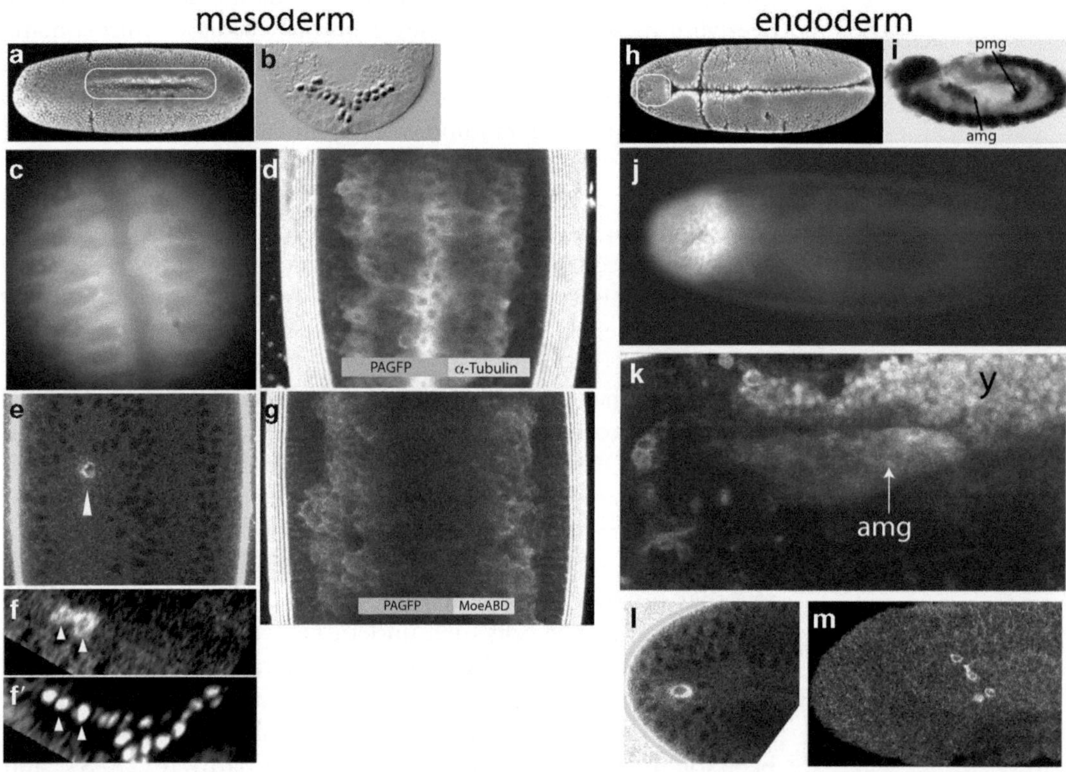

Fig. 1. Application of PAGFP for live imaging and cell fate mapping of the mesoderm (**a–h**) and endoderm (**i–m**). (**a**, **b**) Presumptive mesodermal cells in the ventral epidermis invaginate to form a furrow, undergo an EMT, and then migrate over the ectoderm. (**c**, **d**) Photoactivation of PAGFP-αTubulin in a group of cells labels the entire mesoderm over several segments. (**e**, **f**) Photoactivation of a single cell (**e**; *arrowhead*) results in a clone of two cells (**f**; *arrowheads*), 45 min later, that can be clearly visualised after embryos have been fixed and immunostained – in this case for the mesoderm-specific transcription factor Twist (**f′**; *arrowheads*). (**g**) PAGFP-MoeABD, which binds F-Actin filaments, highlights the protrusions of migrating cells. (**h**, **i**) Endodermal midgut cells arise at both the anterior (*circle*) and posterior (not shown) ends of the embryo, where they invaginate, undergo an EMT and, later, migrate towards each other along the visceral mesoderm (**i**). (**j**, **k**) Photoactivation of cells surrounding the anterior end of the ventral furrow allows the migration of the anterior midgut (AMG) to be visualised. Yolk particle autofluorescence is also visible (y). (**l**, **m**) Single cell labelling in the AMG anlagen results in a clone of four cells approximately 3.5 hr later. Images **a** and **i** are from FlyBase (http://www.flybase.org). Images **b** and **j**, are from FlyMove (http://flymove.uni-muenster.de).

endodermal cells during migration is also largely unknown, since live imaging of endodermal cells is beset by similar problems to the mesoderm. However, as explained in the next section, we are beginning to address this gap in our knowledge with the use of PAGFP.

1.3. Photoactivatable GFP as a Tool for Live Imaging

Photoactivatable molecules have the remarkable property that their absorption and emission characteristics are modified by exposure to short wavelength light (40). In the case of PAGFP, the response to 488 nm light increases by over 100-fold (41, 42). For photoconvertible molecules, such as Kaede, Kikume and EosFP, photoactivation changes the fluorescence from green to red (40). By expressing such FPs ubiquitously one can "turn on" a subset of cells by spatially restricted exposure to the activating light. Photoactivation can be performed with single photon (e.g. this work) or multiphoton excitation (43, 44). Multiphoton excitation permits cells in multilayered tissues to be photoactivated without labelling cells above and below the focal plane (43). This is not a problem in the examples described in this chapter since early cellular blastoderm stage embryos have only a single cell layer.

To use PAGFP in the early embryo we ubiquitously express PAGFP-αTubulin using a set of maternal GAL4 drivers. Presumptive mesodermal cells can then be photoactivated prior to internalisation. By photoactivating a large patch of cells using an epifluorescent microscope (Subheading 3.3) it is possible to visualise the entire mesoderm over several segments as it migrates out over the underlying ectoderm (Fig. 1). Alternatively, individual cells can be labelled by scanning polygonal regions with a confocal laser (Subheading 3.4), and then assayed after several hours of development by which time they may have produced a clone of labelled cells. This cell fate mapping approach is particularly useful since photoactivated fluorescence is able to survive fixation, allowing clones of photolabelled cells to be clearly imaged in embryos that have been immunostained to provide a cellular context (Fig. 1).

Although our studies have focused on mesodermal cells, the approach outlined here is applicable to other cell types in the cellular blastoderm. For example, we have successfully labelled cells in the AMG anlagen and then visualised the AMG cells migrating posteriorly (Fig. 1). We have also photolabelled individual AMG cells and imaged the subsequent clone of cells several hours later (Fig. 1).

Like other FPs, PAGFP can be easily fused to different proteins. For the mesoderm, PAGFP-Tubulin was chosen to highlight both cell morphology and mitoses. To highlight the protrusive morphology of migrating cells, we also fused PAGFP to the Actin-binding domain of *Drosophila* moesin (45), which targets actin filaments. PAGFP-Moe^{ABD} highlights F-actin-rich structures, such as lamellipodia and filopodia (Fig. 1).

2. Materials

2.1. Embryo Collection and Preparation

1. Applejuice-agar egg collection plates: Dissolve 12.5 g of sucrose in 250 ml of dH_2O, and 20 g of agar in 500 ml of dH_2O in a microwave oven. When the agar solution bubbles, add the sucrose to the agar, then add 250 ml of apple juice, and dispense approximately 10 ml into 60 mm plastic dishes and allow to set.
2. Fly stocks: General methods for keeping *Drosophila*, and for setting up crosses are described elsewhere (46). For details on fly lines used to obtain PAGFP-αTubulin expressing embryos consult (35).
3. Plastic egg-laying chamber: 100-ml plastic tripour beakers (ProSciTech). Punch air holes with a needle (e.g. 23 gauge) in the base of the beakers.
4. Dry bakers yeast made into a paste with water.
5. Bleaching chamber: we cut off and use the top third of a 50 ml falcon tube. We use the lid (with a large circular hole cut into it) to hold in place, a piece of nylon gauze.
6. Bleach (available chlorine 4% *m/v*) diluted 50% with water.

2.2. Mounting Pre-gastrulation Embryos for Imaging

1. 22 mm × 40 mm coverslips.
2. Aluminium slides with a circular hole (Fig. 3).
3. Stereo microscope, preferably with backlighting.
4. Gas-permeable oil. We use liquid paraffin, but Voltalef, and Halocarbon oil 7100 are also common.
5. Paintbrush.
6. Glue: We use a Japanese rubber bicycle tube cement (Chiaro). Other glues may work but should be tested for (a) adhesion to embryos; (b) non-toxicity; (c) ability to dislodge embryos after a period in the oil.

2.3. Embryo Fixation Following Microscopy

1. 10× PBS solution: dissolve 3.33 g $NaH_2PO_4 \cdot 2H_2O$ and 56.6 g $Na_2HPO_4 \cdot 12H_2O$ in 800 ml of dH_2O, adjust to pH 7.4. Add 102.2 g NaCl and adjust to 1 l.
2. Fixative: 4% formaldehyde in 1× PBS.
3. Heptane.
4. Methanol.
5. Fixation bottles: Glass McCartney bottles are ideal as embryos float at the centre of the liquid interface, away from the glass surfaces.

2.4. Immunostaining of Embryos

1. Microcentrifuge tubes.
2. PBST = 0.1% Triton X in PBS.

3. Primary antibody: Rabbit-anti-Twist (M. Leptin) at 1:200 in PBST.
4. Secondary antibody: Goat-anti-Rabbit-Cy5.
5. 70% Glycerol in PBS.

3. Methods

3.1. Embryo Collection and Preparation

1. Place flies into the egging chamber covered by a collection plate with a dab of yeast paste and place at 29°C overnight. This temperature, which is higher than 25°C standard, increases GAL4 expression levels.
2. Use a paintbrush and water to wash embryos off the plate and into the bleaching chamber.
3. Add a few millilitre of bleach and gently agitate for 3 min.
4. Wash embryos thoroughly in tepid water to completely remove bleach.
5. Transfer embryos by paintbrush to a new plate. A drop of water helps spread embryos and aids visualisation.

3.2. Mounting Pre-gastrulation Embryos for Imaging

1. Within the overnight collection embryos that are approaching gastrulation (i.e. stage 5 according to (47)) can be distinguished from older embryos as they have no large internal structures, are darker, and have a "mottled" appearance (Fig. 2). Stage 5 embryos submerged in water or oil, show a margin of clearer cytoplasm (Fig. 2, *inset*), which corresponds to the ingressing front of cellularisation. Embryos already gastrulating (i.e. stage 6–7 embryos) have a ventral furrow (VF) (Fig. 2).
2. Draw a line of glue across the coverslip with a 200-μl micropipettor tip (Fig. 3a).
3. With a moistened paintbrush, transfer a few embryos, together with some water, next to the glue (Fig. 3b). Using the paintbrush, pick up embryos and place them gently, posterior side down, next to the glue, and then gently lay them down horizontally onto the glue such that the cells to be photolabelled face the coverslip (Fig. 3c) (see Note 1).
4. Immediately (see Note 2) cover the embryos with a drop of oil (~10 μl) (Fig. 3d), and adhere the coverslip to the underside of the aluminium slide via surface tension using a small amount of oil (Fig. 3e).

3.3. Photolabeling and Live Imaging Groups of Cells

To photolabel the entire mesoderm (over several segments), we use mercury arc lamp light, a 405/20 nm excitation filter, and a high power objective (e.g. 60× or 100× oil immersion).

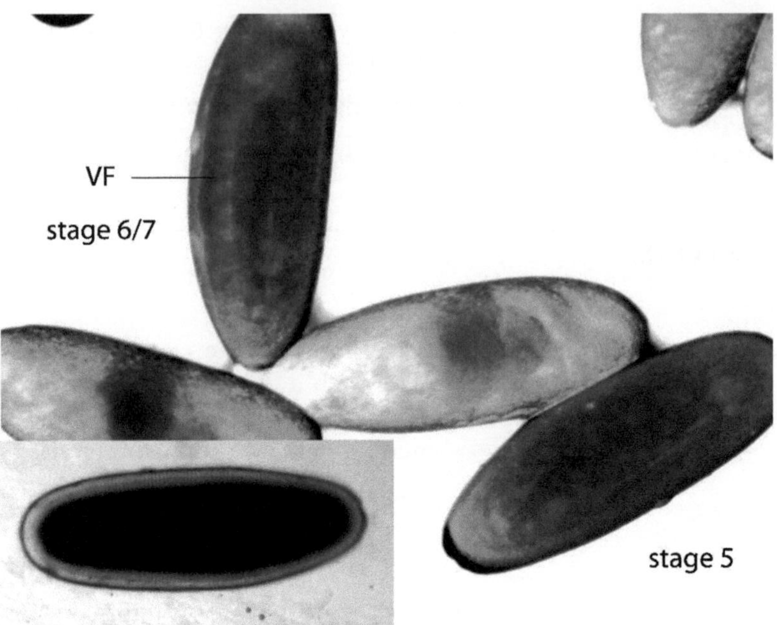

Fig. 2. Pre-gastrulation (i.e. stage 5) embryos are darker than older embryos, show no large internal structures, have a mottled appearance and exhibit a margin when submerged (*inset*). Embryos that are already gastrulating (i.e. stage 6/7) have a ventral furrow. *VF* ventral furrow.

High magnification helps to both increase the concentration of light and spatially restrict the illumination region.

1. The progression of embryos towards gastrulation can be monitored using brightfield illumination and a medium power (e.g. 40×) objective. Just prior to furrowing, the cellularisation front will have extended about 4 cell diameters inwards and a sharp demarcation between the clearer peripheral region and the inner yolk is visible (Fig. 4a).
2. Focus on the ventral epidermal surface and periodically (e.g. 1–2 min intervals) image the embryo using a FITC filter set. Furrowing is first detected by cells on either side of the developing furrow moving towards each other, and cells at the midline constricting laterally. If the FITC fluorescence is too weak, expose the embryo to a short pulse (e.g. 0.5 s) of 405 nm light.
3. Switch to a high power objective and expose cells in the developing furrow to the activating light. When using a DeltaVision deconvolution microscope system we also utilise "critical illumination," which focuses the Hg lamp light, and an aperture diaphragm to create a sharp circular area of illumination. The optimal period of exposure should be empirically determined to give maximum photoactivated fluorescence, without bleaching. In our case this is approximately 30 s.

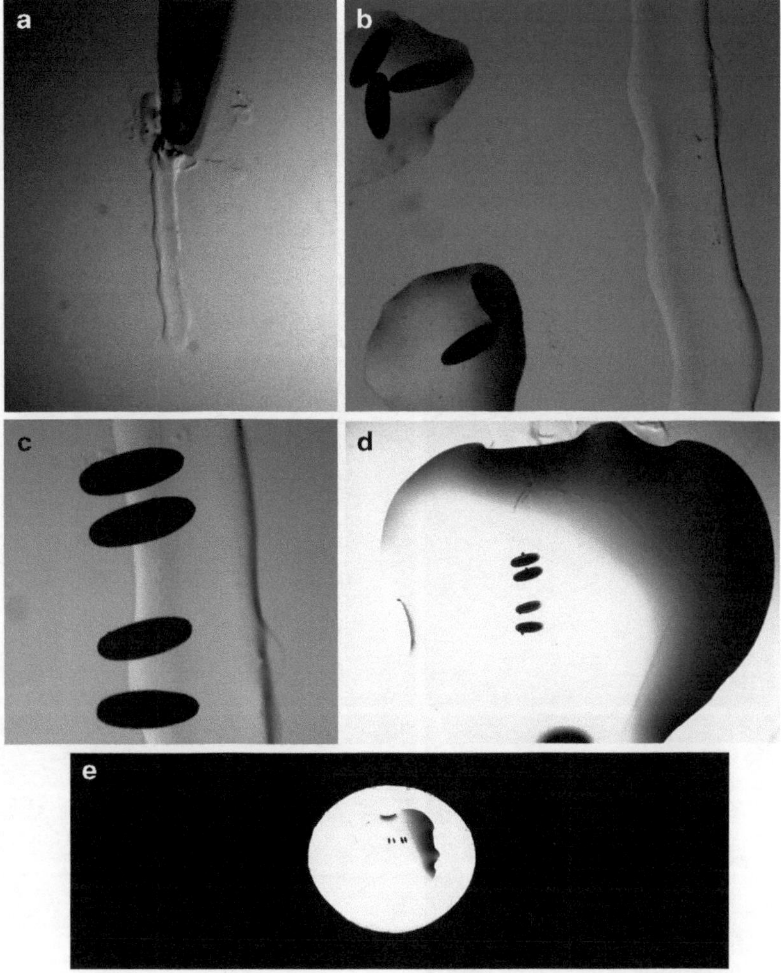

Fig. 3. Mounting embryos for live microscopy. (**a**) A stripe of glue is deposited on a coverslip. Stage 5 embryos are then placed next to the glue in small pools of water (**b**), positioned ventral side down onto the glue (**c**), and then immediately immersed in oil (**d**). Finally, the coverslip is adhered to the aluminium slide (**e**).

4. To capture the subsequent migration of the mesoderm collect confocal Z-stacks at periodic intervals. Live imaging of fluorescently labelled cells involves a trade-off between several conflicting requirements (see (48) for a thorough discussion). For example, high spatial resolution (both in the *x–y* and *z* axes) reduces temporal resolution. Similarly high image quality, usually obtained by high intensity laser light or multiple scans at lower intensity, comes at the expense of photobleaching the fluorophore, or phototoxicity to the specimen. We find the best compromise on our confocal (a Leica SP2) was achieved with medium/low laser intensity, medium speed (e.g. 400 Hz) scanning, 512 × 512 images with 3× averaging, and 3 µm z-steps, taken at 2-min intervals.

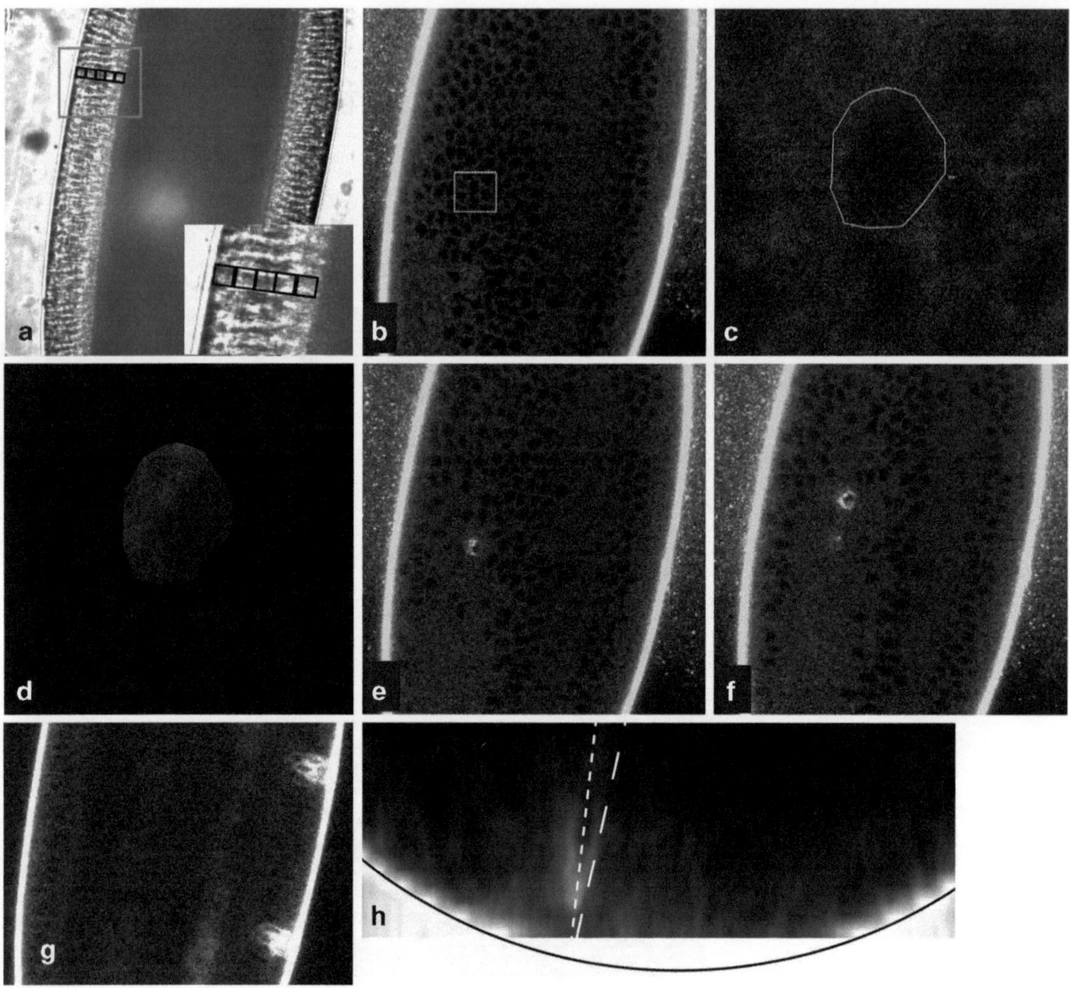

Fig. 4. Photoactivation of individual mesodermal cells at gastrulation. Just prior to furrowing the cellularisation front has moved ~4 cell diameters inwards (**a**). Focussing on the epidermis, a *zoom box* is drawn around the target cell (**b**). After scanning once with low-intensity 488 nm light, a *polygon* is drawn around the nucleus (**c**), and scanned with the photo-activating frequency (**d**), resulting in a labelled cell (**e**). Additional cells may be photolabelled (**f**), and the embryo given identifying marks (**g**). The position of the labelled cell (*dotted line*) with respect to the centre of the furrow (*dashed line*) can be determined by analysing a Z-stack taken just after photolabelling (**h**).

3.4. Photolabelling Individual Cells

For cell fate mapping we photolabel individual cells using region-of-interest confocal laser scanning, and then allow development to proceed for several hours, before fixation and immunolabelling. On a confocal, the high concentration of light required for effective photoactivation is achieved with a high zoom factor. In our hands a 63× oil immersion objective and zoom level of between 12× and 15× is most effective.

1. Use brightfield illumination to monitor embryos approaching gastrulation (Fig. 4a).

2. Focus on the epidermal ventral surface of the embryo, and periodically (e.g. 1–2 min intervals) scan with 488 nm light (Fig. 4b).
3. Draw a zoom box around a cell in the developing furrow (Fig. 4b).
4. Scan with a low intensity (e.g. 10% of maximum) 488 nm laser setting (high intensity 488 nm light can, itself, photoactivate) (Fig. 4c).
5. Draw a polygon around the nucleus (Fig. 4c). One can incorporate a small extra margin but too much will result in photolabelling of adjacent cells.
6. Scan the polygon with the activation (e.g. 405 nm or UV) laser (Fig. 4d) (see Note 3). Do tests to determine a good trade-off between brightness of the photoactivation, potential photodamage to the cell (usually manifested as a delay in mitosis), and bleaching. In our case, we use two 200 Hz, 1024 × 1024 scans with a UV laser at approximately 12 mW.
7. Reimage the embryo at zoom 1× with the 488 nm laser. A single cell should be labelled (Fig. 4e). It is usually possible to repeat this procedure three to five times before the invaginating furrow becomes too folded (Fig. 4f).
8. Document the position of the photolabelled cells with respect to the furrow centre by capturing a Z-stack (e.g. 512 × 512, 400 Hz, 2× averaging, 3 μm z-steps), prior to the furrow aperture closing over.
9. Each embryo can also be given a unique combination of identifying marks, via photoactivation so that they can be immunostained together. We typically mark the lateral sides of the embryo well away from the ventral furrow region (Fig. 4g).
10. Reconstructed cross-sections of the image stacks can subsequently be created with an image processing package (e.g. ImageJ (49)) to visualise the precise location of cells within the furrow. To determine the midline of the furrow, draw a circle over the outline of the embryo, and then a line from the centre of this circle to the convergence point of lines drawn down the long axis of the labelled cell and its neighbours (Fig. 4h).

3.5. Embryo Fixation Following Microscopy

Photoactivated cells remain fluorescent after fixation. Consequently, it is possible to photolabel cells and then allow the embryo to develop for several hours before fixation and immunostaining. In Fig. 1, for example, cell clones are visualised with respect to the entire mesoderm that is immunostained for the transcription factor, Twist. Although the method given here is quite standard, the particulars are well suited to processing individual embryos.

1. On a stereomicroscope, use a fine paintbrush to gently dislodge and transfer the photolabelled embryos from the oil to an apple juice plate.
2. Remove both the oil and any attached glue from the embryo by gently moving and rolling the embryo around on the agar (see Note 4).
3. At this point the apple juice plate can be placed in an incubator to allow further embryonic development.
4. Using a paintbrush free of oil, transfer the embryo to the fixation vial containing equal measures of fixative and heptane. The embryo will drop through the heptane and stay at the interface (see Note 5).
5. Fix embryos for 20 min with constant agitation. In this and subsequent immunostaining steps cover vials/tubes with aluminium foil.
6. Remove most of the fixative. Do not allow the embryo to touch the pipette tip, as they can be awkward to dislodge, once adhered.
7. Add 2 ml of heptane, 2 ml of methanol, and shake the vial for several seconds. The vitelline membrane should rupture or "crack" allowing the embryo to drop down into the methanol (see Note 6).
8. Using a 1-ml micropipettor tip, transfer the embryo with some methanol to a 1.5-ml microfuge tube.
9. Remove most of the methanol and rinse with a further 1 ml of methanol.
10. Remove the methanol and add 1 ml of PBST.
11. Replace the PBST four to five times.
12. If several embryos are to be processed together wrap the tube in aluminium foil and set it aside until other embryos have also been fixed and, cracked and place in PBST.

3.6. Immunostaining of Embryos

Although immunohistochemical staining methods for *Drosophila* embryos have been published (7) a description of the particular constraints and difficulties in our case is appropriate. We find that PAGFP fluorescence fades somewhat overnight. Therefore, we photoactivate, immunostain, and reimage the embryos, in the same day. The protocol given here, therefore, involves short incubation times and quick washes, as well as a minimal base solution. This protocol may not work well with all antibodies, which could require longer incubations and washes, and the use of blocking reagents such as BSA and/or some form of serum.

1. For antibody staining, it is convenient to first transfer all embryos with a 1-ml micropipettor tip into a single microfuge

tube so that they may be processed together (see Subheading 3.4, step 9). Remove excess PBST and replace with a small amount (e.g. 50 μl) of primary antibody, and very gently agitate the tube for 45 min.

2. Remove the primary antibody and rinse four to five times in PBST.
3. Add 50 μl of secondary antibody and gently agitate for 45 min.
4. Remove secondary antibody, rinse four to five times with PBST and replace with 1 ml of 70% glycerol in PBS (see Note 7).
5. After a few minutes, transfer the embryos in glycerol to a slide.
6. Locate the embryos, transfer them to a new slide, position them ventral side towards the coverslip and seal with nail polish.
7. Examine embryos on the confocal to find the cell clones. We typically take Z-stacks with a modest resolution (e.g. 512×512×2 μm) and high levels of averaging (e.g. 6×). Avoid excessive 488 nm laser intensity, which can still photoactivate the fixed PAGFP.

4. Notes

1. Avoid moving and rolling embryos after they are in contact with the glue as mechanical stress on embryos can cause them to rotate within the vitelline membrane.
2. It is critical that the embryos do not desiccate. If embryo alignment is taking more than a minute, deposit a small drop of water against each embryo.
3. After the 488-nm "set up" scan, take the UV/405 photoactivation scan without delay. Because gastrulation is rapid, cell movement can result in the polygonal scan region becoming misaligned with the cell.
4. It is critical that virtually all oil is removed, as it interferes with cracking. Keep the paintbrush oil-free by regularly pinching its bristles with tissue paper, and then rewetting it by pushing down into agar.
5. During fixation, cracking, and antibody staining, keep visual contact with the embryo at critical moments. We use a Halogen fibre optic light guide to provide a strong downlight in a darkened room.
6. Embryos that fail to crack (usually due to excess oil), remain at the interface. Typically the vitelline membrane will be ruptured but still attached to the embryo. Remove most of the heptane, add more methanol, and shake the vial. At a high

enough ratio of methanol to heptane, the embryo will fall into the methanol and can be processed normally. If the vitelline membrane is still attached after immunostaining it can be removed with forceps once the embryo is in 70% glycerol, prior to microscopy.

7. After adding glycerol try to keep the embryos suspended in the middle of the glycerol volume, i.e. prevent them from rising to the surface where they are easily lost. Use a 1-ml micropipettor tip to direct a stream of glycerol onto the embryos and keep pushing them back down until they become indiscernable due to clearing.

Acknowledgments

We thank Michael Zavortink and Ursula Wiedemann for construction of the PAGFP-α-Tub84B and PAGFP-Moe^{ABD} constructs, respectively and Maria Leptin for the gift of the anti-Twist antibody. The PAGFP vectors were kindly provided by George Patterson. This work was supported by an NHMRC project grant to M.J.M. and R.S., the ARC Special Research Centre for the Molecular Genetics of Development, and the Institute of Advanced Studies at The Australian National University.

References

1. Matthews, K.A., Kaufman, T.C., and Gelbart, W.M. (2005) Research resources for *Drosophila*: the expanding universe *Nat Rev Genet* **6**,179–93.
2. Venken, K.J., and Bellen, H.J. (2005) Emerging technologies for gene manipulation in *Drosophila melanogaster Nat Rev Genet* **6**, 167–78.
3. Bate, M., and Martinez Arias, A. (1993) The Development of *Drosophila melanogaster Cold Spring Harbor Laboratory Press.*
4. Chalfie, M., Tu, Y., Euskirchen, G., Ward, W.W., and Prasher, D.C. (1994) Green fluorescent protein as a marker for gene expression *Science* **263**, 802–5.
5. Brand, A.H. (1995) GFP in *Drosophila TIG*, 324–5.
6. Mavrakis, M., Rikhy, R., Lilly, M., Lippincott-Schwartz, J. (2008) Fluorescence imaging techniques for studying Drosophila embryo development. *Curr Protoc Cell Biol.* 4.18.1- **43**.
7. McDonald, J.A., and Montell, D.J. (2005) Analysis of cell migration using *Drosophila* as a model system *Methods Mol Biol*, 175–202.
8. Wood, W., and Jacinto, A. (2005) Imaging cell movement during dorsal closure in *Drosophila* embryos *Methods Mol Biol*, 203–10.
9. Ninov, N., and Martín-Blanco, E. (2007) Live imaging of epidermal morphogenesis during the development of the adult abdominal epidermis of *Drosophila Nature protocols* **2**, 3074–80.
10. Prasad, M., and Montell, D.J. (2007) Cellular and molecular mechanisms of border cell migration analyzed using time-lapse live-cell imaging *Dev Cell* **12**, 997–1005.
11. Oda, H., and Tsukita, S. (2001) Real-time imaging of cell-cell adherens junctions reveals that *Drosophila* mesoderm invagination begins with two phases of apical constriction of cells *J Cell Sci* **114**(Pt 3), 493–501.
12. Brand, A.H., and Perrimon, N. (1993) Targeted gene expression as a means of altering cell fates and generating dominant phenotypes *Development* **118**, 401–15.
13. Duffy, J.B. (2002) GAL4 system in *Drosophila*: a fly geneticist's Swiss army knife *Genesis* **34**, 1–15.

14. Lin, H., and Schagat, T. (1997) Neuroblasts: a model for the asymmetric division of stem cells *Trends Genet* **13**, 33–9.
15. Spradling, A., Drummond-Barbosa, D., and Kai, T. (2001) Stem cells find their niche *Nature* **114**, 98–104.
16. Mandal, L., Martinez-Agosto, J.A., Evans, C.J., Hartenstein, V., and Banerjee, U. (2007) A Hedgehog- and Antennapedia-dependent niche maintains *Drosophila* haematopoietic precursors *Nature* **446**, 320–4.
17. Micchelli, C., and Perrimon, N. (2006) Evidence that stem cells reside in the adult *Drosophila* midgut epithelium *Nature* **439**, 475–9.
18. Ohlstein, B., and Spradling, A. (2006) The adult *Drosophila* posterior midgut is maintained by pluripotent stem cells *Nature* **439**,470–4.
19. Takashima, S., Mkrtchyan, M., Younossi-Hartenstein, A., Merriam, J.R., and Hartenstein, V. (2008) The behaviour of *Drosophila* adult hindgut stem cells is controlled by Wnt and Hh signalling *Nature* **454**, 651–5.
20. Singh, S.R., Liu, W., and Hou, S.X. (2007) The adult *Drosophila* malpighian tubules are maintained by multipotent stem cells *Cell Stem Cell* **1**, 191–203.
21. Kaltschmidt, J.A., Davidson, C.M., Brown, N.H., and Brand, A.H. (2000) Rotation and asymmetry of the mitotic spindle direct asymmetric cell division in the developing central nervous system *Nat Cell Biol* **2**, 7–12.
22. Cheng, J., Türkel, N., Hemati, N., Fuller, M., Hunt, A., and Yamashita, Y. (2008) Centrosome misorientation reduces stem cell division during ageing *Nature* **456**, 599–604.
23. Santos, A., and Lehmann, R. (2004) Germ Cell Specification and Migration in *Drosophila* and beyond *Current Biology*, R578–R89.
24. Kunwar, P.S., Starz-Gaiano, M., Bainton, R.J., Heberlein, U., and Lehmann, R. (2003) Tre1, a G protein-coupled receptor, directs transepithelial migration of *Drosophila* germ cells *PLoS Biol* **1**, E80.
25. Kunwar, P., Sano, H., Renault, A., Barbosa, V., Fuse, N., and Lehmann, R. (2008) Tre1 GPCR initiates germ cell transepithelial migration by regulating *Drosophila melanogaster* E-cadherin *The Journal of Cell Biology* **183**, 157–68.
26. Sano, H., Renault, A.D., and Lehmann, R. (2005) Control of lateral migration and germ cell elimination by the *Drosophila melanogaster* lipid phosphate phosphatases Wunen and Wunen 2 *The Journal of Cell Biology* **171**, 675–83.
27. Wilson, R., and Leptin, M. (2000) Fibroblast growth factor receptor-dependent morphogenesis of the *Drosophila* mesoderm *Philos Trans R Soc Lond B Biol Sci* **355**, 891–5.
28. Stathopoulos, A., Tam, B., Ronshaugen, M., Frasch, M., and Levine, M. (2004) Pyramus and thisbe: FGF genes that pattern the mesoderm of *Drosophila* embryos *Genes Dev* **18**, 687–99.
29. Gryzik, T., and Muller, H.A. (2004) FGF8-like1 and FGF8-like2 encode putative ligands of the FGF receptor Htl and are required for mesoderm migration in the *Drosophila* gastrula *Curr Biol* **14**, 659–67.
30. Gisselbrecht, S., Skeath, J.B., Doe, C.Q., and Michelson, A.M. (1996) Heartless encodes a fibroblast growth factor receptor (DFR1/DFGF-R2) involved in the directional migration of early mesodermal cells in the *Drosophila* embryo *Genes Dev* **10**, 3003–17.
31. Beiman, M., Shilo, B.Z., and Volk, T. (1996) Heartless, a *Drosophila* FGF receptor homolog, is essential for cell migration and establishment of several mesodermal lineages *Genes Dev* **10**, 2993–3002.
32. van Impel, A., Schumacher, S., Draga, M., Herz, H., Goshans, J., and Muller, H. (2009) Regulation of the Rac GTPase pathway by the multi- functional Rho GEF Pebble is essential for mesoderm migration in the *Drosophila* gastrula *Development* **136,** 813–22.
33. Smallhorn, M., Murray, M.J., and Saint, R. (2004) The epithelial-mesenchymal transition of the *Drosophila* mesoderm requires the Rho GTP exchange factor Pebble *Development* **131**, 2641–51.
34. Schumacher, S., Gryzik, T., Tannebaum, S., and Muller, H.A. (2004) The RhoGEF Pebble is required for cell shape changes during cell migration triggered by the *Drosophila* FGF receptor Heartless *Development* **131**, 2631–40.
35. Murray, M.J., and Saint, R. (2007) Photoactivatable GFP resolves *Drosophila* mesoderm migration behaviour *Development* **134,** 3975–83.
36. Mcmahon, A., Supatto, W., Fraser, S., and Stathopoulos, A. (2008) Dynamic Analyses of *Drosophila* Gastrulation Provide Insights into Collective Cell Migration *Science* **322**, 1546–50.
37. Reuter, R., Grunewald, B., and Leptin, M. (1993) A role for the mesoderm in endodermal migration and morphogenesis in *Drosophila Development* **11,** 1135–45.
38. Martin-Bermudo, M.D., Alvarez-Garcia, I., and Brown, N.H. (1999) Migration of the *Drosophila* primordial midgut cells requires

coordination of diverse PS integrin functions *Development* **126,** 5161–9.

39. Devenport, D., and Brown, N.H. (2004) Morphogenesis in the absence of integrins: mutation of both *Drosophila* beta subunits prevents midgut migration *Development* **131**, 5405–15.
40. Lukyanov, K.A., Chudakov, D.M., Lukyanov, S., and Verkhusha, V.V. (2005) Innovation: Photoactivatable fluorescent proteins *Nat Rev Mol Cell Biol* **6**, 885–91.
41. Patterson, G.H., and Lippincott-Schwartz, J. (2002) A photoactivatable GFP for selective photolabeling of proteins and cells *Science* **297**, 1873–7.
42. Patterson, G.H., and Lippincott-Schwartz, J. (2002) Selective photolabeling of proteins using photoactivatable GFP *Methods*, 445–50.
43. Pantazis, P., and Gonzalez-Gaitan, M. (2007) Localized multiphoton photoactivation of paGFP in *Drosophila* wing imaginal discs *J Biomed Opt* **12**, 044004.
44. Post, J.N., Lidke, K.A., Rieger, B., and Arndt-Jovin, D.J. (2005) One- and two-photon photoactivation of a paGFP-fusion protein in live *Drosophila* embryos *FEBS Lett* **579**, 325–30.
45. Dutta, D., Bloor, J.W., Ruiz-Gomez ,M., VijayRaghavan, K., and Kiehart, D.P. (2002) Real-time imaging of morphogenetic movements in *Drosophila* using Gal4-UAS-driven expression of GFP fused to the actin-binding domain of moesin *Genesis* **34**, 146–51.
46. Greenspan, R.J. (1997) Fly Pushing: The Theory and Practice of *Drosophila* genetics *Cold Spring Harbor, NY: Cold Spring Harbor Laboratory Press.*
47. Campos-Ortega, J.A., and Hartenstein, V. (1997) The Embryonic Development of *Drosophila melanogaster 2nd ed. Berlin: Springer-Verlag.*
48. Pawley, J.B. (2006) Handbook of Biological Confocal Microscopy *3 ed: Springer.*
49. Collins, T.J. (2007) ImageJ for microscopy *Biotechniques* **43**(1 Suppl), 25–30.

Part IV

Adult Hematopoietic Stem Cell Migration

Chapter 13

Adhesion, Migration, and Homing of Murine Hematopoietic Stem Cells and Progenitors

Jose A. Cancelas

Abstract

Recent advances in cell biology have demonstrated the role of multiple signaling proteins in the transduction of external signals to cytoplasmic and nuclear effectors, controlling the movement and/or retention of hematopoietic stem cells and progenitors (HSC/P) within the bone marrow, with important clinical implications. Multiple assays have become routine in the analysis of adhesion to the microenvironment, migration toward chemoattractant gradients, and homing of HSC/P in the bone marrow in vivo. In this chapter, we analyze some of the most frequently used assays in our laboratory to explore the ability of HSC to migrate, adhere, and home in in vitro and in vivo assays.

Key words: Hematopoietic stem cells, Migration, Homing

1. Introduction

Hematopoietic stem and progenitor cells (HSC/P) reside in the bone marrow (BM) cavity during postnatal life and may be localized to specific "niches" within the hematopoietic microenvironment (HM). A tiny fraction of HSC, of unknown physiological relevance, is to be found in the blood circulation and the number of these cells in circulation can be increased in a process termed "mobilization." When infused into the blood during BM transplantation procedures these HSC engraft in the BM space and lead to the subsequent reconstitution of multilineage hematopoiesis. Poor engraftment of HSC/P has been shown as one cause of the failure of many protocols of cord blood transplantation, ex vivo HSC/P expansion, and retroviral-mediated gene therapy. Defective homing of HSC/P resulting from decreased adhesion

Marie-Dominique Filippi and Hartmut Geiger (eds.), *Stem Cell Migration: Methods and Protocols*, Methods in Molecular Biology, vol. 750, DOI 10.1007/978-1-61779-145-1_13, © Springer Science+Business Media, LLC 2011

to the extracellular matrix and/or loss of retention in the BM microenvironment may result in decreased HSC/P engraftment.

Fibronectin (FN), a major component of the extracellular matrix protein, is involved in the adhesion of HSC/P to the extracellular matrix of the BM through β_1-integrins and CD44 adhesion receptors. CXCL12, the best characterized chemoattractant has also been involved in BM retention and mobilization of HSC/P. Adhesion to FN and migration toward CXCL12 have become the reference surrogate assays to measure the ability of HSC/P to adhere to and/or migrate toward (chemotaxis) a chemoattractant gradient, and the in vivo homing assay of HSC/P has become the gold standard in studying the ability of HSC/P to migrate into and be retained in the BM.

We have used these three assays in our analysis of the role of Rho GTPases in homing, mobilization, and retention, and in controling the fine equilibrium between the marrow and the circulating pools of HSC/P (1–12).

2. Materials

2.1. Animals, Cells, and Cell Culture Reagents

1. C57Bl/6 and B6.SJL-Ptprc[a] Pepc[b]/BoyJ mice (Jackson Labs, Ban Harbor, ME).
2. Bone marrow (BM) cells are obtained after crunching femorae, tibiae, and iliac crest bones from 8- to 10-week-old mice.
3. Iscove's Modified Dulbecco's Medium (IMDM) supplemented with 10% fetal calf serum (FCS), 2 mmol/L l-glutamine, and 100 IU/mL penicillin and 0.1 mg/mL streptomycin.
4. RPMI 1640 (Invitrogen, Carlsbad CA).
5. Protease-free bovine serum albumin (BSA).
6. Histopaque 1083 (Invitrogen).
7. Cell dissociation buffer (PBS based).
8. Phosphate-buffered saline (PBS).
9. Nontissue culture, 5-μm pore, 24-well polycarbonate transwells.
10. Nontissue culture, 24-well plates, polystyrene.
11. 35 × 10-mm, 2 × 2-mm, 3-mL, gridded dishes, polystyrene.
12. 40-μm mesh filters, nylon.
13. 15-mL conical and 4-mL round-bottom polypropylene tubes.
14. Turk's Solution

2.2. Flow Cytometry and Sorting

1. FACS Canto equipped with two lasers (Argon 488 and HeNe 633 nm) and ability to analyze signals from eight detectors (Becton–Dickinson, San Jose, CA).
2. Antibodies: Antimouse lineage antibodies, anti-CD3-FITC (clone 145-2C11), anti-CD4-FITC (clone GK1.5), anti-CD8-FITC (clone 53-6.7), anti-CD45R/B220-FITC (clone RA3-6B2), anti-Ly6C/G-FITC (clone RB6-8C5), anti-CD11b-FITC (clone M1/70), and anti-TER119-FITC (clone Ly-76). Anti-CD117(c-kit)-PE (clone ACK45), anti-Sca-1 (Ly6A/E)-APC (clone D7), anti-CD45.1-PE (clone A20), and anti-CD45.2-FITC (clone 104). Isotype controls were rat isotype-FITC (IgG2b kappa), rat isotype-PE (IgG2a kappa), and rat isotype-APC (IgG2b kappa). (All BD Biosciences, San Jose, CA).
3. Mouse serum.
4. PKH26 dye (Sigma-Aldrich, St. Louis, MO).

2.3. Recombinant Proteins

1. Recombinant human FN carboxyterminal fragment CH-296, corresponding to a protein of ~63 kDa (574 amino acids) containing a central cell-binding domain (type III repeat, 8–10), a high affinity heparin-binding domain II (type III repeat, 12–14), and a CS1 site within the alternatively spliced IIICS region of human FN (RetroNectin Reagent, Takara Bio USA, Madison, WI).
2. Recombinant murine CXCL12 (stromal cell-derived factor (SDF)-1α) (R&D Systems, Minneapolis, MN).

2.4. Colony-Forming Cell (CFU-C) Assay

1. MethoCult® GF M3434 complete methylcellulose medium with recombinant cytokines for colony assays of murine cells, with erythropoietin (StemCell Technologies, Vancouver, Canada).

3. Methods

A description of the method of analysis of adhesion, migration, and homing of BM cells is presented. These cells can be preselected in vivo through 5-fluorouracil (5-FU) administration followed by in vitro low-density BM gradient.

3.1. Low-Density Gradient BM Cells

1. Crunched bones and their BM are sieved through 40-mm mesh filters and suspended in PBS/5% FCS at a concentration of 20–50 × 10^6/mL.
2. 4 mL of Histopaque 1083 solution (room temperature) are deposited into 15-mL conical tubes. A maximum volume per tube of 2 mL of cells is layered onto the Histopaque medium.

3. The tubes are spun down at $500 \times g$ for 30 min at room temperature.
4. Collect interphase cell layer between the upper phase and the Histopaque phase of the gradient.
5. Collected cells (~2 mL) are washed in 40 mL of PBS once at $400 \times g$ for 10 min at room temperature. The supernatant is removed and the cell pellet resuspended in 1 mL of PBS.
6. Cells are counted in a hemocytometer (1:10 dilution with Turk's solution).
7. The cells are resuspended in PBS at a concentration of 20×10^6 cells/mL.

3.2. Input Lineage-Negative/c-kit$^+$/Sca-1$^+$ BM Cells

1. Three aliquots of 1×10^6 cells per specimen are used for flow cytometry analysis. All the specimens will be stained according to manufacturer's instructions, by incubating cells with antibodies in a final volume of 100 μL, containing cells, antibodies, and 2% mouse serum in PBS.
2. The first is stained with rat isotype-FITC, rat isotype-PE, and rat isotype-APC antibodies.
3. The second is stained each with anti-CD3, CD4, CD8, Mac-1, Gr-1, CD45R/B220, Ter119, rat isotype-PE, and rat isotype-APC antibodies.
4. The third is stained with each of the lineage antibodies along with anti-c-kit and Sca-1 antibodies.
5. The thresholds of positivity are determined to 1% of the population. Lineage-negative/c-kit$^+$/Sca-1$^+$ cells appear as a well-defined population in a dot plot (c-kit vs. Sca-1) gated on lineage-negative low-density BM cells.
6. The percentage of lineage-negative/c-kit$^+$/Sca-1$^+$ cells is recorded.
7. Cell selection of lineage-negative/c-kit$^+$/Sca-1$^+$ cells can be performed in a cell sorter, according to established protocols (see under Chapter 3).

3.3. Input CFU Quantitation

1. Input CFUs are enumerated by mixing 80,000 low-density BM cells (in 4 mL of Methocult medium, in a 4-mL round bottom tube).
2. Tubes are vortexed allowed to rest for 10 min, and plated on three 35-mm gridded dishes.
3. Dishes are incubated at 37°C, 5% CO_2 for 10 days.
4. Colony-forming units (CFUs) are identified as clusters with more than 50 cells and counted. All colonies, including BFU-E, CFU-GM, and CFU-Mix are counted. Differential analysis of these three populations is possible but it requires

experience. If inexperwienced, the count, the total number, of CFU per dish is recommended.

5. The average number of CFU per 20,000 low-density BM cells plated in each of the three dishes is recorded. The average number of CFU per 100,000 plated cells is calculated by multiplying the resulting number by 5 (*input CFU*).

3.4. Adhesion to Fibronectin

1. Nontissue culture, 24-well plates are coated with CH-296 at 8 μg/cm^2 or BSA (as control) and kept overnight at 4°C. Assays are performed by triplicate.
2. The plates are subsequently blocked with 2% BSA for 30 min at room temperature.
3. A total of 50,000 low-density BM cells, suspended in RPMI 1640 medium containing 10% FBS, and allowed to adhere to the coated plates for 1 h at 37°C.
4. After incubation, nonadherent cells are collected by carefully rinsing the plates three times with medium. Adherent cells are harvested by vigorously rinsing the plates with PBS twice and cell dissociation buffer once. Finally, wells are rinsed with PBS and the content is collected into the adherent fraction tube.
5. The adherent fraction is centrifuged at 400 × *g* for 10 min and the supernatant removed.
6. The cell pellet is resuspended in 200 μL of PBS/5% FCS and mixed with 3.8 mL of CFU assay medium. 3 mL of CFU medium with cells is plated, incubated, and scored as for the input CFU assay. The output CFU is the average of the three dishes scored on day +10 of incubation (*output CFU*).
7. The percentage of CFU adhesion is calculated as follows:

   Adhesion of CFU-C to CH-296 (%) = 100 × [(output CFU × 2)/(input CFU)].

   The average of the three adhesion assays is the final result.
8. Expected result: The percentage of CFU-C adhering to CH-296 after 1 h for low-density BM cells from C57Bl/6 mice ranges from approximately 8 to 15%, averaging ~10%.
9. Alternative flow cytometry analysis: Alternatively, the overall volume of the triplicate assay can be pooled, spun down at 400 × *g* for 10 min, resuspended in 100 μL of PBS/5% FCS, and stained with lineage antibodies, anti-c-kit, and anti-Sca-1 as mentioned for the third flow cytometry tube in steps 1, 4, and 6 in Subheading 3.2. The acquisition and analysis of this population will be performed using the same gating analysis as in Subheading 3.2.

3.5. Chemotaxis Toward a CXCL12 Gradient

1. Cells from step 4 in Subheading 3.1 are suspended in a cell solution containing RPMI 1640/0.5% BSA at a concentration of 1×10^6 cells/mL.
2. 600 μL of chemotaxis buffer containing 100 ng/mL CXCL12 is added to the bottom well of the transwell plate.
3. 100 μL of cell solution is added to upper chamber (100,000 cells).
4. The plate is incubated for 4 h at 37°C/5% CO_2.
5. The upper chamber is carefully removed and discarded.
6. The solution from lower chamber is removed and placed into a labeled, sterile 4-mL round-bottomed polypropylene tube.
7. The wells are washed with PBS, the wash added to the tube, and then washed again with cell dissociation buffer once.
8. The percentage of migrating cells are calculated as in steps 5–7 in Subheading 3.4 (see Note 1).
9. Expected result: The percentage of CFU-C from low-density BM of C57Bl/6 mice, adhering to CH-296 after 4 h, ranges from 20 to 40%, averaging ~30%.
10. Alternative flow cytometry analysis: Alternatively, the overall volume of the triplicate assay can be pooled, spun down at $400 \times g$ for 10 min, resuspended in 100 μL of PBS/5% FCS, and stained with lineage antibodies, anti-c-kit, and anti-Sca-1 as mentioned for the third flow cytometry tube in steps 1, 4, and 6 in Subheading 3.2. The acquisition and analysis of this population will be performed using the same gating analysis as in Subheading 3.2.

3.6. Progenitor Homing Assay

1. We recommend that a minimum of three mice (pooled) are used per group to compare different groups of BM cells. BM nucleated cells suspended in PBS/5% FCS will be the starting cell population.
2. Input CFU analysis is done as mentioned earlier (see step 5 in Subheading 3.3).
3. A minimum of eight recipient mice (C57Bl/6 females, ages 8–12 weeks) per group is included.
4. Recipient mice are lethally irradiated at a validated dose able to kill all the endogenous CFUs and CFU-S-Day 12. In our institution, this dose is 700 cGy in a first dose followed by a second dose of 475 cGy 3 h apart. The dose rate is 50–100 cGy/min. If the dose that completely destroys any remaining endogenous CFU is unknown, set up a control group with five irradiated mice just transplanted with 0.2 mL of PBS. Find the minimal dose since an excessive irradiation may induce severe mucositis and animal death (13, 14).

5. One hour after irradiation, a total of 50×10^6 donor BM cells are intravenously injected into the tail vein of each of the recipient mice. This is usually done at 5.00 pm.
6. On the following day (9.00 am), mice are sacrificed by CO_2 inhalation and cervical dislocation and their two femurs and tibiae, and each spleen are harvested (see Note 2).
7. All the harvested bones from each mouse are crunched and processed as mentioned in step 1 in Subheading 3.1. Splenocytes are obtained by shearing each individual spleen between two glass slides and flushing them with PBS/5% FCS into 35-mm dishes. Cell suspensions are sieved and processed as described in step 1 in Subheading 3.1.
8. BM and spleen cells are each resuspended in 200 μL of IMDM + 10% FCS and counted.
9. 0.2 mL of BM or spleen cell suspensions are added to 4 mL of CFU medium. Assays are set as indicated in the CFU assay protocol (see Subheading 3.3).
10. Colonies are scored on day +10. Specific CFU subset (CFU-GM or BFU-E) homing can be calculated if required. In our experience, it is very difficult to address CFU-Mix homing.
11. Homing in BM (%) is calculated as follows (considering that two femurs and two tibiae contain approximately 20% of the total BM of a mouse)(15):

    Homing per BM (%) = 100 × 5 × [(output CFU per femora and tibiae)/(input CFU × 1,000)].
12. Homing per spleen (%) is calculated as follows:

    Homing per spleen (%) = 100 × [(output CFU per spleen)/(input CFU × 1,000)].
13. Expected results: Homing of normal CFU in C57Bl/6 mice (donors and recipients) is around 10% (5–15%).

3.7. Stem Cell BM Homing Assay

This assay is very similar to the previous one but it uses a competitive repopulation readout on 3-h homing cells. It intends to measure the frequency of homed repopulating stem cells in a cell inoculum.

1. Proceed as in steps 1, 2, 4, and 5 in Subheading 3.6.
2. After 3 h, recipient mice are sacrificed and their bones harvested and processed as in step 1 in Subheading 3.1.
3. BM and spleen cells are each resuspended in 1.1 mL of IMDM + 10% FCS and counted.
4. BM specimens are mixed with 1.1 mL of a cell suspension of freshly harvested B6.SJL-Ptprc^a Pepc^b/BoyJ BM (prepared as in step 1 in Subheading 3.1) at a concentration of 5×10^6 cells/mL.

5. Fractions of 0.2 mL of mixed BM (corresponding to 1/6 of bone cell suspensions and 5×10^5 competitor BM cells) are transplanted into ten lethally irradiated (same irradiation dose as in primary recipients) secondary B6.SJL-Ptprc^a Pepc^b/BoyJ recipient mice intravenously via the tail vein.
6. An input control of repopulation ability of BM from CD45.2+ cells can be performed in parallel if two groups of transgenic mice are compared. Low-density BM are harvested and processed as indicated in Subheading 3.1. A cell suspension made by mixing 15×10^6 BM cells/mL of test (C57Bl/6) mice and 15×10^6 BM cells/mL of B6.SJL-Ptprc^a Pepc^b/BoyJ mice BM is made.
7. Every month after transplantation, the chimerism of recipient mice in peripheral blood is determined by flow cytometry. This determination is performed by staining peripheral blood with anti-CD45.1 and anti-CD45.2 antibodies, according to manufacturer's instructions.
8. The content of competitive repopulation units (CRU) in the inoculums of homing specimens and input controls are calculated as follows (Harrison's formula):

   CRU = 5 × [(percentage of CD45.2+ cells)/(1 − percentage of CD45.1+ cells)], which assumes that there is 1 CRU per 100,000 BM cells.
9. Homing in BM (%) is calculated as follows (considering that two femurs and two tibiae contain approximately 20% of the total BM of a mouse) (Boggs et al.):

   Homing per BM (%) = 100 × 6 × 11 × (homed CRU/500).
10. If a control of CRU content in BM is performed, this formula can be corrected:

    Homing per BM (%) = 100 × 6 × 11 × [homed CRU/(control CRU × 500)] (see Note 3).

4. Notes

1. This experiment can be performed at different doses of CXCL12 if a dose–response analysis of chemotaxis is required.
2. Other time points can also be analyzed and the progenitor homing kinetics may differ in specific protein deficiencies. In our experience, the time points that can be used are 3, 6, 12, 16, and 24 h. Later time points may be affected by cell division.

3. We have only validated this assay for BM homing and not for spleen homing, however, it may be applied if a large number of cells are to be transplanted into the primary recipients. We have not done linearity studies beyond 50×10^6 cells transplanted.

Acknowledgments

The author wishes to acknowledge the contributions to these protocols by all the co-authors of his publications, and thanks Margaret O'Leary for her editing assistance. The author wishes to thank Margaret O'Leary for helpful edition. This work has been supported by National Institutes of Health (1R01HL08-7159 and supplement 3R01HL087159), Department of Defense (CM-064050), Heimlich Institute Foundation, Alex's Lemonade Stand Foundation, and National Blood Foundation.

References

1. Yang, L., Wang, L., Kalfa, T.A., Cancelas, J.A., Shang, X., Pushkaran, S., Mo, J., Williams, D.A., and Zheng, Y. (2007) Cdc42 critically regulates the balance between myelopoiesis and erythropoiesis *Blood* **110**, 3853–61.
2. Yang, L., Wang, L., Geiger, H., Cancelas, J.A., Mo, J., and Zheng, Y. (2007) Rho GTPase Cdc42 coordinates hematopoietic stem cell quiescence and niche interaction in the bone marrow *Proc Natl Acad Sci USA* **104**, 5091–96.
3. Williams, D.A., Zheng, Y., and Cancelas, J.A. (2008) Rho GTPases and regulation of hematopoietic stem cell localization *Methods Enzymol* **439**, 365–93.
4. van Hennik, P.B., Verstegen, M.M., Bierhuizen, M.F., Limon, A., Wognum, A.W., Cancelas, J.A., Barquinero, J., Ploemacher, R.E., and Wagemaker, G. (1998) Highly efficient transduction of the green fluorescent protein gene in human umbilical cord blood stem cells capable of cobblestone formation in long-term cultures and multilineage engraftment of immunodeficient mice *Blood* **92**, 4013–22.
5. Thomas, E.K., Cancelas, J.A., Chae, H.D., Cox, A.D., Keller, P.J., Perrotti, D., Neviani, P., Druker, B.J., Setchell, K.D., Zheng, Y., et al. (2007) Rac guanosine triphosphatases represent integrating molecular therapeutic targets for BCR-ABL-induced myeloproliferative disease *Cancer Cell* **12**, 467–78.
6. Gu, Y., Filippi, M.D., Cancelas, J.A., Siefring, J.E., Williams, E.P., Jasti, A.C., Harris, C.E., Lee, A.W., Prabhakar, R., Atkinson, S.J., et al. (2003) Hematopoietic cell regulation by Rac1 and Rac2 guanosine triphosphatases *Science* **302**, 445–9.
7. Ghiaur, G., Lee, A., Bailey, J., Cancelas, J.A., Zheng, Y., and Williams, D.A. (2006) Inhibition of RhoA GTPase activity enhan-ces hematopoietic stem and progenitor cell proliferation and engraftment *Blood* **108**, 2087–94.
8. Ghiaur, G., Ferkowicz, M.J., Milsom, M.D., Bailey, J., Witte, D., Cancelas, J.A., Yoder, M.C., and Williams, D.A. (2008) Rac1 is essential for intraembryonic hematopoiesis and for the initial seeding of fetal liver with definitive hematopoietic progenitor cells *Blood* **111**, 3313–21.
9. Cancelas, J.A., Lee, A.W., Prabhakar, R., Stringer, K.F., Zheng, Y., and Williams, D.A. (2005) Rac GTPases differentially integrate signals regulating hematopoietic stem cell localization *Nat Med* **11**, 886–91.
10. Cancelas, J.A., Koevoet, W.L., de Koning, A.E., Mayen, A.E., Rombouts, E.J., and Ploemacher, R.E. (2000) Connexin-43 gap junctions are involved in multiconnexin-expressing stromal support of hemopoietic progenitors and stem cells. *Blood* **96**,498–505.
11. Cancelas, J.A., Jansen, M., and Williams, D.A. 2006. The role of chemokine activation of Rac GTPases in hematopoietic stem cell marrow homing, retention, and peripheral mobilization *Exp Hematol* **34**, 976–85.
12. Jansen, M., Yang, F.C., Cancelas, J.A., Bailey, J.R., and Williams, D.A. (2005) Rac2-deficient hematopoietic stem cells show defective interaction

with the hematopoietic microenvironment and long-term engraftment failure *Stem Cells* **23**, 335–46.

13. Bierkens, J.G., Hendry, J.H., and Testa, N.G. (1989) The radiation response and recovery of bone marrow stroma with particular reference to long-term bone marrow cultures *Eur J Haematol* **43**, 95–107.
14. Grande, T., and Bueren, J.A. (1994) Involvement of the bone marrow stroma in the residual hematopoietic damage induced by irradiation of adult and young mice *Exp Hematol* **22**, 1283–87.
15. Boggs, D.R. (1984) Experimental hematology and bone marrow transplantation *Exp Hematol* **12**, 147–51.

Chapter 14

Methods to Analyze the Homing Efficiency and Spatial Distribution of Hematopoietic Stem and Progenitor Cells and Their Relationship to the Bone Marrow Endosteum and Vascular Endothelium

Jochen Grassinger and Susie K. Nilsson

Abstract

The tracking of immunofluorescent labeled hematopoietic stem and progenitor cells (HSC/HPC) within the bone marrow (BM) cavity allows the assessment of the regulatory processes involved in transendothelial migration, trans-marrow migration, and finally lodgement into the HSC niche. This is of interest as the extracellular and cellular components involved in the regulation of HSC quiescence and differentiation are still not completely understood. Homing of transplanted HSC is the first critical step in the interaction between HSC and the microenvironment of the BM. As a consequence, murine models allowing the evaluation of the structural relationship between migrating HSC, the endosteal bone surface, and the vascular components of the BM enhance our understanding of hematopoietic regulation.

Key words: Hematopoietic stem cell, Niche, Homing, Bone marrow, Vasculature

1. Introduction

The existence of a "niche" in which HSC reside within the bone marrow cavity was proposed more than 30 years ago (1). Recent data supports the theory that the interaction of HSC with cellular and extracellular components within the endosteal bone marrow region is critical for HSC regulation (2–7). Other data suggests that BM vascular cells are also involved in HSC regulation (8). As the identification of single HSC/HPC exhibiting defined phenotypes in situ is incredibly difficult, tracking of transplanted immunofluorescent-labeled HSC/HPC at the single cell level in steady state or ablated recipients provides a valuable method to investigate the transendothelial and trans-marrow migration as well as

Marie-Dominique Filippi and Hartmut Geiger (eds.), *Stem Cell Migration: Methods and Protocols*, Methods in Molecular Biology, vol. 750, DOI 10.1007/978-1-61779-145-1_14, © Springer Science+Business Media, LLC 2011

the lodgement of HSC/HPC to their niche. In vivo real time imaging techniques that do not disrupt the bone integrity (9, 10) are currently restricted to the calvarium BM and therefore, single cell tracking in fixed long bones provides a valid method to investigate HSC location in these primary hematopoietic compartments. Using this technique the modification of the microenvironment of the recipient or the surface protein expression of the donor HSC in transgenic animal models allows insight into components influencing the homing and engraftment process. In this context, we recently demonstrated that HSC/HPC harvested from the endosteal BM region have superior homing and reconstitution abilities than their phenotypically equal counterparts from the central BM region (11).

Herein, we describe a methodology to analyze the homing efficiency and spatial distribution of candidate HSC/HPC harvested from endosteal and central BM regions in a competitive homing assay. This allows direct comparison of the engraftment ability of different cell types into the same recipient devoid of intra-experimental variability. Furthermore, we demonstrate a method to label BM vasculature in situ with endothelial cell markers allowing the analysis of the relation between HSC/HPC, the endosteal surface, and the vasculature components of the BM.

2. Materials

2.1. Isolation of Bone Marrow and Preparation of HSC/HPC

1. Adult C57BL/6J (Ly5.2) mice, 6- to 8-week-old (see Note 1).
2. Sterile #11 surgical blade and #3 handle.
3. 50-ml polypropylene conical tubes.
4. Phosphate-buffered saline with 2% serum (PBS 2% Se): PBS 310 mOsm, pH 7.2, supplemented with 2% defined bovine calf serum, iron supplemented (Hyclone).
5. 1-ml syringes attached to 23-gauge and 21-gauge needles to flush marrow from bones.
6. Murine stem cell isolation kit (Chemicon) or alternatively the following components that are included in this kit: sterile porcelain mortar and pestle, 40-μm nylon cell strainer, 4 mg/ml dispase II (Chemicon), and 3 mg/ml collagenase I (Chemicon).
7. 37°C orbital shaker, for example, Eppendorf Thermomixer comfort model (Eppendorf).
8. Hemocytometer and microscope equipped with phase contrast or an automated cell counter, for example, Sysmex model KX-21N (Sysmex).
9. Nycoprep™ 1.077 Animal (Axis-Shield).

10. Cannulas, for example, Unomedical (Unomedical) attached to 20-ml syringes.
11. Lineage depletion antibody cocktail: a mixture of purified rat-anti-mouse antibodies recognizing the cell surface antigens: B220 (lymphoid), GR-1 and MAC-1 (myeloid), and TER119 (erythroid, all antibodies BD Pharmingen, see Note 2).
12. PBS supplemented with 2 mM EDTA and 0.1% (w/v) fraction V bovine serum albumin (BSA, Sigma–Aldrich), pH 7.4 (PBS–EDTA 0.1% BSA).
13. Dynabeads for magnetic labeling of the cells: Sheep anti-rat-IgG beads – 4.5 μm diameter, 4×10^8 beads/ml (Dynal Biotech ASA).
14. Magnets: Dynal MPC-MPC-S for 20 μl to 2 ml samples, MPC-L for a 1–8-ml sample (Dynal Biotech).
15. Suspension mixer: allowing both tilting and rotation at 4–8°C for Dynabead incubation step, for example, Ratek suspension mixer (RSM6, Ratek) used in cold room.
16. 5- and 14-ml polypropylene round bottom tubes.
17. 5-ml polystyrene round bottom tube with cell strainer caps (BD Falcon).
18. Rat-anti-mouse-Sca-1-FITC (Ly-6A/E, clone E13-161.7) and rat-anti-mouse-c-kit-APC (CD117, clone 2B8) and rat-anti-mouse-IgG2a-FITC and rat-anti-mouse-IgG2b-APC (isotype control) conjugated antibodies (BD Pharmingen, see Note 3).

2.2. CFDA-SE and SNARF-1 Labeling of Target Cells

1. CFDA-SE (carboxyfluorescein diacetate, succinimidyl ester – 25 mg, molecular weight 557.47, molecular probes). Prepare a 44.8-M stock solution of CFDA-SE, by adding 1 ml of anhydrous DMSO to the 25-mg vial of CFDA-SE. Then, in a glass vial, prepare a working solution of 5 mM CFDA-SE by adding a 100-μl aliquot of the 44.8 M stock solution of CFDA-SE to 796 μl of DMSO (see Note 4). Prepare a 5-μM CFDA-SE working solution by diluting 1 μl of 5 mM CFDA-SE stock in 999 μl of PBS. Keep working solutions light protected on ice.
2. SNARF-1 (Seminaphtorhodafluor-1 carboxylic acid, acetate, succinimidyl ester) – 50 μg, molecular weight 592.56, molecular probes (see Note 5). Prepare a 1-mM stock solution of SNARF-1, add 84.4 μl of anhydrous DMSO to the 50 μg vial (see Note 4). Prepare a 10-μM SNARF-1 working solution by adding 1 μl of 1 mM SNARF-1 stock in 99 μl of PBS. Keep working solutions light protected on ice.
3. Dimethyl sulfoxide (DMSO) AnalaR (Merck).
4. PBS and PBS supplemented with 0.5, 2, and 20% serum.

5. Hemocytometer and microscope equipped with phase contrast.
6. Water bath maintained at 37°C.
7. 0.4% Trypan blue.
8. 1.5-ml Eppendorf tube, 5-ml polypropylene round bottom tube.
9. Fluorescence microscope equipped with dual filter for FITC and Texas Red, for example, Olympus BX51 (Olympus, Watford Hertfordshire, UK).

2.3. HSC/HPC Transplantation

1. 1-ml syringe.
2. Heating lamp.
3. 75% Ethanol made in distilled water.
4. Kleenex tissue.
5. Apparatus to immobilize mouse during injection.

2.4. Perfusion of Mouse and Preparation of Sections

1. 16% Paraformaldehyde (Electron Microscopy Science).
2. 25% Glutaraldehyde (ProSciTech).
3. 0.2 M Sorensen's phosphate buffer. This buffer is made in two parts. Solution A = 0.2 M Na_2HPO_4 (28.3 g/l) and solution B = 0.2 M KH_2PO_4 (27.2 g/l). Both are dissolved in distilled water. Lower pH of solution A to pH 7.4 by slowly adding solution B and mixing well.
4. Perfusion apparatus, for example, Braun compact S (Braun).
5. Perfusion tubing (Microtube Extrusions).
6. 26-gauge needles, 10-ml syringes.
7. 10% EDTA (ethylenedinitrilo tetraacetic acid disodium salt dehydrate, Titriplex III) made in distilled water, adjust to pH 7.0.
8. Ratek suspension mixer.
9. Anesthetic (for example, isoflurane).

2.5. Section Analysis

1. Citrus agent (Histo-Pure, limonene-D BP solvent, Australian Biostain).
2. 100 and 70% ethanol made in distilled water.
3. Distilled water, PBS, cover slips.
4. Vectashield mounting media (Vector Laboratories).
5. Fluorescence microscope, for example, Olympus BX51.

2.6. Homing Analysis

1. Sterile #11 surgical blade and #3 handle.
2. PBS 2% Se (see Subheading 2.1).
3. 50-ml polypropylene conical tubes and 5-ml polypropylene tubes.

4. Murine stem cell isolation kit.
5. 37°C orbital shaker.
6. Flow cytometer capable of detecting CFSE and SNARF-1 fluorescence, for example, Becton & Dickinson LSR II (BD).

2.7. Intramedullar Staining of Vasculature

1. PBS and PBS 0.5% BSA.
2. Purified rat-anti-mouse-CD31 (clone MEC13.3) or purified rat-anti-mouse-CD102 (clone 3C4; both BD Pharmingen, see Note 6).
3. 10 mM citrate buffer pH 6.0 (citric acid, tri-sodium citrate, Merck).
4. Oven with temperature range from 37°C to 90°C.
5. 50 mM glycine in PBS pH 3.5.
6. PBS 0.3% Triton X-100.
7. PBS 0.3% hydrogen peroxide.
8. PBS 0.05% Tween-20.
9. 20× SSC buffer. Dissolve 175.3 g of sodium chloride and 88.2 g of trisodium citrate dehydrate in 1 l of distilled water, adjust to pH 7.0 with 1 M HCl and Autoclave.
10. Blocking buffer: 5% BSA, 5% skim milk powder, and 0.05% Triton X-100 in 4× SSC (see Note 7) supplemented with 10 μg/ml donkey IgG (Jackson ImmunoResearch).
11. Tyramide Signal Amplification (TSA™ Biotin) system (Perkin Elmer, NEL700). Components of the kit include: streptavidin–HRP, blocking reagent, amplification diluent, biotinyl tyramide.
12. TNB buffer (0.1 M Tris–HCl, pH 7.5, 0.15 M NaCl, 0.5% blocking reagent).
13. Streptavidin-Alexa Fluor® 488 antibody conjugate (molecular probes).
14. DAPI 1 mg/ml (4′,6-diamidino-2-phenylindole dihydrochloride, molecular probes).

3. Methods

3.1. Isolation of Bone Marrow and Preparation of HSC/HPC

3.1.1. Central Bone Marrow

1. Kill mice by cervical dislocation and dissect iliac crests, femurs, and tibias (see Note 8).
2. Remove the epiphyseal and metaphyseal region of femurs and tibias using the scalpel blade (see Note 9, and Fig. 1). Store these bone fragments in 40 ml of PBS 2% Se in a 50-ml Falcon tube.

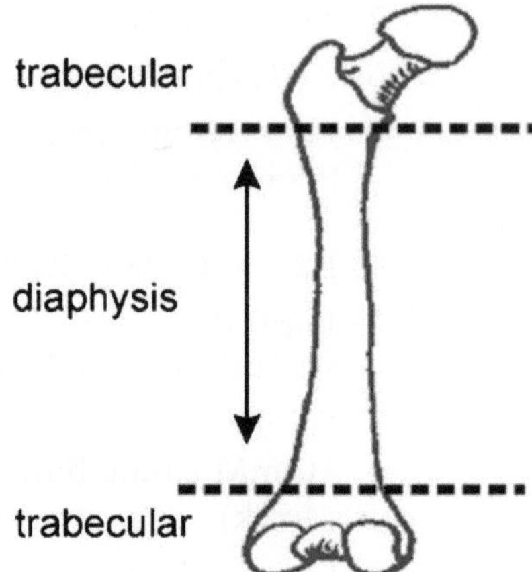

Fig. 1. Trabecular bone regions from femur and tibia are cut (*dot line*) and added to the endosteal bone fraction (see Note 8).

3. Using a 1-ml syringe containing PBS 2% Se attached to a 21-gauge needle insert the needle into each epiphysis of the femoral shaft and the knee epiphysis of the tibia, repeatedly (thee to four times) flush the marrow contents with intermediate pressure into a 50-ml centrifuge tube containing 40 ml of PBS 2% Se. To flush the ileum, use a 23-gauge needle attached to a 1-ml syringe containing PBS 2% Se and flush the bone marrow from the acetabular notch into the 50-ml centrifuge tube. Total marrow from ten donor animals should be flushed equally across two collection tubes. Place the flushed bones in the 50-ml Falcon tube containing the trabecular bone fragments for sampling of endosteal marrow (see Subheading 3.1.2).
4. Wash the flushed cells by centrifuging at 400 × *g* for 5 min at 4°C.
5. Decant supernatant and resuspend the cell pellets in 20 ml of PBS 2% Se.
6. Filter the cell suspension through a 40-μm nylon cell strainer into a fresh 50-ml conical tube.
7. Dilute cells to 40 ml with PBS 2% Se, perform a cell count, and store on ice for density gradient separation (see Subheading 3.2).

3.1.2. Sampling of Endosteal Marrow

1. Kill mice, harvest bones, and flush marrow as for central marrow harvest (see Subheading 3.1.1) or use already flushed bones and epi- and metaphyseal fragments from Subheading 3.1.1.
2. Decant the bones in buffer into a sterile mortar.

3. Grind the bones with the pestle until marrow cavity is open, do not pulverize the bones.
4. Thoroughly mix cell and bone solution by pipetting up and down the supernatant, then remove cell supernatant and filter through a 40-μm nylon cell strainer into a 50-ml conical tube (prelabeled with the code to identify central and endosteal cells from the same recipient).
5. Rinse the crushed bone fragments and filter as in step 4 to collect a total of 100 ml PBS 2% Se. Set tubes aside on ice until step 10.
6. Transfer crushed bones into a 50-ml conical tube (usually the one the bones were collected after flushing) containing 10 ml of 4 mg/ml dispase II and 3 mg/ml collagenase I from the stem cell isolation kit and agitate for 5 min at 37°C in an orbital shaker at 750 rpm.
7. Add 15 ml of PBS to the bone fragments and shake vigorously for 10 s.
8. Filter the cell suspension through a 40-μm nylon cell strainer into another 50 ml prelabeled conical tube. Repeat this step by washing the bone fragments with an additional 15 ml of PBS and vigorously shaking for 10 s.
9. Filter the cell suspension through the cell strainer into the 50-ml conical tube and top up to 50 ml volume with PBS 2% Se.
10. Centrifuge all tubes of cell suspensions (4 × 50-ml tubes) at 400 × *g*, 5 min, 4°C.
11. Decant supernatant, resuspend and pool cell pellets in 50 ml of PBS 2% Se, perform a cell count and store cells on ice for density gradient separation (see Subheading 3.2).

3.2. Density Gradient Separation

1. Centrifuge cell suspension at 400 × *g*, 5 min, 4°C and resuspend cell pellet to approximately 2×10^8 cells/20 ml with PBS 2% Se.
2. Divide 20 ml aliquots of cell suspension over an even number of 50-ml centrifuge tubes.
3. Underlay each gradient with 10 ml of Nycoprep 1.077A using a cannula attached to a 20-ml syringe.
4. Centrifuge the gradients at 600 × *g* for 20 min at room temperature (RT) with no de-acceleration.
5. Collect the mononuclear cells of two gradients from the interface between the PBS layer and the Nycoprep solution into a 50-ml centrifuge tube using a cannula attached to a 10-ml syringe.
6. Centrifuge the tubes at 400 × *g* for 5 min, 4°C, decant the supernatant, and resuspend the pooled cell pellets in 50 ml of PBS 2% Se and perform a cell count.

3.3. HSC Pre-enrichment Immunomagnetic Cell Separation

3.3.1. Immunolabeling Cells with a Cocktail of Lineage Antibodies

1. Centrifuge cells at $400 \times g$ for 5 min, 4°C, decant supernatants.
2. Resuspend cell pellets at 1×10^7 cells/100 µl in a pretitred lineage antibody cocktail. Concentrations between 1 and 2.5 µg/ml usually give good staining results and can be used as starting point for titrations.
3. Incubate cells for 15 min on ice and wash labeled cells in PBS 2% Se by centrifuging at $400 \times g$ for 5 min, 4°C to remove unbound antibody.
4. Remove supernatant completely and resuspend the cell pellets in PBS–EDTA 0.1% BSA at a concentration of between 10^7 and 10^8 cells/ml (see Note 10).

3.3.2. Dynabeads Washing Procedure

1. The optimal Dynabead to cell ratio used in this protocol has been established as half a bead per cell, with a second depletion repeated with the same number of beads. Before use, thoroughly mix Dynabeads. Then Dispense beads for both steps into individual 1.5-ml tubes and follow the washing procedure as outlined below:
   (a) Add 1 ml of PBS–EDTA 0.1% BSA to each tube and mix.
   (b) Place the tubes in the magnet for 1 min, remove, and discard the supernatant.
   (c) Remove the tube from magnet and resuspend the Dynabeads in 1.0 ml of PBS–EDTA 0.1% BSA.
   (d) Repeat step 1b.
2. Remove the tube from the magnet and resuspend the Dynabeads in 0.25 ml of PBS–EDTA 0.1% BSA.

3.3.3. Immunomagnetic Separation

1. Add washed Dynabeads to the cell suspensions.
2. Incubate for 5 min at 2–8°C with gentle tilting and rotation.
3. Place the tube in the magnet for 2 min.
4. Transfer supernatant containing the unbound cells to a fresh 5-ml collection tube.
5. While still in the magnet, rinse the bead bound cells with 1 ml buffer.
6. Transfer the supernatant containing any residual unbound cells to the collection tube.
7. Add the second aliquot of washed Dynabeads to the cell suspension in the collection tube.
8. Incubate for 10 min at 2–8°C with gentle tilting and rotation.
9. Place tubes in magnet for 2 min.

10. Transfer the supernatant containing the unbound cells to new 5-ml collection tube.
11. Place the supernatant in magnet for 2 min to remove any residual beads.
12. Transfer the supernatant containing unbound cells to a 10-ml polypropylene collection tube and wash beads several times with 1 ml of buffer while still placed in the magnet to maximize recovery.
13. Make up the volume of the unbound, lineage negative cell suspension to 10 ml and count.

3.4. HSC Fluorescence Activated Cell Sorting

3.4.1. Labeling of Lineage Depleted Cells

1. Centrifuge lineage negative cells and aspirate supernatant.
2. Resuspend cell pellet at 1×10^7 cells/100 μl in an optimally pretitred antibody cocktail of rat-anti-mouse-Sca-1-FITC and rat-anti-mouse-c-kit-APC, we use both antibodies at 1 in 250 dilution.
3. Incubate light protected on ice for 20 min and wash the cells afterward twice in a tenfold volume of PBS 2% Se.
4. Resuspend the cells in PBS 2% Se and transfer solution into 5-ml collection tube with cell strainer cap prior to fluorescence-activated cell sorting. Adjust cell concentration according to the guidelines of your Flow department (depending on nozzle size and flow rate), we use 10×10^6 cells/ml with a Cytopeia Influx 516SH cell sorter with a 70-μm nozzle.

3.4.2. HSC/HPC Sorting

1. In order to select the instrument settings and fluorescence compensation, prepare an unstained sample tube and positive control tubes for each fluorochrome using whole bone marrow (WBM) cells in 5-ml polystyrene or polypropylene tubes (depending on cell sorter used). Aliquots of the following samples will be required (see Note 11): unstained bone marrow for basic setup of forward side scatter (FSC), side angle scatter (SSC) and lymphoblastoid profile; CD45-FITC for compensation control; CD45-APC for compensation control; Sca-1-FITC and c-kit-APC stained endosteal lineage depleted cells ; Sca-1-FITC and c-kit-APC stained central lineage depleted cells; IgG-FITC and IgG-APC isotype control to set the LSK gate.
2. Set gating strategy according to Fig. 2: exclude doublets by gating singe cells (A); gate lymphoblastoid cell region (B); apply LSK gate (C). By back-gating the LSK population into B, the lymphoblastoid region should be corrected until LSK cells fall perfectly into this region.
3. Sort cells into either 5- or 1.5-ml tubes filled with PBS 2% Se or an equivalent buffer.

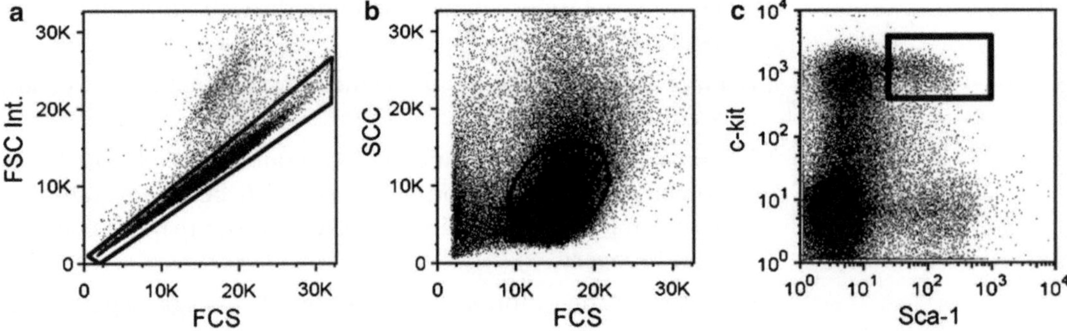

Fig. 2. Sequential sorting strategy for lineage negative, Sca-1⁺c-kit⁺ HSC/HPC. (**a**) single cells; (**b**) lymphoblastoid cell region; (**c**) LSK cells for sorting.

3.5. Transplantation of CFDA-SE and SNARF-1 Labeled HSC/HPC

3.5.1. CFDA-SE and SNARF-1 Labeling of Cells

1. Resuspend each HSC/HPC population at 5×10^6 cells/ml PBS 0.5% Se in a 1.5-ml Eppendorf for CFDA-SE and SNARF-1 staining (see Note 12).
2. Prewarm cell suspensions for 1 min at 37°C.
3. Add 111 μl of the 5 μM CFDA-SE or 10 μM SNARF-1 solution to each 1 ml of cell suspension to give a final concentration of 0.5 μM CFDA-SE and 1.0 μM SNARF-1, respectively (see Note 13).
4. Mix the cell solutions thoroughly to maintain a single cell suspension.
5. Incubate cell suspensions with gentle agitation for 10 min at 37°C in the dark.
6. Add 1 ml of ice cold PBS 20% Se to the cells and mix thoroughly.
7. Centrifuge cells for 5 min at $400\times g$ at 4°C, dry pellet cells, and repeat wash step with PBS 20% Se.
8. Wash cells with cold PBS 2% Se and perform cell counts by excluding nonviable cells using Trypan blue.
9. Use cells counted in hemocytometer to check staining intensity with a fluorescence microscope.
10. Pool SNARF-1 and CFDA-SE stained cells in equal numbers, add 2×10^5 irradiated (15 Gy) WBM filler cells per recipient and centrifuge for 5 min at $400\times g$ at 4°C. Resuspend decanted cell pellet with PBS to a final volume depending on the number of transplant recipients (200 μl per recipient) (see Note 14).
11. Positive controls are required for flow cytometric analysis. For this, stain aliquots of 1×10^5 WBM cells with CFDA-SE or SNARF-1 using the above staining protocol. Store cells in PBS 20% Se in an incubator at 37°C until needed (see Subheading 3.5.4).

3.5.2. Cell Transplantation (see Note 15)

1. Place recipient animals under a heating lamp to dilate the tail vein.
2. Fill 1-ml syringe attached to a 26-gauge needle with well mixed cell suspension.
3. Place recipient into mouse immobilization apparatus and wipe tail with 70% ethanol.
4. Inject cell suspension into recipient via the lateral tail vein (see Note 16).
5. Release mouse and house in appropriate box with chow and water ad libitum.
6. Allow the cells to home for the time period of interest, here 15 h (see Note 17).

3.5.3. Transplantation Analysis: Spatial Distribution

1. Prepare fixative: 4% paraformaldehyde supplemented with 0.1% glutaraldehyde in Sorensen buffer (see Note 18).
2. Anesthetize mice, open the abdominal cavity, and isolate the descending aorta from surrounding muscle and fat. Insert perfusion needle into descending aorta and perfuse at physiological pressure (20.4 ml/h, see Note 19). Once perfusion is started, cut the vena cava above the kidney to avoid a build up of pressure.
3. After 5–6 min perfusion, excise the femurs and immerse in fixative for an additional 4 h at RT while continuously rotating on a suspension mixer.
4. Remove fixative, rinse bones in 10% EDTA, and decalcify in 10 ml of 10% EDTA on a suspension mixer, at 4°C for 14 days. Replace the 10% EDTA daily (decalcification times can take up to 3 weeks if changes of the 10% EDTA is weekly).
5. Dehydrate the bones in graded ethanol and embedded in paraffin.
6. Cut longitudinal sections (3.5 μm) of the femur and mount every alternate section on poly-l-lysine slides as serial sections.
7. Dewax femurs by washing in citrus agent, 100% ethanol, 70% ethanol, and distilled H_2O for 5 min twice each. If an injection of antibodies was done before sacrifice of the recipient, proceed to Subheading 3.6.2, otherwise wash sections in PBS and mount in antifade mounting media (Vectashield).
8. Analyze sections using a fluorescence microscope.
9. The spatial distribution of the transplanted cells is determined by analyzing the location of each CFSE or SNARF-1 labeled cell from at least six longitudinal sections per transplant recipient. The locations of fluorescent cells are designated as either endosteally (arbitrarily defined as within 12 cells of the endosteum and all cells homed to the trabecular bone region) or centrally located (see Note 20).

3.5.4. Transplantation Analysis: Homing

1. After the time period of interest, transplanted mice are euthanized and the marrow harvested from femurs, tibias, and ileum using an "enhanced" method. Bones are dissected and cleaned as described in Subheading 3.1.1, but the epiphysis and metaphysis are not removed. Intact bones are placed into the mortar and crushed using the pestle. Process crushed bone fragments as described in Subheading 3.1.2 and perform a white blood cell count.
2. $10–20 \times 10^6$ events are run through the flow analyzer to assess statistically relevant numbers of CFSE or SNARF-1 positive cells homed to the marrow (Fig. 3a–d, see Note 21). The percentage of white blood cells is determined for each sample and the denominator (D) is mathematically calculated as the total number of WBC cells analyzed (D=%WBC/100×measured events).
3. The proportion donor (% do) of analyzed BM is then calculated using the number of $CFSE^+$ (or $SNARF\text{-}1^+$) events detected (% do=#$CFSE^+$/D×100).
4. The total number of donor cells detected in the bone marrow of each recipient is calculated using the proportion of donor cells, the total number of BM cells harvested as determined in Subheading 3.5.4 (#BM) and the assumption that one femur, tibia, and ileum represent 15% of the total number of cells in the mouse [total # do=(%do/100×#BM)/15×100].

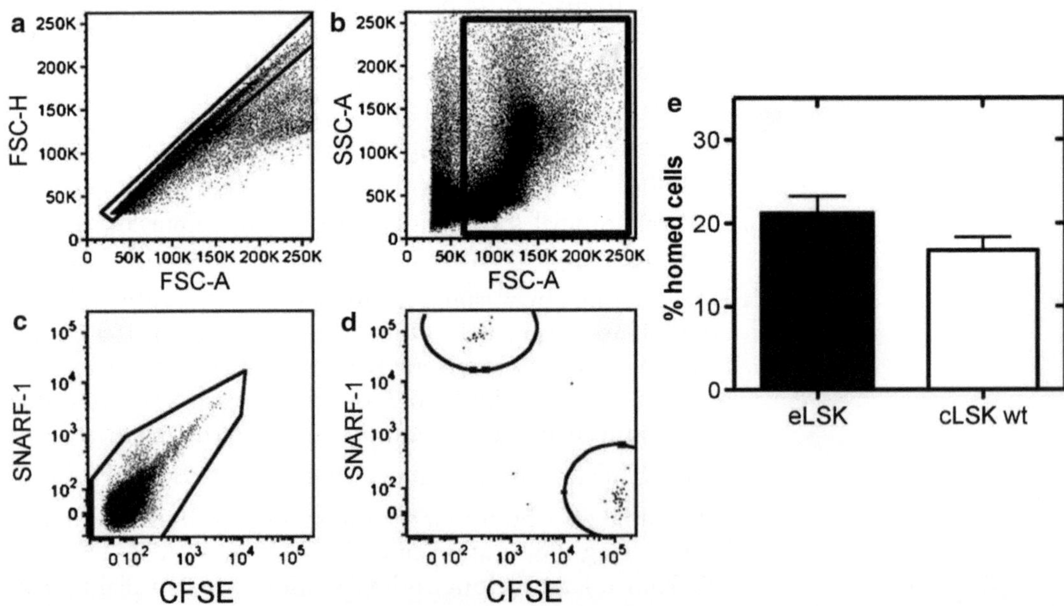

Fig. 3. Analysis of homed CFSE or SNARF-1 stained LSK after 15 h. (**a**) Single cells; (**b**) white blood cells gate; (**c**) unspecific and auto-fluorescent cell fraction as an inverted gate; (**d**) $CFSE^+$ and $SNARF\text{-}1^+$ events from 10×10^6 measured total events; and (**e**) homing efficiency of LSK harvested from the endosteal bone marrow region (eLSK=21%) compared to the central region (cLSK=16%). Data shows mean+SEM of n=9 individual recipients.

5. Finally, the proportion of CFSE+ (or SNARF-1+) cells from each transplant (% CFSE+ tplt) detected in each recipient is calculated using the total number of donor cells and the number of CFSE+ cells transplanted (# cells tplt) into each recipient (% CFSE+ tplt = total # do/# cells tplt × 100). See Fig. 3e for homing efficiency of wt eLSK and cLSK into nonablated C57BL/6 recipients.

3.6. Identification of Marrow Vasculature

3.6.1. Antibody Injection for Vasculature Staining

1. Preheat the recipient using the heat lamp as described in Subheading 3.1.1 (see Note 22).
2. Prepare antibody solution in 200 μl PBS per recipient (see Note 23).
3. Place recipient into mouse immobilization apparatus and wipe tail with 70% ethanol.
4. Inject antibody solution via lateral tail vein.
5. Allow the antibody to circulate for approximately 15 min to ensure robust staining of inner endothelial wall.
6. Process recipient bones and cut and dewax sections as described in Subheading 3.5.3.

3.6.2. Analysis of Sections Injected with Endothelial Antibodies

1. Wash sections twice in PBS for 5 min (see Note 24).
2. Antigen retrieval is done by heating slides in 10 mM citrate buffer pH 6.0 in an oven by placing the slides in the buffer in the bottom of a staining box in the oven at 37°C, then heating the oven to 90°C. When the oven is at 90°C, retrieval is done for 20 min (see Note 25).
3. Remove the boxes and cool to RT – this will take approximately 20 min.
4. Wash sections twice in PBS for 5 min. Encircle sections with Pap pen or glass silicone sealant between washes.
5. Incubate sections in 50 mM glycine in PBS (pH 3.5) for 5 min.
6. Wash sections in PBS 0.3% Triton-X for 15 min.
7. Wash sections twice in PBS for 5 min.
8. Treat sections with hydrogen peroxide in PBS (3%) for 15 min at RT.
9. Wash sections in PBS 0.05% Tween-20 three times for 5 min.
10. Incubate sections in blocking buffer 60 min at RT.
11. Tip excess blocking buffer off sections.
12. Dilute biotinylated antibody in PBS 0.5% BSA containing 10 μg/ml donkey IgG. Incubate sections in diluted antibody overnight at RT.
13. Wash sections in PBS 0.05% Tween-20 three times for 5 min.

14. Block sections in TNB buffer for 30 min at RT.
15. Incubate sections with streptavidin–HRP diluted 1/100 in TNB buffer for 30 min at RT.
16. Wash sections in PBS 0.05% Tween-20 three times for 5 min.
17. Incubate sections in biotinyl tyramide (Amplification Reagent) working solution for 6 min at RT. Stock made of 1/50 using 1× amplification diluent.
18. Wash sections in PBS 0.05% Tween-20 three times for 5 min.
19. Incubate slides with streptavidin-Alexa 488 (diluted in PBS 0.5% BSA) for 30 min at RT light protected.
20. Wash sections in PBS 0.05% Tween-20 for 5 min.
21. Wash slides in DAPI (0.5 μg/ml) in PBS for 10 min.
22. Wash sections twice in PBS 0.05% Tween-20 for 5 min.
23. Mount section in Vectashield Mounting Medium (see Note 26).
24. Analyze using fluorescence microscopy as described in Subheading 3.5.3 (Fig. 4).

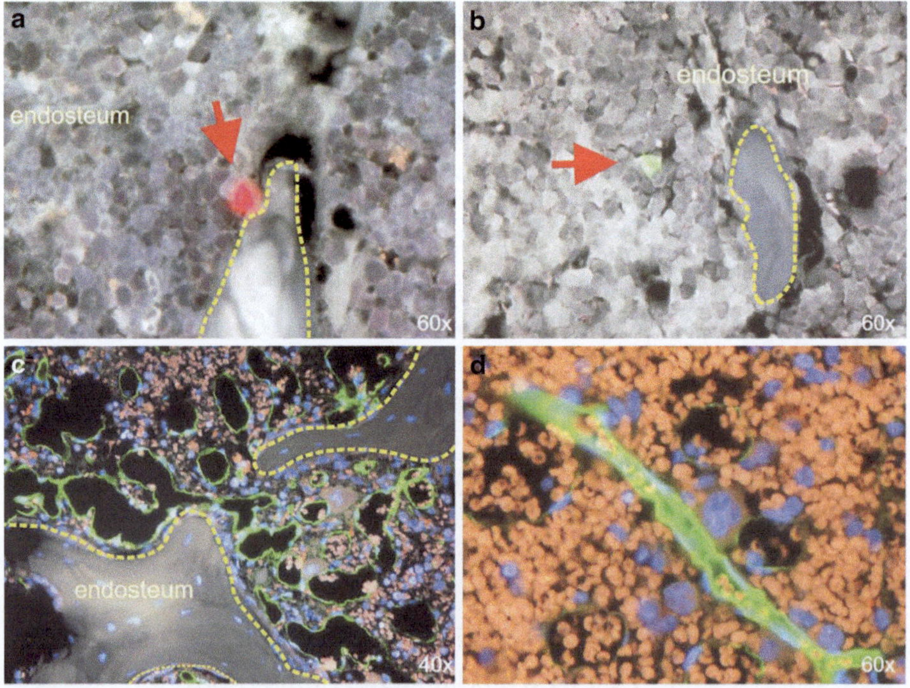

Fig. 4. Analysis of the spatial distribution of SNARF-1 stained eLSK (*arrow*, **a**) and CFSE stained cLSK (*arrow*, **b**). Rat-anti-mouse CD31 (**c**) and rat-anti-mouse CD102 (**d**) antibodies were injected into ablated recipients 15 min prior to perfusion-fixation of the bone marrow and an Alexa 488 conjugated goat-anti-rat secondary antibody was used to visualize the vascular endothelium.

4. Notes

1. This method is for ten donor animals. Volumes can be altered for more or less donors. We use the same method with ablated C57/BL6 mice (irradiated: 10.5 Gy split over two 5.25 Gy doses separated by 4 h and chemo-ablated: 200 mg/kg 5-fluorouracil i.v.), nonablated C57BL/6 mice and a range of transgenic mice.
2. The use of this limited antibody cocktail results in the removal of approximately 70% of WBM cells. To gain higher purity additional antibodies can be added, for example, T-cell markers like CD3, CD4, CD5, and CD8.
3. Other conjugates can be used. Please note that Sca-1-FITC staining is still detectable on the cell surface following homing experiments for short time periods.
4. To dissolve CFDA-SE or SNARF-1 in DMSO prewarm supplied powder. Both stock solutions have to be stored light protected at −20°C in a secondary container with desiccant to avoid hydrolysis and should be kept in small aliquots to avoid frequent freeze–thawing. As the nonfluorescent CFDA-SE is cleaved intracellular by esterase's to the fluorescent CFSE no change in color of the buffer can be observed. In contrast, dissolving SNARF-1 results in a red colored buffer.
5. Seminaphtorhodafluor-1 carboxylic acid, acetate, succinimidyl ester (SNARF-1) is a pH-sensitive fluorescent dye that can be excited at 488 nm. We use a 610/30 nm band pass filter for SNARF-1 detection.
6. Both, anti-CD31 (PECAM), and anti-CD102 (ICAM-2) antibodies result in a good staining of the inner endothelia wall, however, other antibodies of interest can be used. Use low endotoxin, azide-free antibodies for injection into the mice.
7. To make 50 ml blocking buffer add 2.5 g skim milk powder and 2.5 g BSA to 10 ml 20× SSC, allow mix completely (may need over night in cold room). Add 37.3 ml distilled H_2O and 25 μl Triton X-100. Filter solution sterile. Store aliquots at −20°C.
8. Cut the legs from the spinal cord using scissors. Clean the bones from muscle by carefully scraping the bones with the scalpel blade. Dislocate the femur from the knee and scrape the area around the head of the femur as clean as possible. Excise the tibia by pulling the foot and peeling the muscle away. Dislocate the knee. When the muscle is removed from the ileum, a flat triangular piece of cartilage is exposed. Remove this cartilage by cutting it through the acetabular notch.

9. For extended method of cleaning bones refer to (12). Trabecular sections of the long bones are cut to avoid a contamination of flushed cells with endosteal located HSC/HPC.
10. Typically, we resuspend cells as follows: endosteal cells ($<3\times10^8$) in 3 ml buffer in 5 ml polypropylene tube, central cells ($>3\times10^8$) in 6 ml buffer and place 3 ml in each of 2.5-ml polypropylene tubes.
11. This protocol describes sorting for an enriched HSC/HPC cell population with the phenotype Sca-1$^+$c-kit$^+$lineage$^-$ (LSK); however, any other HSC/HPC phenotype can be used.
12. In this protocol, small numbers of HSC and HPC are used (<200,000 cells). It is recommended that a minimum volume of 200 μl is used for cell labeling. If higher cell numbers are to be stained, resuspend cells in a 5-ml polypropylene tube.
13. The protocol is optimized for murine HSC/HPC. Different final concentrations of CFDA-SE and SNARF-1 might be required for other cell types and should be determined by the user in regard to the detection method used. Please note that high concentrations of the intracellular dyes can cause cell death. Described solutions of CFDA-SE and SNARF-1 in this protocol are used to obtain an at least 1:1,000 dilution of DMSO in the final staining solution.
14. In order to detect sufficient numbers of cells in the bone marrow typically $30–40\times10^3$ cells are transplanted per recipient for the evaluation of homing and lodgement. Filler cells are required to decrease the loss of HSC/HPC cells posttransplant due to nonspecific entrapment in organs, such as the lung or spleen, and immunological responses of the recipient.
15. Ablated or not ablated recipients can be used in this assay.
16. To avoid back flushing of the cell suspension, wait for 20 s before removing the needle after injecting the cells. Only blood should be visible at the injection site after removing the needle.
17. We have analyzed homing and spatial distribution from 15 min to 20 h. For spatial distribution analysis, a time point less than 24 h is recommended to ensure the analysis of homed cells and not their progeny (13). Analysis of short homing time points (<5 h) need to consider labeled cells still in the peripheral circulation.
18. Fixative solution should be made and perfusions performed in a fume hood to avoid exposure to toxic paraformaldehyde. For 40 ml add 10 ml paraformaldehyde (16%) to 9.84 ml of sterile water and dissolve in 20 ml of Sorensen buffer (0.2 M). Add 160 μl of glutaraldehyde and adjust to pH 7.
19. If homing is to be concurrently assessed using the same transplant recipient, isolate both femoral arteries and ligate one

using a piece of surgical silk thread. Remove iliac crest, femur, and tibia distal to the silk thread, then insert the perfusion needle into the other femoral artery.

20. Using micro-CT and vascular cast techniques we showed that after 15 h no HSC/HSP homed to the trabecular area of the long bones was outside 12 cell diameters of bone endosteal surface. For this we defined cells located in the trabecular region as endosteal (14).
21. To minimize each samples file size, we collect two files from each sample. The first file is used to determine the WBC count, for this we safe 100,000 total events. The second file of $10–20 \times 10^6$ events (depending on the incidence of transplanted cells in the bone marrow) excludes unstained and autofluorescent events (bone marrow cells) not positive for CFSE or SNARF (see Fig. 2).
22. Depending on the experimental design steady-state recipients, ablated recipients, or recipients that have received CFSE or SNARF-1 labeled cells can be used.
23. An antibody concentration of 15 mg/ml results in robust staining with rat-anti-mouse-CD31 and rat-anti-mouse-CD102 antibody.
24. For the washing steps a horizontal rocker is used and the slides are kept vertically in glass boxes.
25. In order to prevent sections lifting off, slides need to be laid flat for antibody retrieval. Microwave retrieval is also not recommended.
26. For analysis of the spatial distribution of homed cells we usually do not use a DAPI staining as this decreases the visibility of CFSE or SNARF-1 stained cells. However, for the analysis of antibody injected sections DAPI is usually included.

Acknowledgment

German Cancer Aid, NHMRC.

References

1. Schofield, R. (1978) The relationship between the spleen colony-forming cell and the haemopoietic stem cell *Blood Cells* **4**, 7–25.
2. Adams, G.B., et al. (2006) Stem cell engraftment at the endosteal niche is specified by the calcium-sensing receptor *Nature* **439**, 599–603.
3. Calvi, L.M., et al. (2003) Osteoblastic cells regulate the haematopoietic stem cell niche *Nature* **425**, 841–6.
4. Haylock, D.N. and S.K. (2007) Nilsson, Expansion of umbilical cord blood for clinical transplantation *Curr Stem Cell Res Ther* **2**, 324–35.

5. Nilsson, S.K., et al. (2003) Hyaluronan is synthesized by primitive haemopoietic cells, participates in their lodgement at the endosteum following transplantation, and is involved in the regulation of their proliferation and differentiation in vitro *Blood* **101**, 856–62.
6. Nilsson, S.K., et al. (2005) Osteopontin, a key component of the hematopoietic stem cell niche and regulator of primitive hematopoietic progenitor cells. *Blood* **106**, 1232–9.
7. Zhang, J., et al. (2003) Identification of the haematopoietic stem cell niche and control of the niche size *Nature* **425**, 836–41.
8. Kiel, M.J., et al. (2005) SLAM family receptors distinguish hematopoietic stem and progenitor cells and reveal endothelial niches for stem cells *Cell* **121**, 1109–21.
9. Lo Celso, C., et al. (2009) Live-animal tracking of individual haematopoietic stem/progenitor cells in their niche *Nature* **457**, 92–6.
10. Xie, Y., et al. (2009) Detection of functional haematopoietic stem cell niche using real-time imaging *Nature* **457**, 97–101.
11. Haylock, D.N., et al. (2007) Hemopoietic stem cells with higher hemopoietic potential reside at the bone marrow endosteum *Stem Cells* **25**, 1062–9.
12. Williams, B. and S.K. Nilsson. (2009) Investigating the interactions between haemopoietic stem cells and their niche: methods for the analysis of stem cell homing and distribution within the marrow following transplantation *Methods Mol Biol* **482**, 93–107.
13. Nilsson, S.K., et al. (1997) Potential and distribution of transplanted hematopoietic stem cells in a nonablated mouse model *Blood* **89**, 4013–20.
14. Ellis SL, et al. (2008) Defining the HSC Niche: The relationship between bone, HSC and vasculature *Exp Hem* **36**, 61.

Chapter 15

Imaging Hematopoietic Stem Cells in the Marrow of Long Bones In Vivo

Anja Köhler, Hartmut Geiger, and Matthias Gunzer

Abstract

Hematopoietic stem and progenitor cells (HSPCs) are located in the bone marrow in zones of residence specialized in supporting them which are referred to as niches. It is in such a specialized niche that normal HSPCs are maintained to perform their self-renewal and differentiation duties in a highly controlled manner. One challenge in dissecting the functional significance of the complex cellular and molecular interactions in the niche is to link the types and qualities of cell–cell contacts to the intracellular signaling components involved in cell regulation. Attempts to study the interactions of HSPC with their niche eventually have to be performed in their natural location in vivo, as isolation of the cells from bone marrow will disrupt the HSPC–niche interactions and thus not reveal functionally critical cell–cell contacts. Intravital imaging of individual cells in the bone marrow has just recently been introduced, almost exclusively focusing on imaging inside the marrow of the calvaria. However, calvarial marrow is functionally distinct from marrow of long bones, the major source of HSPC for both physiology and study. To overcome these limitations, we developed a novel method for multiphoton intravital imaging of HSPC in the marrow of long bones.

Key words: Hematopoietic stem cells, In vivo imaging, Bone marrow niche

1. Introduction

The biology of hematopoietic stem and progenitor cells (HSPCs) depends on their interaction with local stromal elements in the bone marrow (1). These stromal elements have been termed "niche" and are thought to control both the retention of HSCs as well as their future fate leading to daughter cells that maintain their pluripotency or develop into mature effector cells by either symmetric or asymmetric division (2, 3). How this control is exerted on the single cell level has remained elusive for a long time, most likely due to experimental inaccessibility of the bone marrow niche and an inability to regenerate functional stem cell

Marie-Dominique Filippi and Hartmut Geiger (eds.), *Stem Cell Migration: Methods and Protocols*, Methods in Molecular Biology, vol. 750, DOI 10.1007/978-1-61779-145-1_15, © Springer Science+Business Media, LLC 2011

niches in vitro. The type and nature of cell–cell interactions in the niche has only recently been clarified more precisely thanks to the advent of modern intravital imaging approaches (4, 5).

Previous approaches into this area of research have used imaging in a small stretch of bone marrow located in the calvaria of mice (4–6). Marrow in the calvarium is generated by a process called intramembraneous ossification, while the marrow of long bones develops in a distinct process termed endochondral ossification (7). In addition, only marrow of fetal long bones harbors a very rare cell type that has the ability to ectopically generate bones with functional niches when transplanted under the kidney capsule of mice (7). These differences between marrow in the calvarium and in long bones seem to have consequences for the "niche quality" of the marrow. An example is the fact that the activity of a HSC-specific promoter in long-term repopulation experiments in vivo was found to be mostly located in the hind legs and thus in long bones of mice (8).

In light of these findings, we have established a novel protocol that allows imaging of individual HSPC within their bone marrow niche in long bones of mice in vivo (9). This approach allows direct imaging of single cells in vivo and thus the analysis of HSPC motility/migration and their entry into blood vessels, the determination of the distance to the endosteum and the dynamics of cell to cell interactions.

2. Materials

2.1. Mice

C57BL/6J mice (8–12 weeks) have to be housed under SPF conditions.

2.2. Isolation of Early Hematopoietic Progenitor Cells

1. Iscove's Modified Dulbecco's Medium (IMDM, Invitrogen) + 2% Fetal Calf Serum (FCS): 500 ml IMDM (1×) and add 10 ml of FCS.
2. IMDM + 10% FCS: Take 23 ml of IMDM + 2% FCS and add 2 ml of FCS.
3. Hank's Balanced Salt Solution (HBSS) + 2% FCS: 500 ml of HBSS, and add 10 ml of FCS.
4. Ficoll 1083.
5. Trypanblue-Solution (0.4%).
6. Magnet for cell separation (Dynal).
7. Dynabeads Sheep anti-Rat IgG (Dynal).
8. Blocking-Antibody 2.4G2 (BD-Biosciences), 0.5 mg/ml.
9. Streptavidin FITC, PE or APC labeled (BD-Biosciences, 0.5 mg/ml).

10. c-Kit (CD117) APC, IgG2b, clone 2B8 (BD-Biosciences), 0.5 mg/ml.
11. Sca-1 (Ly6A/E) PE, IgG2a, clone E13-161.7 (BD-Biosciences), 0.5 mg/ml.
12. 7-Aminoactinomycin (7-AAD, Molecular Probes) 1 mg/ml.
13. Modified Neubauer chamber.

Linage (LIN) cocktail:

Name	Supplier	Dilution	Add to LIN cocktail (μl)
CD5 (Ly-1) Biotin, IgG2a, clone 53-7.3	BD-Biosciences	1:200	25.0
B220 (CD45R) Biotin, IgG2a, clone RA3-6B2	BD-Biosciences	1:300	16.7
Mac-1 (CD11b) Biotin, IgG2b, clone M1/70	BD-Biosciences	1:320	15.7
CD8a (Ly-2) Biotin, IgG2a, clone 53-6.7	BD-Biosciences	1:200	25.0
Gr-1 (Ly-76) Biotin, IgG2b, clone RB6-8C5	BD-Biosciences	1:350	14.3
TER-119 (Ly-76) Biotin, IgG2b, clone TER-119	BD-Biosciences	1:320	15.7
			Total: 112.4

2.3. Early Hematopoietic Progenitor Cell Staining

1. Phosphate-buffered saline (PBS): Prepare 10× stock with 1.37 M NaCl, 27 mM KCl, 100 mM Na_2HPO_4, 18 mM KH_2PO_4 (adjust to pH 7.4 with HCl if necessary) and autoclave before storage at room temperature. Prepare working solution by dilution of one part with nine parts distilled, autoclaved water.
2. Carboxy Fluorescein Succinimidyl Ester-Dye (CFSE, Molecular Probes). Prepare stock of 5 mM in DMSO, store in 5 μl aliquots at −20°C.
3. Cell Tracker Orange-Dye (CTO, Molecular Probes). Prepare stock of 5 mM in DMSO, store in 5 μl aliquots at −20°C.

2.4. Mouse Anesthesia

1. Ketamine-Rompun-narcosis: 3.5 ml NaCl (0.9%), 1 ml Ketamine (Inresa Arzneimittel GmbH), 0.5 ml Rompun (Braun).
2. Isofluran (Delta Select, Actavis).
3. 22G (0.9×25 mm) permanent venous catheter (Braun).
4. Mini Vent (Hugo Sachs Elektronik).
5. Univentor (Univentor Limited).

2.5. Preparation of the Tibial Bone

1. Cutter (105 mm, Omnilab) and forceps (115 mm, Omnilab).
2. Scalpel.
3. Electric drill (model 780, Dremel) with flexible wave (model 225, Dremel).
4. Aluminium oxide grindstone (model 997, Dremel).
5. Syringe (5 ml, Becton Dickinson).
6. Needles (21G, 23G, Becton Dickinson).
7. Cell strainer (100 μm pores, Becton Dickinson).

2.6. Imaging and Analysis

1. A suitable intravital two-photon microscopy setup with at least three independent detectors as well as conventional widefield fluorescence illumination as described (10).
2. A suitable setup for analysis of three-dimensional (3D) and 4D imaging data (e.g., Volocity®, Improvision/Perkin Elmer, UK; Imaris®, Bitplane AG, Switzerland or Arivis Browser, Arivis, Rostock, Germany).

3. Methods

3.1. Purification of Early Hematopoietic Progenitor Cells (eHPCs) (see Note 1)

1. Get tibia and femur from both legs of ten mice, open the bones on either side with scissors or a scalpel and put them into IMDM/2.0% FCS (medium).
2. Using a 3-ml syringe equipped with a 21 or 22G needle flush bone marrow cylinders into a 50-ml falcon tube within a total volume of 10 ml, wash one time in medium (400 g, 1,400 rpm with Sorvall Super 21, 5 min).
3. Bring cells to a total volume of 24 ml in HBSS with 2% FCS, RT, pass through cell strainer.
4. Count cells: take 10 μl of cell suspension, dilute with 90 μl of medium and take 10 μl, add 10 μl of Trypan Blue and determine cell number in a modified Neubauer chamber.
5. Carefully layer on 16 ml of Ficoll at room temperature (RT).
6. Spin for 25 min at 600 g (1,650 rpm with Sorvall Super 21, RT, w/o brake).
7. Carefully remove white blood cell layer on top and first parts of Ficoll and transfer into a new 50-ml falcon tube (Fig. 1).

All following steps on ice (cool down centrifuge if necessary)!

8. Wash cells in 50 ml of IMDM, 2% FCS (from now on use only IMDM). Spin at 400 g (1,400 rpm on Sorvall Super 21, 4°C).
9. Wash again in 25 ml of Medium (see step 8).
10. Resuspend cells in 3 ml of IMDM/2.0% FCS.

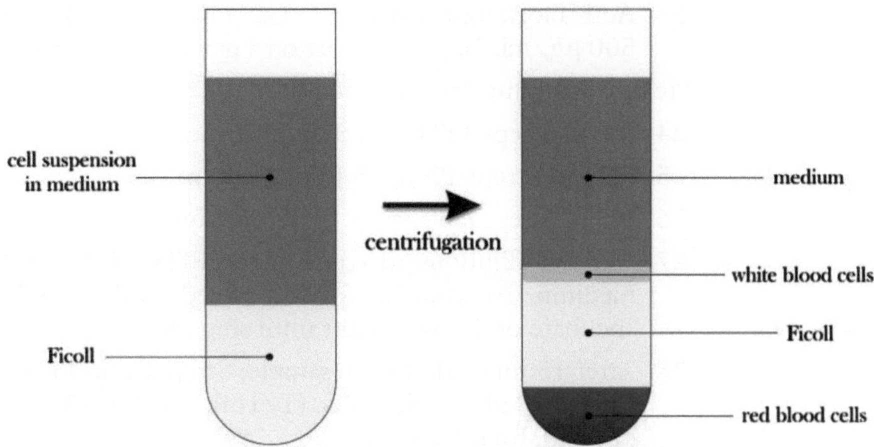

Fig. 1. Ficoll gradient before and after centrifugation.

11. Determine cell count (see step 4).
12. Add LIN antibodies: according to the antibody table: per 1×10^8 cells add 112.4 μl of LIN cocktail.
13. Incubate 25 min on a gentle shaker, 4°C, cold room.
14. Wash two times with 25 ml of Medium, each (see step 8).
15. Resuspend in 2 ml of medium.
16. Aliquot 10 μl of cells each into five Polystyrene FACS tubes to obtain single color controls for FACS compensation (see below for description of control staining). Label tubes: unstained, FITC, PE, 7AAD, APC.
17. Dynabeads: Resuspend beads stock carefully to obtain a homogenous emulsion (attention: beads sink down rapidly). Add from stock an amount of four beads per cell to the cell suspension (usually 1 ml original stock (4×10^8 beads/ml) per 10^8 cells).
18. Add beads (total volume with beads less than 8 ml) to cells in 50 ml tube.
19. Incubate cells on a gentle shaker, 25 min, 4°C, cold room.
20. Transfer cell suspension with beads into 15 ml tube, fill up with medium to 8 ml.
21. Separate with magnet and transfer supernatant into a new tube. Wash beads once with 7 ml medium to increase cell yield, also transfer this supernatant into a new 15 ml tube. Separate both supernatants at least three times with magnet (to get rid of remaining beads).
22. Combine and spin both supernatants in 50 ml tube 400 g, 5 min, 4°C. Resuspend pellet in 1 ml of medium. Determine cell count.

23. Add blocking-antibody 2.4G2 (1/100 stock solution of 500 μg/ml, 10 μl), incubate on a gentle shaker, 15 min at 4°C.

Prepare AB-dilutions for controls:

24. 0.5 μl Strept-FITC in 75 μl of medium.
25. 0.25 μl Strept-PE in 75 μl of medium.
26. 0.5 μl Strept APC in 75 μl of medium.
27. Add AB-dilutions to controls for FITC, PE and APC and medium to control for unstained and 7AAD-staining. Incubate on ice w/o light until step 35.
28. After 15 min, add to cell sample: Streptavidin FITC (1/150, total 6.6 μl), c-kit APC (1/160, total 6.25 μl), Sca-PE (1/160, total 6.25 μl).
29. Incubate on a gentle shaker, 20 min at 4°C.
30. Get a 15 ml falcon tube for each population to sort, fill with 1 ml of IMDM + 10% FCS.
31. Mix 3 ml of Medium and 15 μl of 7-AAD (1 mg/ml), final concentration 5 μg/ml: this is the resuspension medium.
32. Wash cells and controls once (25 ml of cells, 1.5 ml of controls, see step 8).
33. Resuspend in 700 μl of resuspension medium, transfer into new FACS tube.
34. Resuspend 7-AAD control in 300 μl of 7-AAD-medium, the other controls in 300 μl of medium.
35. Purify eHPCs by cell sorting (see Note 2). As we have predepleted the cells for LIN+ cells, the major task is to gate the LIN− population correctly. The way we address that is as follows (a) We include a LIN+ (not-depleted control) to unambiguously determine how to define a LIN− population and (b) we look at lineage signal over the c-kit signal (Fig. 2c), with the majority of the c-Kit positive cells being LIN− and thus gate them accordingly.

3.2. Cell Staining for Two-Photon Visualization After Adoptive Transfer

1. After the sort wash cells two times with 15 ml of PBS (see Subheading 3.1, step 8).
2. Resuspend in 1 ml of PBS.
3. Add 1 μl of CFSE (5 mM) or 5 μl of CTO (5 mM), incubate 15 min at RT, w/o light.
4. Wash once with 15 ml of PBS (see step 1).
5. Resuspend the cells in PBS and transfer by retro-orbital injection into the recipient (in a volume of 100 μl per animal).

3.3. Mouse Anesthesia

1. Inject 100–150 μl of ketamine-rompun i.p.
2. Intubate the animal with a permanent venous catheter (remove the sharp top of the catheter with a cutter, so that

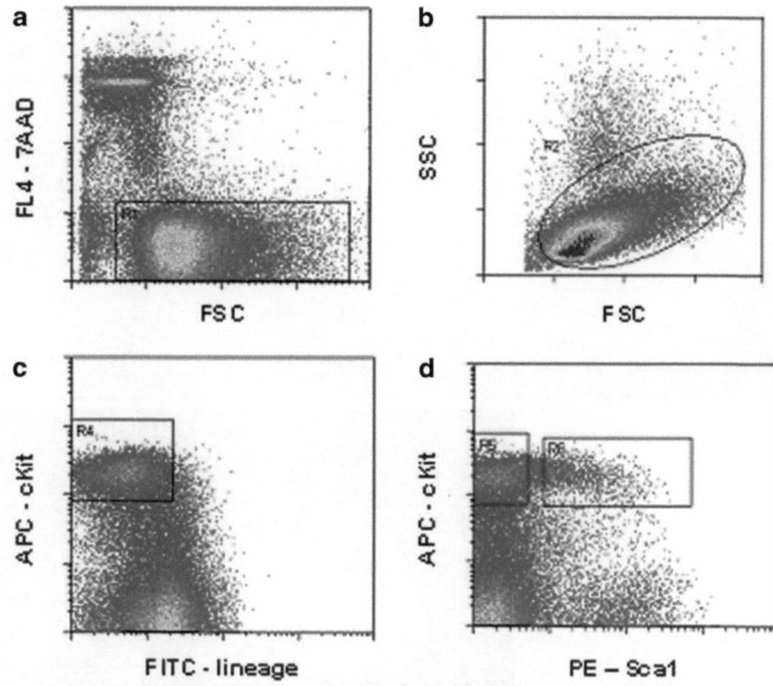

Fig. 2. Setup and gating strategy for hematopoietic progenitor and early hematopoietic progenitor sorting. First we gate on living cells based on negativity of 7-AAD-staining and correct size by forward-scatter (**a**) and for the right cell morphology (**b**). In the next step, c-Kit positive and lineage negative cells are selected (**c**). Gating on c-Kit and Sca-1 allows the isolation of c-Kit positive and Sca-1 negative hematopoietic progenitor cells (**d**, gate R5) as well as c-Kit and Sca-1 positive hematopoietic stem cells (see **d**, gate R6).

the metal part is shorter than the plastic part to prevent the animal for tracheal injuries. Use the metal part for stabilization of the plastic part but remove it immediately, when the mouse is intubated).

3. Ventilate by a mechanical small animal respirator (Mini Vent) at rates of 250×/min and a volume of 250 μl.
4. The animals receive a mixture of O_2 and 1% isoflurane (mixed within the Univentor).
5. Transfer animals into a chamber and fix them with needles.
6. Start tibia preparation.

3.4. Preparation of the Tibia

1. Expose the tibia by removing the skin on top of the bone, and further removing muscle tissue with a scalpel.
2. Carefully remove the calcified bone with an electric drill to obtain a very thin (30–50 μm) remaining layer of bone tissue covering the BM cavity to permit a better light penetration. Clean and cool the bone in between with PBS (see Note 3).
3. Control the procedure permanently by stereomicroscopy to maintain the necessary thickness of the remaining bone tissue (Fig. 3).

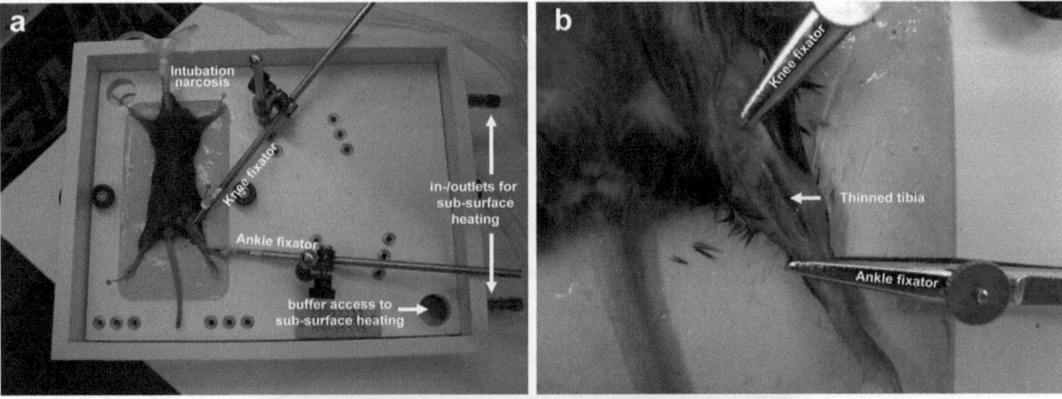

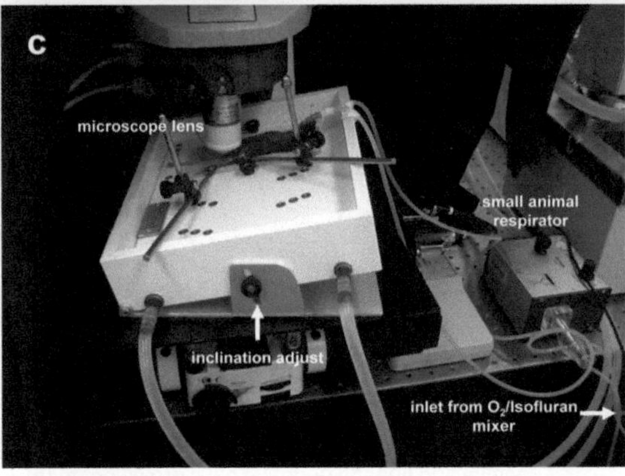

Fig. 3. Setup for intra-tibial two-photon imaging. (**a**) Overview of an anesthetized animal within a custom-made imaging chamber. Note: to allow better images the PBS that usually fills the part of the imaging chamber containing the prepared leg has been omitted. (**b**) Higher magnification showing the fixed and thinned tibia of the left leg. (**c**) The complete chamber as positioned under the two-photon microscope ready for imaging. Note: the Univentor (O_2/Isofluran mixer) is outside of the field of view (not visible).

3.5. Two-Photon Microscopy Inside the Tibia

1. Fill the chamber with PBS (prewarmed to 37°C) and keep animals at 37°C while experiments are performed.
2. Under the microscope search for stained cells in the bone marrow using the conventional fluorescent light illumination (choosing a green, red, or red/green filter set, depending on how the cells were stained, (see Note 4)).
3. Switch to two-photon observation mode. The bone tissue can be identified by using its second-harmonic (SHG) signal (PMT).
4. Use an illumination wavelength of either 800 or 880 nm to detect green (530 nm) and red (580 nm) fluorescence, as well as SHG (at 400 or 440 nm emission).
5. For imaging, normally a 300-μm area is scanned in 30 steps of 4 μm down to 120 μm depth (see Note 5 and 6).
6. Repeat the sequence every 60 s for a total of up to 60 min.

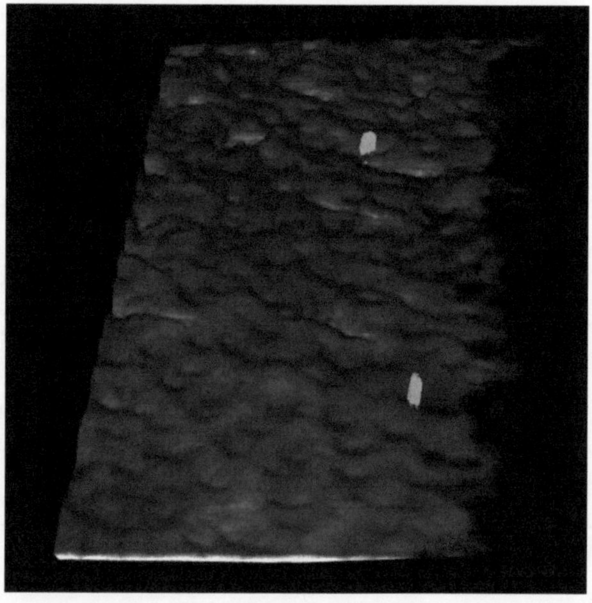

Fig. 4. Visualization of bone marrow cells (highlighted) by intravital microscopy.

3.6. Generation of Movies for Analysis

1. Import primary data from the microscope into the available software package for further processing and conversion into 3D movies.
2. Adjust colors and brightness of the channels (precisely described in (9)) (Example in Fig. 4).
3. For cell tracking we export the movies into the .avi format and analyze these by computer-assisted cell tracking. We also have developed a novel tool to analyze the typical wigling of HSPC (9). Depending on the used software cell tracking can also be done automatically or semiautomatically. Exported movies are also suitable for presentation.

4. Notes

1. The protocol describes the isolation of early hematopoietic progenitor cells, but any type of cell that can be fluorescently labeled and is able to home to the endosteal part of the bone marrow can be used for this intravital imaging procedure. Additional protocols for purification of distinct populations of hematopoietic stem cells are provided elsewhere in this book.
2. Cell sorting: We initially used a Moflo system, but get similar results with a FACSAria or FACSDiva system.
3. Take extreme care not to completely open the bone marrow channel. This will result in immediate leakage of BM

constituents into the medium. In this case, imaging is no longer possible on this bone.

4. Take care to keep the illumination with fluorescent light as short as possible to avoid bleaching.
5. If the motility of HSC is investigated, it might be sufficient to obtain one z-stack per 5 min, since the movements of these cells are very minute.
6. It is possible to also visualize the blood flow by injection of Rhodamine-labeled Dextran (40 kD) or suitable fluorescent Quantum dots and visualization of their signal in the relevant fluorescence channel of the microscope. This gives a direct proof of the physiological imaging conditions (intact blood flow!) and allows observing extravasations of cells, if they do occur.

Acknowledgments

This work was supported by grants from the German Research foundation (DFG, GU 769/1-3 and GU 769/2-1) and from the European Union (NEST, Mamocell) to M.G and grants from the National Institute of Health, HL076604 and DK077762 to H.G. H.G is a New Investigator in Aging of The Ellison Medical Foundation.

References

1. Geiger, H., Köhler, A., and Gunzer, M. (2007) Stem Cells, Aging, Niche, Adhesion and Cdc42: A Model for Changes in Cell-Cell Interactions and Hematopoietic Stem Cell Aging *Cell Cycle* **6**, 884–87
2. Scadden, D. T. (2006) The stem-cell niche as an entity of action *Nature* **441**, 1075–79
3. Wilson, A., Trumpp, A. (2006) Bone-marrow haematopoietic stem-cell niches *Nat Rev Immunol* **6**, 93–106
4. Lo Celso, C., Fleming, H. E., Wu, J. W., Zhao, C. X., Miake-Lye, S., Fujisaki, J.et al. (2009) Live-animal tracking of individual haematopoietic stem/progenitor cells in their niche *Nature* **457**, 92–6
5. Colmone, A., Amorin, M., Pontier, A. L., Wang, S., Jablonski, J., and Sipkins, D. A. (2008) Leukemic Cells Create Bone Marrow Niches That Disrupt the Behavior of Normal Hematopoietic Progenitor Cell *Science* **322**, 1861–65
6. Mazo, I. B., Gutierrez-Ramos, J. C., Frenette, P. S., Hynes, R. O., Wagner, D. D., and von Andrian, U. H. (1998) Hematopoietic progenitor cell rolling in bone marrow microvessels: parallel contributions by endothelial selectins and vascular cell adhesion molecule 1 *J. Exp. Med* **188**, 465–74
7. Chan, C. K., Chen, C. C., Luppen, C. A., Kim, J. B., Deboer, A. T., Wei, K.et al. (2009) Endochondral ossification is required for haematopoietic stem-cell niche formation *Nature* **457**, 490–94
8. Xie, Y., Yin, T., Wiegraebe, W., He, X. C., Miller, D., Stark, D.et al. (2009) Detection of functional haematopoietic stem cell niche using real-time imaging *Nature* **457**, 97–101
9. Köhler, A., Schmithorst, V., Filippi, M. D., Ryan, M., Daria, D., Gunzer, M.et al. (2009) Altered cellular dynamics and endosteal location of aged early hematopoietic progenitor cells revealed by time-lapse intravital imaging in long bones *Blood* **114**, 290–98
10. Nitschke, C., Garin, A., Kosco-Vilbois, M., and Gunzer, M. (2008) 3-D and 4-D imaging of immune cells in vitro and in vivo *Histochem. Cell Biol* **130**, 1053–62

Part V

Nonhematopoietic Stem Cell Migration

Chapter 16

Studies of Adult Neural Stem Cell Migration

Adam C. Puche and Serena Bovetti

Abstract

Adult neural progenitors originating is the subventricular zone (SVZ) proceed toward the olfactory bulb (OB) along a peculiar path of migration known as the rostral migratory stream. Once reaching the inner core of the bulb, neuroblasts migrate radially to various layers of the adult OB where they differentiate into local interneurons. We recently demonstrated that this latter form of migration is guided by blood vessels that serve as a scaffold for neural progenitor migration to their respective OB layers.

Migration of progenitors along blood vessels can be visualized by time-lapse confocal imaging, which requires labeling of both migrating cells and blood vessels. In this article, we describe procedures to label neural progenitors in adult through in vivo stereotaxic cell tracer injection into the SVZ, and to visualize blood vessels through luminal filling with stabilized fluorescent dyes. Moreover, we illustrate the methods for acute brain slices preparation for imaging and procedures used during confocal time-lapse imaging to follow the movement of adult neural progenitors along blood vessels.

Key words: Neural stem cells, Migration, Brain slices, Imaging

1. Introduction

Neuronal migration in the central nervous system (CNS) is typically discussed in the context of developmental processes occurring during embryogenesis and to a lesser degree during postnatal periods. However, discoveries made in the last decade have highlighted active neurogenesis and long range neuronal migration occurring in adult brain. The microenvironment through which adult neural progenitors migrate is very different to that present in embryonic development. During the development of brain structures, such as cortex, cerebellum, and hippocampus, migrating neural progenitors use radial glia processes as scaffolds to reach their final destination in a classic process (1, 2). This radial glia-dependent migration, referred to as gliophilic migration

Marie-Dominique Filippi and Hartmut Geiger (eds.), *Stem Cell Migration: Methods and Protocols*,
Methods in Molecular Biology, vol. 750, DOI 10.1007/978-1-61779-145-1_16, © Springer Science+Business Media, LLC 2011

due to the tight molecular interactions between the glial cell and migrating neuron, is prominent in the embryonic and early postnatal CNS. Neurophilic migration has been used to describe a process of neural progenitor migration along axons of already differentiated neurons to reach their final location. Homophilic neural migration involves the movement of neuroblasts using each other as a substrate forming long chains of cells, sometimes referred to as chain migration, which was first described in the rostral migratory stream (RMS) of adult rodent. Neural progenitors originating is the subventricular zone (SVZ) migrate tangentially in chains surrounded tubes of astrocytic processes along the RMS into the center of the olfactory bulb (OB) (3). Upon reaching the bulb, these cells migrate radially to various layers of the adult OB, where they differentiate into the different classes of interneurons present in the bulb. We demonstrated that SVZ progenitors migrate radially into the OB along blood vessels, hypothesizing that these vessels serve as a scaffold and called this migration modality "vasophilic" (4). Neuronal progenitors migrating in the granule cell layer of the OB are closely associated with blood vessels, have leading/trailing processes, and extend the leading process along or, in same cases, around a vessel. Time-lapse confocal microscopy demonstrated that neuroblasts migrate directly along the vessels in a saltatory fashion altering periods of slow and high speed. Recent reports suggest that this interaction between migratory neurons and blood vessels in this region involve calcium and the induction of BDNF/TrkB systems (5). Neural migration in adult into damaged brain regions, e.g., stroke, and exogenous transplants also appear to tightly follow vascular elements (6, 7). The molecular environment of the perivascular space permitting neural migration in an adult brain is a topic of intense interest due to the clear clinical implications of this vasophilic process.

In this article, we describe procedures to label neural progenitors in adult and follow their migration along blood vessels. This approach combines dye labeling of neural progenitors through in vivo stereotaxic cell tracer injection into the SVZ, visualization of blood vessels through luminal filling with stabilized fluorescent dyes, and time-lapse confocal imaging of acute brain slices to follow the movement of the neural progenitors.

2. Materials

2.1. Basic Solutions

1. 0.9% NaCl.
2. Prepare 4% paraformaldehyde in 0.1 M phosphate buffer: Add 4 g of paraformaldehyde to 80 ml of phosphate buffer, 1 ml

of 10 N NaOH, mix till dissolved, pH 7.4 with HCl, add phosphate buffer to 100 ml, filter through Whatman Q8 filter paper. Contrary to many PFA preparation protocols in the literature, it is recommended not to heat, or at most use of very low warming, when preparing the reagent for fluorescence microscopy as high heat during preparation will increase fluorescence background; stable at 4°C for 1 week.

3. Artificial cerebrospinal fluid (ACSF): mix 120 mM NaCl, 8 mM KCl, 1.3 mM $CaCl_2$, 1.3 mM $MgSO_4$, 10 mM glucose, 25 mM $NaHCO_3$, 5 mM 5 *N,N*-bis(2-hydroxyethyl)-2-aminoethanesulfonic acid (BES). The final osmolarity should be 290–310 mOsm and when oxygenated with 95% O_2/5% CO_2 the pH should be 7.3–7.5, stable at 4°C for several weeks.
4. Phosphate Buffer (PB): 0.1 M sodium phosphate.
5. Phosphate-Buffered Saline (PBS): 0.1 M sodium phosphate, 0.9% NaCl.

2.2. Chemicals

1. Bovine serum albumin fluorescein isothiocyanate (BSA-fluorescein).
2. Bovine serum albumin tetramethyl-rhodamine isothiocyanate (BSA-rhodamine, Sigma).
3. Various cell trackers: Cell Tracker Green CMFDA (Molecular Probes), Cell Tracker Orange (Molecular Probes), Cell Tracker Red (Molecular Probes).
4. Cholera toxin β (List Biologicals).
5. Fluorogold (Fluorochrome Incorporated).
6. Leibovitz's L15, minus phenol red (Invitrogen).
7. Low-gelling temperature tissue culture-tested agarose.
8. Nembutal (sodium pentobarbital).
9. Porcine type-A gelatin.
10. Superglue (Loctite 401 Prism General Purpose Instant Cyanoacrylate, Small Parts).

2.3. Materials

1. Calibrated microcapillary tubes (Drummond 1–5 μl, part number 2-000-001).
2. Carbide steel blades (Ted Pella).
3. Loctite 401 Prism General Purpose Instant Cyanoacrylate adhesive (Small Parts).
4. Micropipette holder part number (World Precision Instruments).
5. Whatman Q8 filter paper.

3. Methods

Adult neural migration occurs in several brain regions; here we focus on several techniques to investigate adult brain "vasophilic" migration though time-lapse imaging. Time-lapse confocal imaging of "vasophilic" migration requires labeling of both migrating cells and blood vessels. In this chapter, we will describe a method for labeling SVZ neural progenitors by stereotaxic injection of cell tracer dyes (Subheading 3.1), blood vessel labeling for immunohistochemistry or time-lapse studies (Subheading 3.2), preparing adult brain slices for imaging (Subheading 3.3), and the methods used during confocal time-lapse imaging (Subheading 3.4).

3.1. SVZ Progenitors by Stereotaxic Injection of Tracer Dyes

Stereotaxic surgery is a well-established technique for targeting specific structures within the brains of living animals for delivery of substances (8). In this section, we describe the procedure to label subventricular zone (SVZ) progenitors in adult mice by stereotaxic injection of dyes (i.e., Cell Tracker Green, CTG, Fluorogold etc.; (9)). We have found that the most effective tracers for labeling adult neural progenitors are molecules typically considered retrograde tracers, the anterograde tracers (e.g., Dextran) were ineffectively absorbed by neural progenitors (9). Labeled progenitors will migrate tangentially from the SVZ to the olfactory bulb along the rostral migratory stream (at 2–4 days postinjection), and upon reaching the center of the bulb they migrate radially to their final destination (at 6–7 days postinjection) and begin to differentiate (at 10–30 days) (10).

1. Prepare dye at the final concentration. Several dye options are suitable for labeling adult progenitors including: (1) cholera toxin β (CTb; 1% in 0.9% NaCl), (2) cholera toxin β FITC conjugate (CTb-FITC; 1% in 0.9% NaCl), (3) Fluorogold (FG; hydroxystilbamidine, methanesulfonate; 1% in 0.9% NaCl), and (4) Cell Tracker Green CMFDA (10 mM in dimetheylsulfoxide; Cell Tracker Orange and Cell Tracker Red are also effective dyes).
2. Load pulled glass micropipettes with the dye solution chosen. These micropipettes can be prepared from microcapillary tubes with any standard capillary pipette puller (e.g., Narishige PE21 model puller) from a wide range of capillary glass. We favor glass containing printed calibration scales such that delivery volume can be verified (e.g., Drummond 1–5 μl calibrated microcapillary tubes). Ensure during loading that air bubbles are not present near or at the tip as these will disrupt the injection.
3. Anesthetize animals with intraperitoneal anesthetic injection (e.g., Nembutal sodium pentobarbital 50 mg/kg body weight in 0.9% saline solution). After the animal has reached full

anesthesia, remove the fur from the top of the head with a small electric shaver.

4. Place animal on a thermo-regulated heating pad (to maintain body temperature) in the stereotaxic apparatus with the head approximately level to the ear bars. For accurate stereotaxic injections it is important that the animal be positioned correctly in the apparatus. Move the animals head to guide the tip of the first clamped ear bar into the external auditory meatus and hold firmly against the skull; slide the other ear bar into the opposite ear against the skull and clamp when in position. The bars should rest in small depressions in the skull (immediately dorsal to the external skull tympanic bulla and slightly rostral to the post-tympanic hook). The ears should be resting flat against the bars and the top of the skull level. The skull should freely rotate vertically around the axis of the ear bars without any lateral side-to-die movement. If gentle lateral pressure moves the snout >1 mm in either direction, the ear bars are likely improperly seated within the skull depressions. The anterior position of the head is secured in the nose assembly by resting the "bite bar" against the hard pallet with most apparatus designed for the front incisors to rest through a hole in the mouthpiece and clamping the nose bar into position against the bridge of the nose.

   Placing ear bars properly can be difficult, particularly in mouse, and incorrect placement will affect the accuracy of any injection and can compromise the health of the animal. Incorrect placement/insertion of the ear bars (e.g., into the space formed between the skull and mandible) can cause respiratory distress in mouse and lead to death. The presence of lateral movement of the snout and/or a total distance between the bars of <5 mm in an adult mouse likely indicates placement into the mandiblar space. Standard ear bars have an 18° taper at the tip resulting in penetration into the ear canal and a more secure hold, but puncture the tympanic membrane. Ear bars with a 45° taper do not puncture the tympanic membrane, but offer a slightly less stable hold. Refer to Athos and Storm (8) for more detailed information about stereotaxic surgery.

5. Surgical tools/materials should be sterilized and incision areas prepared with alternate wipes of Betadine and alcohol strictly in accordance with NIH and local IACUC guidelines on aseptic surgery in rodents. Make an incision along the midline with a scalpel and use skin clamps to keep the skull surface exposed. Cleaning/wiping the skull surface with 100% ethanol will make the cranial sutures (plate joins) more visible for identification of the bregma/lambda reference points.

6. Fit a drill to the stereotaxic frame, set to the appropriate target position (the anterior SVZ in adult is at antero-posterior,

AP, +0.5 mm and medio-lateral, ML, ±1.0 mm from Bregma) and drill a hole through the skull at a rate of approximately 25–50 µm/s.

7. Place the micropipette loaded with dye into a holder fitted to the stereotaxic frame (e.g., World Precision Instruments micropipette holder) and connect to a microinjector or Picospritzer. Lower the pipette through the skull opening at the correct AP/ML position at a rate of approximately 25 µm/s (higher than this rate can be used distal to the target site slowing as the target structure is neared) down to the correct dorsoventral (DV) co-ordinate (e.g., 2 mm from brain surface for the anterior SVZ). Once the correct position is reached wait ~20 s for the brain to stabilize before beginning dye injection. Injection into a solid brain tissue should not exceed 100–200 nl/min up to a maximum volume of ~250 nl (high volumes and/or rapid delivery can result in pressure lesions of the brain tissue, if high volumes are used substantially longer infusion times should be used). Following injection leave the pipette in place for at least 30 s so the brain can stabilize before beginning removal of the pipette. Similar to insertion, the initial pipette removal rate should be slow (e.g., 25 µm/s) with a higher rate possible once distant to the tarter structure. The number of adult stem cells labeled will depend on the dye, concentration, and injection volume (100–200 nl of 10 mM Cell Tracker Green CMFDA labels a suitable number of migrating SVZ cells for analysis).
8. Suture or staple the skin closing incision. Remove the mouse from the stereotaxic frame and place into a thermo-regulated recovery area (~30°C, thermal depression is common in anesthetized mouse and a warm environment should be maintained until recovery). Preemptive analgesic treatment following any craniotomy is recommended (e.g., 2.5 mg/kg body weight subcutaneous injection of flunixin meglumine). Consult guidelines for surgery and local IACUC regulations regarding postoperative care (see Note 1).

3.2. Blood Vessels Labeling

Identification of blood vessels can be performed by immunohistochemistry for endothelial cell antigens; however, we found immunohistochemical approaches to be unreliable due to heterogeneity in expression of many of the markers in smaller vessels in brain. Therefore, we updated the concepts from an old technique of perfusion with Indian ink for use with fluorescent confocal microscopy. This approach involves perfusion of a fluorescent molecule with a stabilizing matrix. The type of stabilizing matrix depends on whether the tissue will be used for immunohistochemistry (protocol A below) or time-lapse microscopy (protocol B below). Although the technique is similar, the differences between the two protocols reside in solutions and concentrations and have been optimized for the requirements of the two different tasks (fixation stability for immunohistochemistry vs. temperature stability for time-lapse imaging).

3.2.1. Blood Vessels Labeling for Immunohistochemistry

1. Prepare and filter 0.9% NaCl solution through a Q8 Whatman filter to remove any particulate and warm to 40°C for use.
2. Prepare two solutions, one containing 1% porcine type-A gelatin and the other 5% gelatin. To dissolve gelatin in pH 7.4 phosphate-buffered saline (PBS) heat to 42–45°C while stirring until the gelatin is completely dissolved. Cool the gelatin solution and maintain at 40 ± 2°C under continuous slow stirring (do not exceed 50°C at any point during preparation). Add 1 mg of BSA-rhodamine or BSA-fluorescein per 5 ml of gelatin solution; the choice of fluorophore, rhodamine, or fluorescein, should be complementary to the fluorescent dye used to label the cells (e.g., Cell Tracker Green CMFDA with BSA-rhodamine). Load a 5 ml syringe with each solution and mount a 24-gauge needle (a typical adult mouse requires ~5 ml of each solution; however, higher volumes, 10 ml, can provide more complete vessel filling should a suboptimal perfusion technique occur). Minimize bubbles in the solution during loading of the syringes as bubbles introduced during perfusion will block blood vessels leading to poor vascular filling. To prevent the gelatin setting inside the syringes they need to be maintained at 40°C (e.g., place the loaded syringes in an oven, on a heating plate, or in a container of saline solution held at 40°C).
3. Anesthetize the animal with an intraperitoneal injection of an appropriate dose of anesthetic (e.g., sodium pentobarbital, 50–100 mg/kg body weight in 0.9% saline). The animal should be secured on their back with limbs spread against a warming platform to maintain body temperature at 37°C during the procedure (a chemical heating pad works well).
4. To perfuse adult animals, open the chest cavity by a lateral incision through the abdominal wall below the rib cage followed by longitudinal incisions through the rib cage up to each armpit on the lateral aspects of the chest thus allowing the breastplate and most of the ribcage to be hinged upward exposing the heart. Use a syringe or tubing connected to a pump loaded with 0.9% saline at 37–40°C attached to a 24-gauge needle (either sharp or blunt points work). Insert the needle into the left ventricle and clamp the heart across the needle insertion point with a pair of haemostatic forceps to prevent any fluid leakage out the insertion point. Then quickly cut the right atrium, and start infusion. Perfuse the animal with 0.9% saline at a rate approximately 2–3 ml/min for 1 or 2 min (or until the liver becomes pale and blood stops flowing from the atrium opening). Remove the saline needle and insert (into the same opening) the needle/syringe

containing 1% gelatin mix. Clamp the heart across the needle insertion point with a pair of hemostatic forceps as above and slowly infuse the gelatin mix through the vasculature (~2.5 ml/min). Quickly remove the expended 1% syringe and insert the 5% syringe. Clamp and infuse the 5% solution as above. Toward the end of the 5% infusion the resistance to injection will increase if the temperature drops under 37°C due to rising viscosity as the gelatin sets, if this occurs, it is better to stop prior to all of the 5% solution being used as excessive pressure can damage/swell vessels.

5. After infusion of the 5% solution immerse the animal in ice cold water for 3–5 min. This temperature drop causes the gelatin within the vessel lumen to solidify.
6. Remove the brain and immersion fix in 4% paraformaldehyde in 0.1 M phosphate buffer for 16–24 h at 4°C (see Note 2).

3.2.2. Blood Vessels Labeling for Time-Lapse Confocal Imaging

1. Prepare 1.5% low-gelling temperature (LGT) tissue culture-tested agarose in Leibovitz's L15 culture media (minus phenol red). Regular electrophoresis-grade LGT agarose contains toxic contaminants that dramatically reduce tissue health and should not be substituted. To dissolve LGT agarose the solution must be heated to >65°C (a microwave with intermittent agitation is suitable). Once completely dissolved, the LGT agarose must be cooled back to 40°C. Upon reaching 40°C add 1 mg of BSA-rhodamine or BSA-fluorescein per 5 ml of LGT agarose solution and dissolve. Load a 10 ml syringe with the LGT mix and mount a 24-gauge needle (a typical adult mouse requires ~10 ml for this protocol). To prevent the LGT agarose setting inside the syringe it should be maintained at 40°C (e.g., place the loaded syringes in an oven, on a heating plate, or in a container of saline solution held at 40°C).
2. Anesthetize the animal with an intraperitoneal injection of an appropriate dose of anesthetic (e.g., sodium pentobarbital, 50–100 mg/kg body weight in 0.9% saline). The animal should be secured on their back with limbs spread against a warming platform to maintain body temperature at 37°C during the procedure (a chemical heating pad works well).
3. To perfuse adult animals, open the chest cavity by a lateral incision through the abdominal wall below the rib cage followed by longitudinal incisions through the rib cage up to each armpit on the lateral aspects of the chest thus allowing the breastplate and most of the ribcage to be hinged upward exposing the heart. Use a syringe or tubing connected to a pump loaded with Leibovitz's L15 minus phenol red that has been preheated to 37–40°C attached to a 24-gauge needle (sharp or blunt point). Insert the needle into the left ventricle and clamp the heart across the needle insertion point with a pair of hemostatic forceps to prevent any fluid leakage out the

insertion point. Then quickly cut the right atrium and start infusion. Perfuse the animal with Leibovitz's L15 at a rate approximately 2–3 ml/min for 1 min (or until the liver becomes pale and blood stops flowing from the atrium opening). Remove the saline needle and insert (into the same opening) the needle/syringes containing LGT agarose mix. Clamp the heart across the needle insertion point with a pair of hemostatic forceps as above and slowly infuse the LGT agarose mix through the vasculature (~2–3 ml/min).

4. After infusion of the LGT solution immerse the animal ice cold water for 2 min. This temperature drop causes the agarose within the vessels to solidify in the luminal spaces of all the blood vessels.
5. Remove the brain following the protocol for time-lapse slices preparation below (see Note 3).

3.3. Acute Brain Slices for Time-Lapse Imaging

The preparation of acute brain slices for time-lapse imaging follows the same procedures as preparation of slices for electrophysiology. Preventing mechanical damage to the tissue during dissection/slicing (compression or torsional stresses) and rapid processing to minimize hypoxic tissue degeneration are key factors. This slice preparation protocol should begin immediately after the end of the LGT agarose mix perfusion and all materials and solutions for brain slicing need to be prepared in advance.

1. Prepare oxygenated ACSF solution at 4°C. The ACSF should be saturated with 95% O_2/5% CO_2 by means of bubbling the gas through a diffuser (e.g., 10 µm pore HPLC gas diffuser) in the solution for at least 30 min prior to starting slicing.
2. Rapidly remove the brain from the skull taking care to avoid compression or other mechanical damage to the brain. Trim with a scalpel/razor blade to approximate the region of interest (e.g., for olfactory bulbs cut coronal at the level of the prefrontal and piriform cortices). Reducing the size of the block being used for slicing improves oxygenation of the tissue and reduces cutting time.
3. Securing the tissue to the vibratome chuck can take place in one of two methods depending on the region of interest. For cutting in a conventional plane (coronal, horizontal, saggital) with a small tissue slice (approximately 1 cm cube) the tissue can be secured directly with glue (see following, step 4A). If the tissue needs to be positioned at a specific plane or the tissue block is very large then a support matrix needs to be used during cutting (see below, step 4B).

4A. Place a small square of Q8 Whatman filter paper onto the vibrating microtome chuck. Apply a small drop of superglue Loctite 401 adhesive onto the filter paper, this glue will spread through the paper providing good contact with the chuck.

Pick the brain piece up on the end of a spatula, wick away excess liquid with a small square of fresh filter paper (careful not to touch the brain directly with the filter paper), and use forceps to gently slide the brain piece in the correct orientation off the spatula onto the glue spot in the filter paper on the vibrating microtome chuck. There should be less than 30 s between placement of the glue droplet and placement of the brain into the glue. The Loctite 401 adhesive will solidify almost immediately upon contact with an aqueous object, thus, the moment the brain touches the glue it will adhere, thus, orientation needs to be accurate since reorienting/repositioning is not possible.

4B. For precision brain orientation prepare a 3% agar solution in Leibovitz's L15 without phenol red. To dissolve agar the solution must be heated to above 65°C (a microwave with intermittent agitation is sufficient) and then maintained at 40°C until needed. Pour the liquid agar into a 35 mm diameter Petri dish and quickly place the brain piece into the agar. Invert the brain several times to allow complete washing of the brain crevasses with the liquid agar solution. Position the brain according to your cutting slice orientation (the cutting plane will be parallel to the Petri dish), place the Petri dish onto a 4°C surface, and adjust brain orientation at need while the agar solidities (approximately 90 s). This allows the brain to be positioned at any orientation. Once set, cut the agar/brain into a pyramidal-shaped block (such that the base is wider than the top to improve stability during cutting) and glue to the vibratome as in step 4A.

5. Submerge the brain in the vibrating microtome chamber with oxygenated 4°C ACSF and place a small gas diffuser bubbling 95% O_2/5% CO_2 in the corner of the chamber to continue oxygenation of the solution during cutting. If available, the temperature regulation controls of the vibrating microtome chamber should be set at 4°C, alternatively pack around the cutting chamber or place a small freezer bag of ice in a corner of the chamber if there is room. Cut brain slices at 300–400 μm thickness and collect into a holding chamber containing oxygenated ACSF (again, include a diffuser line into the solution bubbling 95% O_2/5% CO_2, but avoid direct bubble contact with the slices). For living brain slices the vibrating microtome should be set to a slow advance rate (~0.2–0.3 mm/s resulting in approximately 20 s to cut each slice) with moderate oscillation frequency (if adjustable, 50–70 Hz) or moderate amplitude (if adjustable, 0.8–1.2 mm). The exact parameters will vary greatly with the brain region, animal age, model of instrument (some models are of fixed frequency variable amplitude, other models are of

variable frequency fixed amplitude), and type of blade (we recommend carbide steel "feather" blades from Ted Pella). The general principle is to set the speed to oscillation ratio such as to minimize cutting time while avoiding compression of the brain in front of the blade (indicating too rapid an advance) or tearing of the slice as it cuts (indicating too high oscillation) (see Note 4).

3.4. Time-Lapse Confocal Imaging and Analysis

Acute brain slices have the advantage of preserving the local cellular environment through which a cell migrates, but that is offset of the difficulties in resolving fluorescent cells in thick (300–400 μm) slices. However, by using a confocal microscope imaging depths up to 100 μm into the slice are possible (up to 250 μm depths with a multiphoton instrument). The operational principles and characteristics of confocal microscopes are well established and covered in detail by other works thus we focus upon the conditions and details specific to imaging stem cell migration in brain slices.

Time-lapse imaging is performed on a confocal microscope fitted with water immersion objectives with the brain slice maintained in a thermo-controlled flow chamber. These chambers are commercially available from several vendors (e.g., Warner Instruments), but all of the designs feature an imaging area which holds the brain slice, inlet/outlet ports to allow constant fluid flow across the slice, and heater elements to maintain a set temperature.

1. Prepare oxygenated ACSF solution at room temperature. The ACSF should be saturated with 95% O_2/5% CO_2 by means of bubbling gas through a diffuser (e.g., 10 μm pore HPLC gas diffuser) in the solution for at least 30 min prior to use.
2. Place the slice in the thermo-controlled chamber and secure the tissue to prevent movement. We recommend a slice holding system consisting of interlocking plastic rings containing a nylon monofilament mesh glued to one ring (nontoxic adhesive such as plexiglass dissolved in chloroform), upon which the slice is placed, and the second ring fitted with parallel monofilament threads (0.5–1.0 mm spacing allows imaging while holding the tissue securely). This allows fluid flow and gas exchange on both the upper and lower surfaces of the slice while still preventing slice movement. Care must be taken not to apply excessive pressure with the threads holding the slice. Excessive pressure will result in the threads squeezing/cutting through the tissue damaging and deforming the slice.
3. Run ACSF at a linear flow of 2.5 ml/min from a reservoir with the chamber temperature set to 34°C (typically 5–10 min is required for a cold chamber to reach temperature depending on the heater elements in the chamber). The ACSF reservoir

should be maintained at room temperature with a gas diffuser bubbling 95% O_2/5% CO_2 to maintain oxygenation during use of the solution).

4. Find the cell of interest and begin confocal imaging of the different wavelength emission molecules (a 40× water immersion objective is a good compromise between image area and magnification/detail). It is important to minimize the light exposure to minimize phototoxicity. The illumination light intensity (epifluorescent light and/or laser) should be reduced to the minimum necessary to detect the cell with the instrument detectors set to high gain. The electronic noise present in high gain detectors results in a slightly "grainy" image but as long as the level of this noise does not obscure details a "prefect" image is not necessary for analysis purposes. The frame rate for imaging needs to be carefully selected for the experimental goal; high frame rates provide good temporal resolution of the movement but expose the cell to high amounts of light and more rapid phototoxic injury, whereas slow frame rates permit long imaging sessions but reduce temporal resolution on migratory dynamics. For long range migration images every 5–15 min allows cells to be followed for many hours. If using the Cell Tracker dyes, these are lost by any cell with compromised membrane integrity, thus, if the imaged cell undergoes apoptosis the dye is lost and the cell disappears.
5. Measuring the speed of the imaged cell is a simple process of plotting the XY (and Z if measured) coordinates of the centre of the imaged cell and calculating the distance travelled between each frame using 2D or 3D Pythagorean Theorems. The average migratory rate will be the summated distances traveled between each frame divided by the total imaging time. Cells typically migrate in a saltatory manner with periods of high speed followed by periods of quiescence. The max/min speeds, oscillation frequency, etc., can also be derived (see Note 5).

4. Notes

1. The entire surgical/injection protocol takes approximately 30 min per animal to complete, thus an anesthesia that will provide surgical level anesthesia for that time frame is required. Nembutal or ketamine/xylazine is suitable; however, mice of different strains, ages, sex, and body weight/fat content can have different sensitivities to the anesthetic. Young-adult mice are highly sensitive to anesthetic overdose, whereas older

animals with higher body fat content tend to require greater than the "standard" anesthesia dose due to fat partitioning of some anesthetics.

2. If poor/incomplete vessel filling or blood clearance is observed the infusion can be assisted by adding to all solutions of 10 units/ml Heparin sodium salt (Sigma H4784), 2 mg/ml Atropine, and 30 μg/ml ketamine. This will reduce blood clot formation and induce vasodilation assisting clearance of blood and infusion of labeling solution. A volume larger than 5 ml is sometimes needed for regions of poor/slow circulation.
3. Adult neurons are sensitive to hypoxia; the length of time from the initial Leibovitz's L15 infusion to immersion of the removed brain into oxygenated artificial cerebrospinal fluid (ACSF; see below) should be minimized, ideally the process should take ~7 min.
4. The overall time for this procedure (from brain to slices in the oxygenated holding chamber) should not exceed ~20 min per animal. Despite the reduced temperature during the procedure extending the time before hypoxic damage occurs, total durations significantly greater than 20–30 min can cause degeneration.
5. Slices in oxygenated ACSF can be maintained at room temperature for up to 6 h, whereas slices at 34°C are healthy for approximately 2–4 h. Therefore, slices in holding chambers should be maintained at room temperature to extend their health prior to imaging in the thermo-regulated flow chambers.

As oxygenated ACSF heats up in the imaging chamber, small bubbles can "seed" and slowly form against the threads and slice. Care should be taken to monitor and remove the bubbles if they form near the imaging area as these can cause optical distortion. These can be easily removed without damaging the slice by a pipette evacuating/replacing the chamber liquid.

Time-lapse imaging is often performed across several hours, thus, long-term positional stability of microscope and apparatus are essential. The recording chamber should be attached securely to the microscope stage to prevent any movement of the chamber (likewise, the slice holding system should be secure inside the chamber). All microscopes exhibit drift in their focus mechanisms, this should be measured (and repaired if needed) to ensure it is less than ~1 μm/h. Excessive axial drift will move the focal plane of the microscope resulting in error in *Z*-axis and even lose of the imaged cell. In cases where the migration direction is perfectly parallel to the imaging plane axial drift may not be important. The flow of ACSF through the chamber can introduce vibration into the system if improperly configured. A pump-based flow system, commonly used for electrophysiology, introduces rhythmic fluid pulsations/vibrations that are problematic for imaging and thus these systems should be avoided. The ideal method involves flow-regulated gravity feed from an elevated ACSF reservoir running through the imaging chamber and out into a waste bottle. This type of system is the subject of other articles and the reader is referred to the white paper by Dr. Leinders-Zuffall available at the Warner Instruments website (http://www.warneronline.com/pdf/whitepapers/perfusion_strategies.pdf), this white paper is an extract of a longer chapter titled "Regulating the perfusion speed and height of microscope mounted chambers" published in "Electrophysiological Methods for the Study of the Mammalian Nervous System" (Kocsis JD, ed; Appleton and Lange 2004).

References

1. Gasser, U.E., and Hatten, M.E. (1990) Central nervous system neurons migrate on astroglial fibers from heterotypic brain regions in vitro *Proc Natl Acad Sci USA* **87,** 4543–47.
2. Wichterle, H, Garcia-Verdugo, J.M., and varez-Buylla, A. (1997) Direct evidence for homotypic, glia-independent neuronal migration *Neuron* 1997 **18,** 779–91.
3. Lois, C., Garcia-Verdugo, J.M., and Alvarez-Buylla, A. (1996) Chain migration of neuronal precursors *Science* 1996 **271,** 978–81.
4. Bovetti, S., Hsieh, Y.C., Bovolin, P., Perroteau, I., Kazunori, T., and Puche, A.C. (2007) Blood vessels form a scaffold for neuroblast migration in the adult olfactory bulb *J Neurosci* **27,** 5976–80.
5. Snapyan, M., Lemasson, M., Brill, M.S., Blais, M., Massouh, M., Ninkovic, J., Gravel, C., Berthod, F., Gotz, M., Barker, P.A., Parent, A., and Saghatelyan, A. (2009)Vasculature guides migrating neuronal precursors in the adult mammalian forebrain via brain-derived neurotrophic factor signaling *J Neurosci* **29,** 4172–88.
6. Honda, S., Toda, K., Tozuka, Y., Yasuzawa, S., Iwabuchi, K., and Tomooka, Y. (2007) Migration and differentiation of neural cell lines transplanted into mouse brains *Neurosci Res.* **59,** 124–35.
7. Ohab, J.J., and Carmichael, S.T. (2008) Poststroke neurogenesis: emerging principles of migration and localization of immature neurons *Neuroscientist* **14,** 369–380.
8. Atho,s J., and Storm, D.R. (2001) High precision stereotaxic surgery in mice *Curr Protoc Neurosci* Appendix 4: Appendix.
9. De Marchis, S., Fasolo, A., Shipley, M., and Puche, A. (2001) Unique neuronal tracers show migration and differentiation of SVZ progenitors in organotypic slices *J Neurobiol* **49,** 326–38.
10. Petreanu, L., and Alvarez-Buylla, A. (2002) Maturation and death of adult-born olfactory bulb granule neurons: role of olfaction *J Neurosci* **22,** 6106–13.

Chapter 17

Dissecting Mesenchymal Stem Cell Movement: Migration Assays for Tracing and Deducing Cell Migration

Erika L. Spaeth and Frank C. Marini

Abstract

Targeted migration is a necessary attribute for any gene delivery vehicle. Mesenchymal stem cells (MSC) have been used as effective delivery vehicles for treatments against cancer, graft versus host disease, arthritis, multiple sclerosis, and many other diseases. MSC migrate toward sites of inflammation, however, the true migratory mechanism has yet to be elucidated. There are several receptors and respective chemokines known to be involved in the migration of the MSC. Further insight to MSC migration will be revealed both in vivo and in vitro through the application of migration assays from the most simple, to the more technologically demanding.

Key words: MSC, Tumor microenvironment, Inflammation, Migration, Transwell assay, Scratch assay, Bioluminescence imaging

1. Introduction

The mesenchymal stem/stromal cell (MSC) is a unique population of multipotent cells defined by plastic adherence, differentiation potential, and cell surface marker expression (1). MSC, or MSC-like cells, have been isolated from nearly every organ or tissue in the body making it challenging to characterize the MSC as a completely homogenous population. MSC contribute to the maintenance and regeneration of connective tissues and have the capacity to differentiate into osteoblasts, adipocytes, chondrocytes, myocytes, and cardiomyocytes. MSC are negative for the hematopoietic markers CD45 and CD34, the endothelial marker CD31, and the monocyte marker CD11b. MSC are positive for the expression markers, including CD29, CD44, CD90, CD140b, CD73 (SH3/4), CD105 (SH2), CD166 (ALCAM), and Stro-1, but the expression of specific combinations of markers appear

Marie-Dominique Filippi and Hartmut Geiger (eds.), *Stem Cell Migration: Methods and Protocols*,
Methods in Molecular Biology, vol. 750, DOI 10.1007/978-1-61779-145-1_17, © Springer Science+Business Media, LLC 2011

microenvironment-dependent, suggesting a strong influence of tissue context on MSC phenotypes. In general, MSC appear to be a nonimmunogenic population of cells; however, a few studies demonstrate immune-suppressive functions of MSC through the induction of peripheral tolerance evident in autoimmune disorders such as multiple sclerosis (2).

The use of bone marrow-derived MSC have been employed in support and engraftment of the transplantation of hematopoietic stem cells following high-dose chemotherapy in an effort to replenish the destroyed bone marrow cell population (3). Additionally, preclinical studies have explored the use of MSC in the reduction of GVHD (graft vs. host disease), for tissue repair; including cerebral injury (4) bone fracture (5), myocardial ischemia/infarction (6), muscular dystrophy (7), and tumor homing (8).

The mechanism and factors responsible for the targeted tropism of MSC to these wounded microenvironments remain to be fully elucidated. MSC are likely to have chemotactic properties similar to other immune cells that respond to injury and sites of inflammation. Thus, the well-described migration models of leukocytes and their progenitor cell line, the hematopoietic stem cell can serve as a reasonable example to facilitate the identification of factors involved in MSC migration.

Inflamed microenvironments establish the migratory destination of MSC. Inflammation is a cellular response that takes place under conditions of cellular injury and in sites of tissue wounding. There are many instigators of "inflammation" and subsequently MSC migration. Degenerate tissues involve states of injury that exhibit an array of inflammatory chemotactic molecules that MSC respond to. Such states include (a) hypoxia or ischemia (9–12), a state of reduced oxygen that often parallels and perpetuates inflammation, (b) radiation, used as therapy or as weapon (13–15), (c) tumor cells, which mimic the phenotype of an "unhealed wound" (16), and (d) cutaneous cuts or punctures that define the layman's definition of inflammation (17).

MSC migration can be measured and quantified both in vitro and in vivo. We review migration methods ranging from the basic standards to the technologically advanced techniques used today including competitive migration (modified Ouchterlony Assay) and migratory acceleration (Scratch and Transwell Assays) in vitro under which several conditions can be manipulated both externally in the medium and internally within the cell itself. Furthermore, the analysis of migration can be pursued in an in vivo model, under which noninvasive techniques have allowed long-term analysis of physiologically relevant cell migration to be possible (bioluminescence, fluorescence, positron emission tomography (PET)/CT, and magnetic resonance imaging (MRI)). Finally, conventional methods such as IHC on tissue sections can be used to validate and corroborated in vivo imaging data that will eventually promote the use of the noninvasive techniques in clinical settings.

2. Materials

2.1. MSC Isolation

1. Alpha-modified Eagle's medium (α-MEM) with supplements (L-glutamine and penicillin–streptomycin mixture and 20% fetal bovine serum (FBS)).
2. Trypsin–EDTA (0.25%).
3. Phosphate-buffered saline (PBS).
4. Ficoll-Hypaque gradient (for human MSC isolation from harvested bone marrow).
5. Sterile mortar and pestle (for murine MSC isolation).
6. 180 cm^2 tissue culture dish.
7. Following initial isolation, size of tissue culture plate/flask can vary by experimental conditions – make sure to keep at a cell density of 2,000–6,000 cell/cm^2.
8. Conjugated CD11b antibody (for murine MSC isolation – during immunodepletion step).
9. See Fig. 1 for list of additional antibodies used for human and/or murine MSC characterization.

2.2. MSC Adipocyte Differentiation

1. Culture media: Prepare DMEM-HG (40%; BioWittaker) and HamF12 (50%; BioWittaker) containing 10% FBS and supplemented with 10^{-8} M dexamethasone, 0.5 Ag/ml insulin, and 2 mM glutamine.
2. Oil red O (Sigma).

2.3. MSC Osteoblast Differentiation

1. Dexamethasone (10^{-7} M).
2. Glycerophosphate (2 mM).
3. Ascorbic acid 2-phosphate (0.05 mM; Wako).
4. HCl (0.5N).
5. Calcium quantification kit (Bi-Tron).

2.4. Modified Ouchterlony-Type Assay

1. Matrigel (normal or reduced growth factor; BD Biosciences).
2. Tissue culture dish or plate (depending on assay size).
3. Chemotactic agent (e.g., antibody, growth factor, serum, peptide).
4. LipofectAmine and PLUS transfection reagents (Life Technologies) (optional).
5. Plasmid-encoding GFP or other markers (optional).
6. CO_2-independent medium (optional).

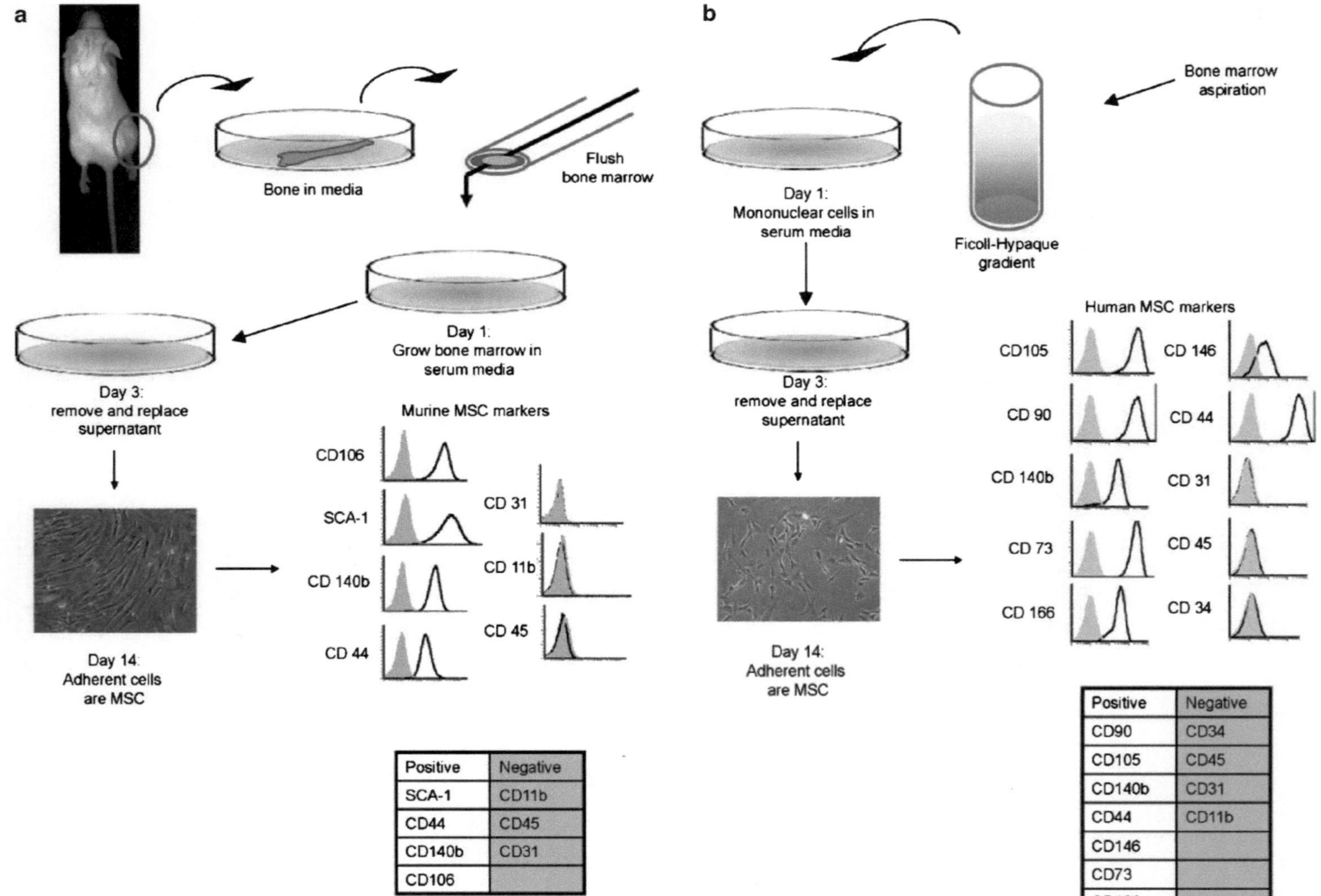

Positive	Negative
SCA-1	CD11b
CD44	CD45
CD140b	CD31
CD106	

Positive	Negative
CD90	CD34
CD105	CD45
CD140b	CD31
CD44	CD11b
CD146	
CD73	
CD166	

Fig. 1. Flow chart describing MSC isolation.

2.5. Scratch Assay

1. α-MEM with supplements (L-glutamine and penicillin–streptomycin mixture and 20% FBS).
2. Trypsin–EDTA (0.25%).
3. PBS.
4. Cell line (MSC, tumor, endothelial, epithelial).
5. LipofectAmine and PLUS transfection reagents (Life Technologies) (optional).
6. Plasmid-encoding GFP or other markers (optional).
7. CO_2-independent medium (optional).
8. Razor.
9. Phase-contrast microscope and fluorescence microscope equipped with stage incubator (hypoxia or normoxia chamber), CO_2 supply, video camera, and charge coupled device camera (CCD camera).
10. Image analysis software.

2.6. Transwell Migration

1. Transwell plate inserts – and transwell plates, number of cells and amount of media are dependent on the size of the well (6, 12, 25).
2. α-MEM with supplements (L-glutamine and penicillin–streptomycin mixture containing 0–2% FBS (for starvation) or 20% FBS (for normal MSC).
3. Trypsin–EDTA (0.25%).
4. PBS.
5. Attractant cells (e.g., tumor cells, "injured" cells can be cultured and the resulting conditioned media can be used as a potential "attractant").
6. Chemotactic agent (attractant or inhibitor – recombinant protein, antibody, serum).
7. LipofectAmine and PLUS transfection reagents (Life Technologies) (optional).
8. Plasmid-encoding GFP or other markers (optional).
9. CO_2-independent medium (optional).
10. Wright–Giemsa staining reagents (Protocol Hema-3, Fisher Scientific).
11. 2% deoxycholic acid (store at RT).

2.7. Bioluminescent Imaging

1. α-MEM with supplements (L-glutamine and penicillin–streptomycin mixture) containing 0–2% FBS (for starvation) or 20% FBS (for normal MSC).
2. Trypsin–EDTA (0.25%).
3. PBS.
4. LipofectAmine and PLUS transfection reagents (optional).

5. Plasmid-encoding GFP or other markers (optional).
6. CO_2-independent medium (optional).
7. Xenogen IVIS bioluminescence/fluorescence optical imaging system (Caliper Life Sciences).
8. IVIS Living Image software (Caliper Life Sciences).
9. Isoflurane (IsoSol.).
10. Insulin syringe (1 cm^3; 28G 1/2 in.).
11. Chemoluminescent substrate reagent depends on the bioluminescent expression system in your cells. Luciferase renilla requires colenterazine (40 mg/ml) (Biotium, Inc.) resuspended in methanol while firefly requires D-luciferin firefly-potassium salt (125 mg/kg) (Biosynth).

2.8. IHC

1. Tissue sections stored in formaldehyde or can be snap-frozen in liquid nitrogen and stored at –80°C.
2. OTC compound (Miles, Inc.) or formaldehyde.
3. Hematoxylin–eosin and immunohistochemical staining kits.
4. Research optical microscope with reflected and transmitted light sources.
5. A high-resolution digital camera.
6. Adobe Photoshop software (Adobe Systems Inc.).
7. Chemoluminescent substrate reagent depends on the bioluminescent expression system in your cells.

3. Methods

While validated MSC cell lines can be purchased from the ATCC, they are also easily isolated from human or mouse (as explained below). Simply, the population of adherent cells that grows out from harvested bone marrow is enriched in the cell population we define as MSC. Phenotypic characterization of MSC can be carried out by flow cytometry (see Fig. 1 for marker reference). Differentiation potential assays to adipocytes and osteoblasts will be briefly discussed although MSC have the potential to differentiate into chondrocytes and myocyte – which will not be discussed in this chapter (18). It is important to note that for all of the ensuing assays, that the MSC should be serum starved 24 h prior to migration.

3.1. MSC Isolation

3.1.1. Human MSC Isolation

1. Collect clinical bone marrow sample (according to institutional protocol).
2. Separate the mononuclear cells by centrifugation over Ficoll-Hypaque gradient.

3. Plate at initial seeding density of 1×10^6 cells/cm^2 in α-MEM with supplements and 20% FBS.
4. After 3 days, remove the nonadherent cells by washing with PBS.
5. Culture adherent cell monolayer until confluency.
6. Trypsinize (0.25% trypsin with 0.1% EDTA) cells and subculture at densities of 5,000–6,000 cells/cm^2.
7. Use cell passages 3–4 for the experiments (see Note 1).

3.1.2. MSC Isolation (Murine)

1. Anesthetize and sacrifice mouse according to institution approved protocol.
2. Remove the two hind limbs (femur and tibia) and the coxae (hip) (see Notes 2 and 3).
3. Place clean bones in warmed αMEM (no serum needed – in a petri dish or conical tube).
4. When all bones are removed, in sterile environment, fill an 18-gauge needle with serum media, cut the ends (both proximal and distal) of the bone, and insert the needle to flush marrow out into a new, sterile tissue culture dish with warmed, serum (20%) αMEM medium.
5. Incubate at 37°C for about 3–5 days.
6. Discard the supernatant and floating cells, culture the adherent cells, which will be the MSC population, until confluent.
7. Trypsinize (0.25% trypsin with 0.1% EDTA) cells and subculture at densities of 5,000–6,000 cells/cm^2.
8. Use cell passages 3–4 for the experiments (see Note 4).

3.2. Adipocyte Differentiation

1. Culture mMSC for 3 weeks in 40% DMEM-HG, 50% HamF12 (both from BioWittaker), 10% FBS (Sigma), 10^{-8} M dexamethasone, 0.5 Ag/ml insulin (Sigma), 2 mM glutamine, and antibiotics.
2. Change the medium twice a week with a volume of 2 ml per well.
3. Assay cells via staining using oil red O (Sigma) (see Note 5).

3.3. Osteogenic Differentiation

1. Grow cells in medium containing 10^{-7} M dexamethasone, 2 mM glycerophosphate, and 0.05 mM ascorbic acid 2-phosphate
2. Change the medium twice a week with a volume of 2 ml of medium per well.
3. Harvest the cell culture between days 21 and 30.
4. Extract the deposited calcium in the cell layer with 0.5N HCl.
5. Quantified with a commercial kit (Bi-Tron) according to the manufacturer's instructions (see Note 5).

3.4. Modified Ouchterlony-Type Assay

The Swedish immunologist Örjan Ouchterlony developed a method of analyzing the antibodies within the patient's serum in an agar gel plate (19, 20). Within the gel are a series of holes – one in the middle and a number of holes surrounding the middle (Fig. 2). If one, or a group of antibodies is placed in the middle well, and antigens place in the outer wells – the plate is allowed to develop for 24–48 h. If the plate contains complementary antibody–antigens, they will form a complex and precipitate out of the gel forming a visible line at the border between the two wells from which the pair originated. The basic structure of this technique can be modified to assess MSC migration. By placing MSC in the middle and potential chemoattractant mediators 10 mm away from center, one can detect and quantitate the migrated MSC that move toward the attractant gradient. This technique works best if the MSC are genetically tagged with a fluorescent protein. We prefer to use a gridded culture plate that facilitates counting cells/grid square.

1. Place MSC in the middle well of a matrigel-filled plate (see Note 6).

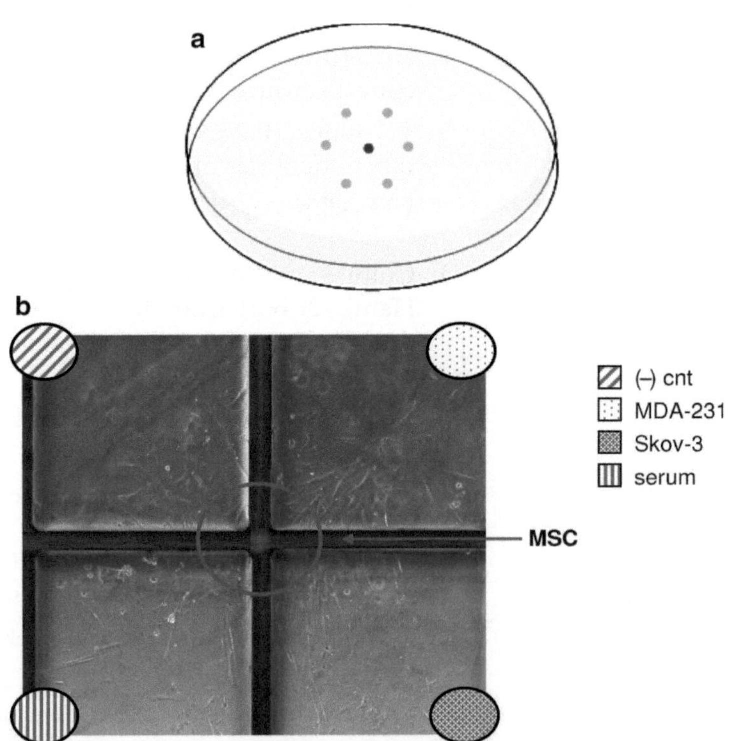

Fig. 2. Diagram of matrigel plated well. (**a**) MSC can be placed in the center puncture with cells, recombinant proteins, or antibodies surrounding the middle puncture-well. (**b**) MSC placed in the middle of a chamber slide. Four opposing corners have a chemoattractant: serum media, Skov-3 or MDA-231 tumor cells, or a negative serum-free control.

2. In the surrounding wells (8–10 mm away from center) place potential chemoattractant mediators (e.g., inflammatory antigens, chemokines, cytokines, tumor conditioned media, infracted cell media).
3. A top layer of matrigel can be added to the well to cover the cells.
4. Add enough medium to lightly dampen the matrigel (see Note 7).
5. Incubate for 24–96 h depending on experimental conditions (see Note 8).
6. Visualize migratory direction by microscope if cells are fluorescently tagged, else a normal light microscope is sufficient for visualizing and counting (see Note 9).

3.5. Scratch Assay

The scratch assay is an easy method to observe migration of cells in an injury-induced environment. The migration will be compared to a standard control cell migration. Various properties of MSC migration can be measured including paracrine stimulation that can be altered by changing the chemotactic media conditions or the molecular genes of the cell that can be manipulated by knock-down, over-expression, or mutant phenotypes. Comparing the gap in the cells left from the scratch mark will give a crude time point of the migration capacity of the cells under the set conditions (Fig. 3).

1. Cells are grown to a confluent monolayer. Using a p200 pipette tip make a “scratch” along the bottom of the dish. It is important to create the same size scratch for every dish. Make sure to reference the location of your scratch on the dish – but outside of the field of view (see Note 10).
2. Gently rinse the cells with serum medium to clean the edges and discard of debris. Add enough reduced serum medium to cover the cells (see Note 11).
3. Incubate the cells at 37°C for at least 24 ± 12 h, as determined by the cell line and the conditions (see Note 12).
4. Place the dish under a fluorescence or phase contrast microscope and acquire images at select time points over the duration of the incubation – every 30–60 min (see Notes 13 and 14).
5. Using image analysis software, the distances between the scratch and the migrating cells can be quantitatively compared. Image Pro-Plus software (Media Cybernetics) or a freeware (http://rsb.info.nih.gov/ij/) (see Note 15).

Variations of the scratch assay can be conducted for native or transfected cells, in normoxic or hypoxic conditions, in the presence of growth serum, conditioned media, etc.

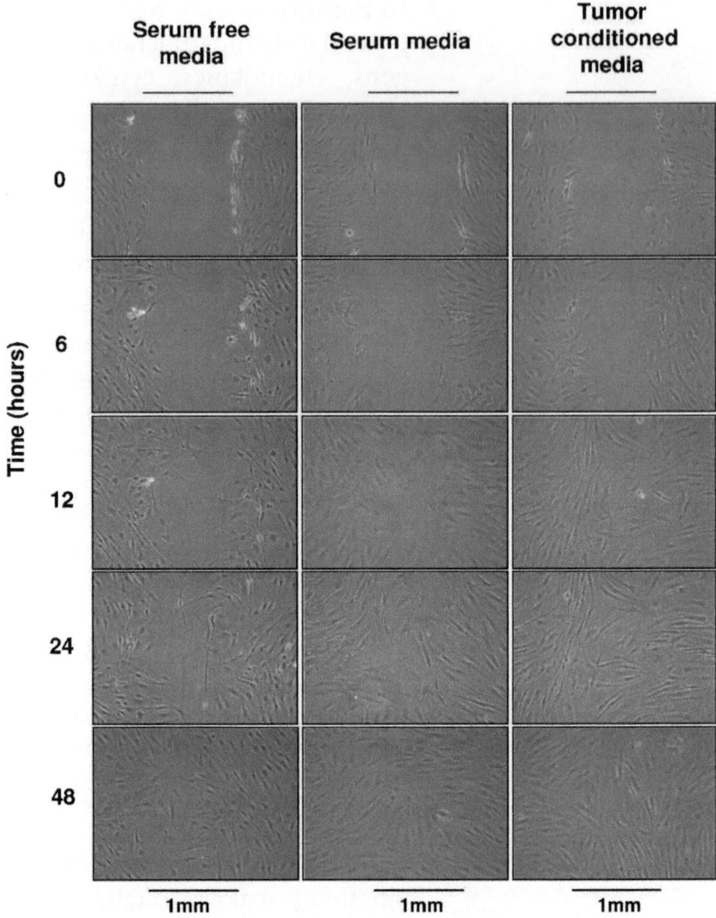

Fig. 3. MSC migration analyzed by scratch assay. The MSC migration images correspond to 0, 6, 12, 24, and 48 h time points. The MSC conditioned with the tumor medium infiltrates the 1-mm gap about 2× faster than the serum media conditioned, and about 4× faster than the serum-free conditioned.

3.6. Transwell Assay

The transwell migration is a modern interpretation on the Boyden chamber assay. Although several variations on this method have been applied to the study of invasion – with the addition of coating factors, such as collagen, fibronectin, or matrigel – we will focus on migration and not invasion, but note that similar methods apply to an invasion assay. The transwell migration assay is performed using a lower chamber, or well, in which the stimulant/attractant is applied. Within this lower chamber, one places a smaller insert containing a porous membrane. Pore size is variable and will depend on the cell line in use. The cells of interest are plated on top of the porous membrane and allowed to migrate to the stimulant within the bottom well. Migration can be quantified by electronic or manual cell counts (Fig. 4).

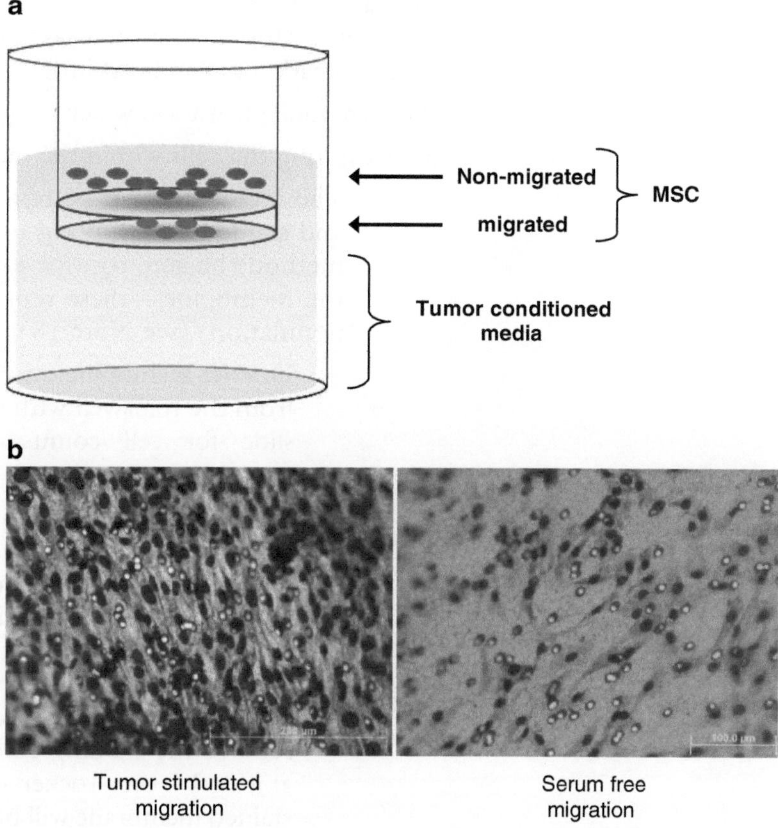

Fig. 4. Transwell migration of MSC toward tumor conditioned media. (**a**) Cartoon depiction of the transwell and the membrane seeded with two layers of MSC, a nonmigratory and a migratory population. (**b**) Transwell membranes are stained for the presence of MSC. Visually, the tumor-stimulated MSC are more abundant in the membrane compared to serum-free stimulated MSC.

1. Cells grown to a subconfluent state of 70–80% are serum starved for 12–24 h prior to migration assay (see Note 16).
2. Serum-starved cells are trypsinized, washed in serum media, and resuspended in small volume of serum-free media. The number of cells per well depends on the cell line in addition to the plate size being used, 6-, 12-, 24- or 96-well plates. Volume will be less than 100 μl per well for 24-well plates and, additional media can be added following the initial plating. 2×10^4 primary MSC per well of a 24-well plate is a reasonable starting off point. And manufacturer's instructions may provide further detail.
3. The bottom well will consist of the stimulant or attractant of interest. Plate bottom well before adding the cells to the transwell membrane (see Note 17).

4. Incubate cells at 37°C for duration of time specified by your conditions. Most migration assays are measured around the 18–24 h mark, but can be as little as 4 h to as great as 96 h.
5. There are multiple ways in which to quantify the migrated cells.
   (a) Counting the cells within the membrane.
      - The membrane can be removed and the cells fixed and stained for counting using the Wright–Giemsa method (be sure to wipe excess cells off the top of the membrane – these represent the nonmigratory population) (see Note 18).
         - Dry the membrane overnight before removing it from the transwell with a razor blade. Mount on slide for cell counts and imaging under a microscope.
         - Alternatively, if further quantification is wanted, place the membrane in between two slides to take images and cell counts by eye or using an interactive cell counting program like Cell Analyst (AssaySoft Inc). Remove the membrane and place it in a 1.5-ml eppendorf tube in 300 μl of deoxycholic acid (or other detergent that will remove the staining color from the cells). After at least 1 h on a rocker at room temperature, the stained membrane will be opaque, and the deoxycholic acid solution can be measured at light wavelength of 595 nm on a spectrophotometer and compared with the cell counts.
   (b) Further analysis of the cell migration can be done by trypsinizing and counting the number of migrated cells in the bottom well compared to the number of cells remaining on top of the transwell membrane (see Note 19).
   (c) The ratio of migrated to nonmigrated cells can be averaged and compared between samples (see Note 20).

Variations on the transwell migration assay include the manipulation of the cell line through stable or transient transfections, neutralizing antibodies and cell conditioning prior to migration, and the manipulation of the migratory attractant, growth factors, serum, cell conditioned media antibodies, and small molecule inhibitors. Additionally, the confluency and passage number of the cell line can also be important factors in the migration of the MSC (21).

3.7. Analysis of In Vivo Migration: Bioluminescent Imaging

Bioluminescent imaging is a noninvasive method to study biological processes such as cell migration in vivo without sacrificing the animal. As MSC are known to home to sites of injury, an in vivo animal model can provide many options to observe injury-induced migration from brain and heart ischemias (11, 22) to bone fractures to puncture wounds or cuts (23) to tumors (8). In this method, we will focus on a solid tumor-induced migration of the MSC to the tumor microenvironment. The application of cell labeled (firefly or renilla) luciferase, an enzyme capable of catalyzing a bioluminescent reaction, enables the visualization of the cells following the SC injection of respective substrate (luciferin or coelenterazine). Light emitted from the enzyme/substrate reaction will be proportional to the amount of enzyme expressed by the cells. For our purposes, we will label our tumor cells with renilla luciferase (rLuc) and MSC with firefly luciferase (ffLuc) but there exist several other commercially available luminescent enzyme and substrate pairs. Labeling the cells with the appropriate enzyme can be done with viral transfection or plasmid transduction and may be transient or stable. For migration of MSC, transient transfection is sufficient because the bulk migration can be visualized within 3 days (Fig. 5a).

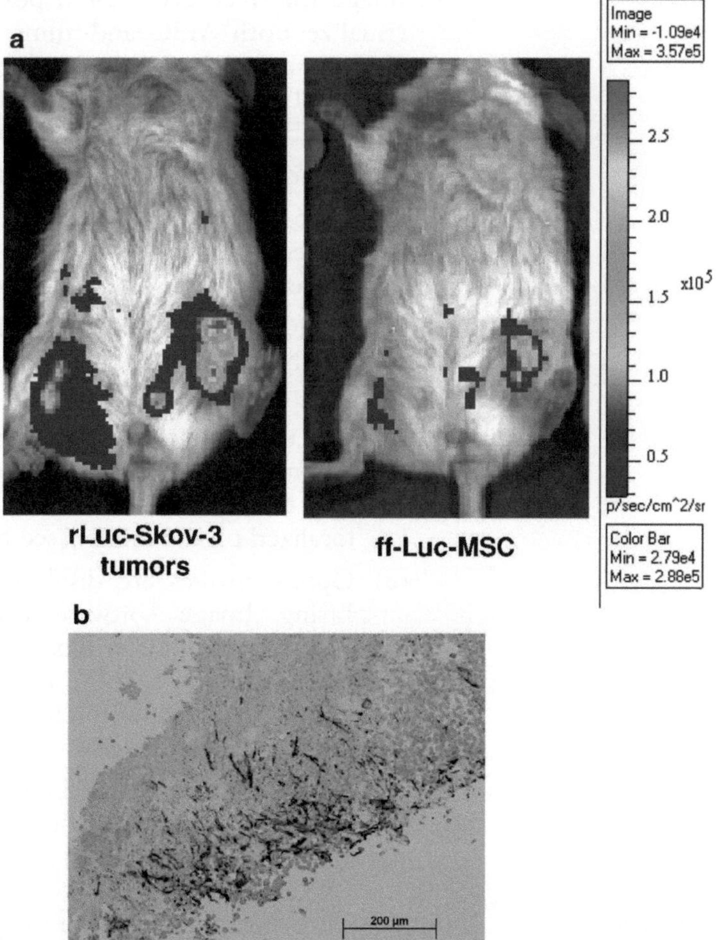

Fig. 5. In vivo imaging of a mouse with bilateral tumors implanted subcutaneously in the hind legs. (**a**) Imaging using coelentrazine to view the rLuc-labeled tumor cells and co-localization of the D-luciferin to view the ffLuc-labeled MSC. (**b**) Immunohistochemical analysis of MSC within a tumor section.

However, if tumor cells are to be labeled, creating a stable line is advisable due to long-term engraftment within the animal. The Xenogen IVIS bioluminescence imaging system is an example of a bioluminescent device capable of measuring both luminescence and fluorescence in sedated, live animals.

1. Tumor injections can be performed subcutaneously (SC), intraperitoneally (IP), or intracranially (IC) according to the type of study being performed. Engraftment can be measured through palpation and/or bioluminescent imaging. In this example, tumor cells (rLuc) injected SC into the limbs of mice are viewed using coelenterazine (see Note 21).
2. For MSC migration, tumor should be palpable and large enough to induce a sufficient migratory signal.
3. MSC (ffLuc) are injected intravenously (*iv*) via tail vein injection (see Note 22).
4. Image the mice every 24 h postinjection. If you want to visualize both MSC and tumor on the same day, MSC (ffLuc) should be imaged prior in the morning and the tumor (rLuc) should be imaged in the evening to ensure the former substrate has been cleared from the animal (see Note 23).
   (a) Induce general anesthesia with 5% isoflurane.
   (b) Place mouse – anterior or posterior – in the light-tight heated chamber with continued anesthesia during the procedure (2% isoflurane introduced via nose cone).
   (c) After acquiring photographic images of each mouse (see Note 24), a luminescent image will be acquired with 1–5-min exposure times (see Note 23).
   (d) The resulting gray scale photographic and pseudocolor luminescent images are automatically superimposed so that the identification of any optical signal is correctly localized on the mouse (see Note 25).
   (e) Optical images are displayed and analyzed with IVIS Living Image software packages. Optical signal is expressed as photon flux, in units of photons/second/centimeter^2 (p/s/cm^2). Direct correlation between cell number and the photon flux can be made if an in vivo cell dilution gradient is made (24).
5. Analysis of the regions of interest (ROI) can be manually drawn around the bodies of the mice – or the specific site of photon flux – to assess signal intensity emitted (see Note 26).

3.8. Immunohistochemistry

Immunohistochemical analysis of the previously described experiment can be done once the experiment is terminated and the mice are sacrificed. The bioluminescent data can be corroborated throughout the individual timepoints if desired; tissue from mice sacrificed at each individual imaging time point can be used to identify cellular localization by IHC. To identify the presence of MSC within a tumor, or any type of tissue, a number of antibodies can be employed. If the MSC are fluorescently labeled, (e.g., GFP) a GFP antibody can be used to detect the cells. If the MSC are unlabeled, commonly used antihuman antibodies may include CD90+ (Thyl), CD166+, CD73+, and CD105+ depending on the resident tissue (Fig. 5b). Antibodies for the detection of murine MSC may include Sca-1+, CD44+, CD140b+, and CD106+.

1. Tissue sections stored in formaldehyde or can be snap-frozen in liquid nitrogen and stored at –80°C until sectioning (6–8 μm in width).
2. Stain – hematoxylin–eosin or immunohistochemical staining (according to manufacturer's protocol).
3. Image with microscope equipped with a CCD camera and Adobe Photoshop software (Adobe Systems Inc., San Jose, CA).

3.9. Additional In Vivo Imaging Techniques

In vivo, noninvasive imaging potential for clinical applications is upon us. There is currently an ongoing preclinical studies in rabbits to image gold-dextran coated particles that are loaded into MSC. As described at the 2008 Radiological Society of North America conference, these MSC can be visualized using X-ray, Raman spectroscopy, computed tomography (CT), or ultrasound (US) modalities. There are numerous possibilities for biocompatible nanoparticle constructs, tracers, or superparamagnetic particles that can be loaded into MSC with properties to enable cell visualization by X-ray, CT, US, PET (25), or MRI (26). The fusion between biology, nanotechnology, and radiology will be important to the future MSC in a clinical setting to ensure the localized delivery to the target tissue.

4. Notes

1. MSC isolation (human): Cells can be sorted by flow cytometry for human MSC markers if a subpopulation is desired (CD44+, CD90+, CD105+, CD73+, CD166+, CD146+, CD140b+ and CD34–, CD45–).
2. Remove the bone immediately after sacrificing the mouse.
3. Remove as much muscle tissue, skin, and fur from the bone using a scapula to scrape the remaining tissue away before placing in warmed α-MEM.
4. Cells can be sorted by flow cytometry for mouse MSC markers if a subpopulation is desired (CD44+, Scal+, NG2+ and CD34–, CD45–).

5. Differentiation assays. Media kits for differentiation (adipocyte, osteoblast) can be purchased. Follow protocol as per manufacturer's instructions.
6. After the matrigel sets to gel form at room temperature (be sure to follow matrigel instructions for plate preparation), you can pierce the wells – best done with a glass pipette tip and a pipette bulb in order to carefully, slowly, and steadily remove the small well of matrigel.
7. Serum content in the media as well as the matrigel serum content can be modified and is left to user discretion. Less serum (2%) will make your migration more specific as opposed to random diffuse migration.
8. Distance apart from the wells is dependent on the duration of the assay. If you want to assay a 24-h timepoint, the wells should be within a few millimeters of each other.
9. This assay is basic and may be less quantifiable than other assays that will be mentioned below if extreme care is not taken to measure both the distance between wells and multiple time points.
10. Cell scratch marks can be marked on the dish with sharpies, or etched with razor blades.
11. The serum media does not need to contain the same amount of serum needed for cell growth, but just enough to maintain cell attachment and avoid apoptosis.
12. If conducting imaging in intervals – it is best to begin imaging in short-time intervals – 3–6 h for the first image – and then successive images should be taken within constant intervals thereafter. Ultimately, the time of incubation should be tailored to the migration time of the fastest cell line. Depending on the conditions and the type of cell, <15 min imaging intervals may be necessary to catch the migratory window.
13. Several modifications to the image-capture process can be made, including the possibility to timelapse video-capture the entire duration of the migration assay. This requires a microscope (confocal or inverted fluorescence with DSU Spinning disk confocal technology and CCD camera attachment), a microincubator (or some type of precision controlled incubator system), and an image capture/analysis software package.
14. If your migration window closes too fast, try a serum-free media to retard migration so that image capture is possible.
15. To achieve statistically significant data, three replicates and a large sample size is necessary. Sample size can be obtained by taking many distance comparisons between the scratch mark and the migrating cell.

16. Cells can be starved in 0–2% serum media depending on the complete duration of the experiment, and your cell viability under stress.
17. For human MSC, both the 8 and 12 μm membranes are acceptable, because immortalized human MSC and murine MSC are smaller than normal human MSC, the 8-μm pore membranes are better.
18. After staining, wash the membrane well with milli-q water and using a Q-tip, wipe off the excess cells on top of the membrane, as their presence will skew your cell counts.
19. Remove the whole transwell and place in a new plate with prewarmed trypsin–EDTA in the bottom wells, after 2–10 min at 37°C, the cells will detach from the membrane and be found in the lower chamber. These cells can then be quantified by hemocytometer, or if fluorescently labeled, they can be quantified with a fluorescence plate reader.
20. This number can be compared against the total number of cell initially present. This ratio is more of an endpoint assay, and assay time points should be taken into account in addition to the cell type that is migrating (e.g., transformed cells replicate faster than primary cells, thus long time points may not be measuring only migration but may be measuring replication).
21. Both coelenterazine and luciferin should be prepared immediately prior to injection. Substrate should be kept on ice and in the dark to maintain efficacy.
22. Tail vein injections require a restraining device, 1 ml tip syringe, 27G 1/2 in. needle, and a heat lamp or 50°C water bath. Heat applied to the tail will allow for better visualization of the two lateral tail veins. Holding the restrainer and the tail at the same time, align the needle (bevel side up) with the vein. If you insert the needle correctly, you will see the needle in the vein and the fluid will be easy to inject. Make sure there are no air bubbles within the needle. Do not force fluid – if the tail swells, you have missed the vein.
23. For an inclusive measurement, several capture times may be required. The bioluminescence of luciferin will peak in about 3–10 min, whereas coelenterazine may peak in 5–15 min. It is vital to make sure that your image capture includes the peak of the substrate activation without saturation of the system. This will require user optimization based on each individual experiment.
24. The imaging system (consisting of a cooled, back-thinned CCD camera) will capture both a visible light photograph of the animal taken with light-emitting diodes and the luminescent image.

25. Fluorescence can also be measured in the same manner, although there are high levels of autofluorescence observed in the fluorescein channel, there are methods to distinguish between GFP and autofluorescence if necessary.
26. It is best to wait until the end of the entire experiment before any analysis is done. This will ensure the global normalization between images in order to produce publishable data.

Acknowledgments

Supported in part by grants from the National Cancer Institute (CA-1094551 and Ca-116199), the Susan G Komen Breast Cancer Foundation, and the Army Department of Defense (BC083397).

References

1. Prockop, D.J. (1997) Marrow stromal cells as stem cells for nonhematopoietic tissues *Science* **276**, 71–4.
2. Zappia, E., Casazza, S., Pedemonte, E., Benvenuto, F., Bonanni, I., Gerdoni, E., et al. (2005) Mesenchymal stem cells ameliorate experimental autoimmune encephalomyelitis inducing T-cell anergy *Blood* **106**, 1755–61.
3. Lopez Ponte, A., Marais, E., Gallay, N., Langonne, A., Delorme, B., Herault, O., et al. (2007) The in vitro migration capacity of human bone marrow mesenchymal stem cells: Comparison of chemokine and growth factor chemotactic activities *Stem Cells.*
4. Li, Y., Chen, J., Wang, L., Zhang, L., Lu, M., and Chopp, M. (2001) Intracerebral transplantation of bone marrow stromal cells in a 1-methyl-4-phenyl-1,2,3,6-tetrahydropyridine mouse model of Parkinson's disease *Neurosci Lett* **316**, 67–70.
5. Niedzwiedzki, T., Dabrowski, Z., Miszta, H., and Pawlikowski, M. (1993) Bone healing after bone marrow stromal cell transplantation to the bone defect *Biomaterials* **14**, 115–21.
6. Kawada, H., Fujita, J., Kinjo, K., Matsuzaki, Y., Tsuma, M., Miyatake, H., et al. (2004) Nonhematopoietic mesenchymal stem cells can be mobilized and differentiate into cardiomyocytes after myocardial infarction *Blood* **104**, 3581–87.
7. Dezawa, M., Ishikawa, H., Itokazu, Y., Yoshihara, T., Hoshino, M., Takeda, S., et al. (2005) Bone marrow stromal cells generate muscle cells and repair muscle degeneration *Science* **309**, 314–17.
8. Studeny, M., Marini, F.C., Champlin, R.E., Zompetta, C., Fidler, I.J., and Andreeff, M. (2002) Bone marrow-derived mesenchymal stem cells as vehicles for interferon-beta delivery into tumors *Cancer Res* **62**, 3603–08.
9. Murdoch, C., Giannoudis, A., and Lewis, C.E. (2004) Mechanisms regulating the recruitment of macrophages into hypoxic areas of tumors and other ischemic tissues *Blood* **104**, 2224–34.
10. Okuyama, H., Krishnamachary, B., Zhou, Y.F. Nagasawa, H. Bosch-Marce, M., and Semenza,G.L. (2006) Expression of vascular endothelial growth factor receptor 1 in bone marrow-derived mesenchymal cells is dependent on hypoxia-inducible factor 1 *J Biol Chem* **281**, 15554–63.
11. Chen, D., Zhang, Z., Wu, X., and Lin, J. H. (2007) Distribution of intravenously grafted bone marrow mesenchymal stem cells in the viscera tissues of rats before and after cerebral ischemia *J. Clin Rehab Tissue Eng. Res* **11**, 10160–64.
12. Spitkovsky, D., and Hescheler, J. (2008) Adult mesenchymal stromal stem cells for therapeutic applications *Minimally Invasive Ther Allied Technol* **17**, 79–90.
13. Klopp, A.H., Spaeth, E.L., Dembinski, J.L., Woodward, W.A., Munshi, A., Meyn, R.E., et al. (2007) Tumor irradiation increases the recruitment of circulating mesenchymal stem

cells into the tumor microenvironment *Cancer Res* **67**, 11687–95.

14. Mouiseddine, M., François, S., Semont, A., Sache, A., Allenet, B., Mathieu, N., et al. (2007) Human mesenchymal stem cells home specifically to radiation-injured tissues in a non-obese diabetes/severe combined immunodeficiency mouse model *Br J Radiol* **80**, Spec No 1:S49–55.
15. Gorin, N., Fliedner, T.M., Gourmelon, P., Ganser, A., Meineke, V., Sirohi, B., et al. (2006) Consensus conference on European preparedness for haematological and other medical management of mass radiation accidents *Ann. Hematol* **85**, 671–79.
16. Dvorak, H.F. (1986) Tumors: wounds that do not heal. Similarities between tumor stroma generation and wound healing *N Engl J Med* **315**, 1650–59.
17. Wu, Y., Wang, J., Scott, P.G., and Tredget, E.E. (2007) Bone marrow-derived stem cells in wound healing: A review *Wound Repair Regen* **15**, Suppl 1:S18–26.
18. Platt, I.D., and El-Sohemy, A. (2008) Regulation of osteoblast and adipocyte differentiation from human mesenchymal stem cells by conjugated linoleic acid *J Nutr Biochem* **20**, 956–64.
19. Ouchterlony, O. (1953) Antigen-antibody reactions in gels. IV. Types of reactions in coordinated systems of diffusion *Acta Pathol. Microbiol.Scand* **32**, 230–240.
20. Ouchterlony, O., Ericsson, H., and Neumuller, C. (1950) Immunological analysis of diphtheria antigens by the gel diffusion method *Acta Med. Scand* **138**, 76–79.
21. De Becker, A., Van Hummelen, P., Bakkus, M., Broek, I.V., De Wever, J., De Waele, M., et al. (2007) Migration of culture-expanded human mesenchymal stem cells through bone marrow endothelium is regulated by matrix metalloproteinase-2 and tissue inhibitor of metalloproteinase-3 *Haematologica* **92**, 440–9.
22. Kim, Y.J., Kim, H.K., Cho, H.K., Bae, Y.C., Suh, K.T., and Jung, J.S. (2007) Direct comparison of human mesenchymal stem cells derived from adipose tissues and bone marrow in mediating neovascularization in response to vascular ischemia *Cell Physiol Biochem.* **20**, 867–76.
23. Sasaki, M., Abe, R., Fujita, Y., Ando, S., Inokuma, D., and Shimizu, H. (2008) Mesenchymal stem cells are recruited into wounded skin and contribute to wound repair by transdifferentiation into multiple skin cell type *J Immunol* **180**, 2581–87.
24. Kidd, S., Spaeth, E., Dembinski, J.L., Dietrich, M., Watson. K., Klopp, A., et al. (2009) Direct evidence of mesenchymal stem cell tropism for tumor and wounding microenvironments using in vivo bioluminescence imaging *Stem Cells* **27**, 2614–23.
25. Hung, S.C., Deng, W.P., Yang, W.K., Liu, R.S., Lee, C.C., Su, T.C., et al. (2005) Mesenchymal stem cell targeting of microscopic tumors and tumor stroma development monitored by noninvasive in vivo positron emission tomography imaging *Clin Cancer Res* **11**, 7749–56.
26. Ju, S., Teng, G., Zhang, Y., Ma, M., Chen, F., and Ni, Y. (2006) In vitro labeling and MRI of mesenchymal stem cells from human umbilical cord blood *Magn Reson Imaging* **24**, 611–7.

Chapter 18

Epithelial Stem Cells

Kyle M. Draheim and Stephen Lyle

Abstract

It is likely that adult epithelial stem cells will be useful in the treatment of diseases, such as ectodermal dysplasias, monilethrix, Netherton syndrome, Menkes disease, hereditary epidermolysis bullosa, and alopecias. Additionally, other skin problems such as burn wounds, chronic wounds, and ulcers will benefit from stem cell-related therapies. However, there are many questions that need to be answered before this goal can be realized. The most important of these questions is what regulates the adhesion of stem cells to the niche versus migration to the site of injury. We have started to identify the mechanisms involved in this decision-making process.

Key words: Hair follicle stem cells, Bulge cells, RhoA, NET1, LEF1, Migration, Adhesion

1. Introduction

A greater understanding of adult epithelial stem cells of the skin will be important for furthering our knowledge and treatment of hereditary cutaneous diseases, as well as conditions such as burns and chronic wounds (1–5). The mammalian skin represents a physical barrier between the body and its external environment receiving most of the damage brought on by physical trauma and mutagenic UV radiation. The skin has a multilayered epithelium that is comprised of sebaceous glands, hair follicles, and the interfollicular epidermis. In order to protect against the accumulation of mutations, the epidermis is characterized by rapid cellular turnover, renewing themselves in humans every 2–3 weeks (6, 7). This regeneration is sustained by many types of epidermal stem cells which additionally participate in the repair of the skin after injuries. These cells are quiescent but upon injury can be mobilized into an extensive and sustained self-renewal capacity. The bulge region of the hair follicle represents the best characterized epidermal stem cell population described to date, but there is

Marie-Dominique Filippi and Hartmut Geiger (eds.), *Stem Cell Migration: Methods and Protocols*, Methods in Molecular Biology, vol. 750, DOI 10.1007/978-1-61779-145-1_18, © Springer Science+Business Media, LLC 2011

evidence of other stem populations in the interfollicular epidermis and the sebaceous glands (8–10).

In human fetal and murine follicles, the bulge is a morphologically prominent outgrowth of epithelial cells below the opening of the sebaceous gland marking the lower end of the permanent portion of the follicle (11). In contrast, this morphological structure is not typically seen in adult human follicles (only at the onset of anagen) and therefore the location of the bulge is estimated to be roughly located at the insertion of the arrector pili muscle. These bulge cells have been demonstrated to possess stem cell properties in vivo and in vitro (12, 13). In addition to the in vitro studies, the bulge cells possess the in vivo proliferative behavior expected of stem cells. While the stem cells are generally slowly cycling, as determined in label-retaining studies, cell proliferation analysis shows that the stem cells of the mouse bulge are activated to transiently proliferate at the onset of anagen (14) starting a cycle characterized in the "bulge activation hypothesis" (13).

There are two main methods used to identify epidermal stem cells. The first makes use of the slow-cycling nature that defines all stem cells in a pulse-chase assay. All actively proliferating cells are pulse-labeled with the introduction of a DNA precursor such as ^{3}H-TdR or BrdU. For in vivo studies, mice are injected at postnatal day 3 or shortly after the engraftment of human skin onto an adult nude mouse. The DNA precursor can be added into the media for in vitro studies. The labeling period is then followed by an extended chase period (4–10 weeks) where all proliferating cells dilute the label by half with every cell division. Over time, the rapidly proliferating cells, such as the transit amplifying cells, will no longer contain detectable levels of label whereas stem cells will (13, 15, 16). Additionally, this method has been adapted to use inducible GFP and β-galactosidase as pulse-labeling reagents (17).

The second method used in the identification of epidermal stem cells involves the use of stem cell markers. Unfortunately, most of the markers identified lack specificity to epidermal stem cells alone. However, these markers in combination with knowledge of the approximate physical location relative to histological markers will allow for accurate isolation through microdissection and laser capture.

Similar to other epithelial tissues (18, 19) the bulge has all the features expected of a stem cell niche and the bulge keratinocytes display the characteristics of stem cells. The bulge represents the permanent portion of the hair follicle; while the lower follicle undergoes apoptosis and degenerates during catagen. The bulge keratinocytes are also tightly adherent to the basement membrane and are protected from accidental loss to plucking (13). Plucking of human follicles, which can remove a majority of the hair follicle epithelium below the level of the bulge, will still result in hair regeneration (20). The basement membrane zone of the bulge also appears specialized for its role in protecting the bulge keratinocytes. K-laminin and type VII collagen are much more highly expressed in the basement membrane zone of the bulge than in the lower portion of the ORS allowing for increased adhesion to the niche (11). Additionally, the high vascularity of the bulge provides nourishment for this important area and the bulge-containing isthmus region is the most richly innervated skin area (21). The bulge area also contains a concentrated number of Langerhans cells that are thought to help maintain the epidermal stem cell niche (22, 23).

Under normal circumstance, epidermal stem cells have very limited motility as it is crucial for the cells to remain in the niche (24). However, hair follicle stem cells can be induced to migrate downward to the hair follicle as well as outward to resurface the epidermis in response to wounding (25). After injury, keratinocytes become activated; simultaneously secreting and responding to growth factors and cytokines (26–28). After mechanical injury, epidermal stem cells will start migrating and proliferating (29, 30). During the process of tissue repair, keratinocytes undergo significant changes of their junctions and adhesion molecules: hemidesmosomes have to be dissolved in order to allow for keratinocyte migration. The migrating keratinocytes produce a different set of membrane molecules, such as vitronectin, fibronectin receptors, and integrin $\alpha5\beta1$, and these replace the collagen receptor to allow for keratinocyte migration (31–33). Many factors active during wound healing, such as EGF, keratinocyte growth factor (KGF) and TGFβ, are potential regulators of these processes (34–36).

To identify cellular pathways which are involved in mediating stem cell properties, such as stem cell migration, expression array analyses comparing the hair follicle stem cell compartment to the transit-amplifying (TA) cell compartment have been performed. These results have revealed a number of interesting genes. Thus the mechanisms anchoring skin stem cells to the niche and controlling cell migration are slowing becoming clearer. RhoA has a constitutive, high expression within the bulge region during all phases of the hair cycle, and thus may play a role in maintaining the steady state of stem cells. Since RhoA is known to regulate the actin cytoskeleton (37), its high level in the stem cells may help maintain the stem cell niche. Additionally, RhoA could be important for anchoring the stem cells within the niche and prevent their loss through migration during their activation. Additionally, NET1 (guanine exchange factor which specifically activates RhoA to induce stress fiber formation) (38) is present in the activated stem cells as the new bulb is developing, but NET1 is not present in any portion of the follicle at any other time of the hair cycle.

2. Materials

2.1. Identification

1. Monoclonal antibodies against ITGα6 (integrin alpha-6), ITGβ1 (integrin beta-1), $CD71^{lo}$, CD34, CD200, KRT15 (cytokeratin 15), KRT19 (cytokeratin 19) are used for immunohistochemistry (IHC), immunofluorescence (IF), and cell sorting.
2. Pulse Chase Labeling uses BrdU or ^{3}H-TdR in neonatal mice or human tissues xenografted onto nude or SCID mice.

2.2. Keratinocyte Stem Cell Medium

1. Combine 670 mL of DMEM (Gibco) with 223.2 mL of Hams F12 (Gibco) at a ratio of 3:1.
2. Put 20 mL of mixture into 50 mL tube containing 24 mg of adenine (Sigma) and adjust pH to 7.5.
3. Add 100 mL of Fetal Bovine Serum (FBS) (final 10%).
4. Add back the medium with adenine (final 24 mg/L).
5. Make Cholera Toxin (Gibco) stock (10 mM): add 1.18 mL of H_2O to 1 mg of cholera toxin; add 0.1 mL of concentrated stock to 10 mL of DMEM with 10% serum, aliquot 1 mL, freeze.
6. Add 1 mL of cholera toxin to keratinocyte stem cell medium (KCM) (final 0.1 nM).
7. Make hydrocortisone (Sigma) stock (5 mg/mL): dissolve 25 mg in 5 mL of 95% ethanol, 4°C storage 400 mg/mL: add 0.8 mL of concentrated stock to 9.2 mL of DMEM, aliquot 1 mL, freeze.
8. Add 2 mL of hydrocortisone to KCM (final 0.4 μg/mL).
9. Make transferrin/T3 stock (transferrin: 5 mg/mL/T3: 2 mM): dissolve 13.6 mg of T3 (Sigma) in a small volume of 0.02 N NaOH, add H_2O to 100 mL; dissolve 250 mg of transferrin (Behring Diagnostics) in 30 mL of PBS; add 0.5 mL of concentrated T3 stock to transferring solution, add H_2O to 50 mL, aliquot 10 mL, freeze.
10. Add 1 mL of transferrin/T3 to KCM (final 5 μg/mL transferrin/2 nM T3).
11. Make insulin (Sigma) stock (5 mg/mL): dissolve 50 mg in 10 mL of 0.005N HCl, aliquot 1.3 mL, freeze.
12. Add 1.3 mL of insulin to KCM (final 6.5 μg/mL).
13. Add 10 mL of Pen/Strep (final 100 u Pen/100 μg Strep/L).
14. Add 2 mL fungizone/amphotericin B (final 0.5 mg/L).
15. Adjust pH to 7.2.
16. Sterile filter 500 mL through 0.22 mm filter.

17. Make epidermal growth factor (EGF) (Sigma) stock (2 mg/mL): add 1 mL of H_2O to the vial, aliquot 0.1 mL, freeze 10 μg/mL: dilute 0.1 mL of 2 mg/mL EGF with 20 mL of H_2O, filter and aliquot 1.0 mL, freeze.
18. Add EGF, 100 μL of 10 μg/mL EGF per 100 mL of medium (use within 2–4 weeks).

2.3. Clonogenic Assay (Colony Formation and Serial Replating)

1. Epidermal stem cell culture materials.
2. Six-well plates.
3. Methanol fixation and washes.
4. Crystal Violet: Dissolve 200 mg in 100 mL of 2% methanol (v/v) for a 0.2% solution.

2.4. Immuno-histochemistry/Immunofluorescence

1. Cell or tissue fixative: acetone or 10% neutral-buffered formalin.
2. Blocking buffer: 5% Goat Serum (Sigma) in phosphate-buffered saline (PBS) with 0.1% Triton X-100.
3. Wash buffer is PBST (0.1% Triton X-100 in PBS).
4. Vectashield mounting media (Vector Labs).

2.5. Adhesion and Migration

1. Collagen IV (Sigma).
2. Heat-inactivated BSA (Sigma).
3. Keratinocyte SFM (Gibco).
4. DPBS (Gibco) or PBS + 1 mM $CaCl_2$ + 1 mM $MgCl_2$.
5. HEPES (Gibco).

3. Methods

3.1. Epidermal Stem Cell Identification

3.1.1. Protein Markers

1. Use antibodies targeting desired marker in FACS, IF, or IHC protocols (see below).

3.1.2. Pulse Chase

1. Subcutaneously inject neonatal mice (age 1–2 days) with either 5 μCi of ^{3}H-TdR or 100 μg of BrdU per gram of body weight once daily for 7–10 days.
2. Sacrifice one mouse 24 h after final dose and test for complete labeling.
3. Sacrifice at desired endpoint (6–8 weeks after final dose) and process tissue for histological analysis.
4. ^{3}H-TdR is detected using autoradiography.
5. BrdU-specific antibodies are used for detection in FACS, IF, or IHC.

3.2. Epidermal Stem Cell Isolation and Culture

3.2.1. Isolation and Primary Culture of Explants or Cell Suspension

1. Incubate fresh human scalp in DMEM/10% FBS/Dispase (4 mg/mL) 14 h at 4°C or 2–4 h at 37°C (see Note 1).
2. Pluck follicles under dissecting microscope and transfer to 50 mL tube (Fig. 1) (see Note 2).
3. Digest the follicles with 1 mL of 1× trypsin–EDTA for 10 min.
4. Add 1.33 mL of versene and 1 mL of 1× trypsin–EDTA.

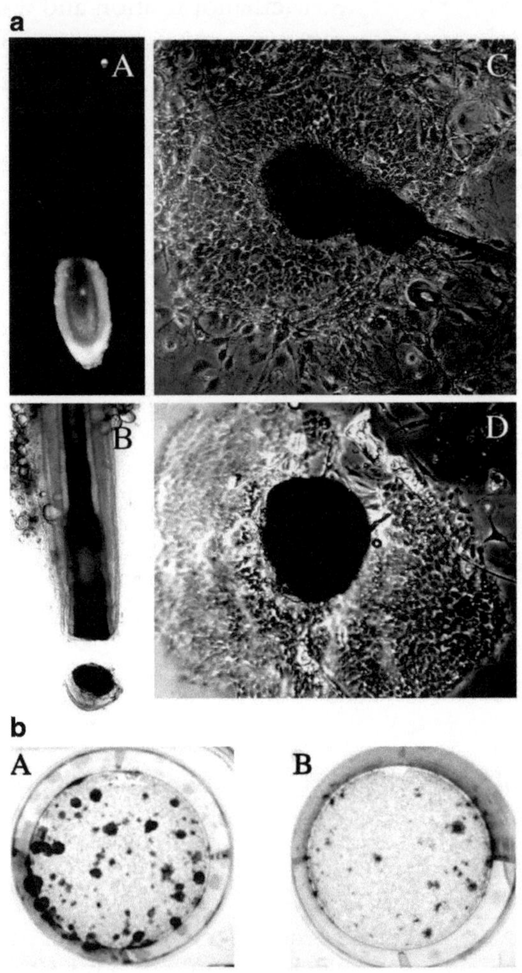

Fig. 1. Panel (**a**): Plucked follicles from dispase-treated human scalp skin. The epithelial sheath is removed along with the hair shaft from telogen follicles (**A**) and anagen follicles (**B**). The follicles can be microdissected to isolate the stem cell enriched telogen club and the hair matrix containing transit-amplifying cells. Cells can be expanded from explant cultures (**C** and **D**) or from single cell suspensions after trypsinization (not shown) (adapted from 24 with permission from Nature Publishing Group). Panel (**b**): Isolated stem cells (**A**) and transit amplifying cells (**B**) grown at clonogenic density show significant differences in colony formation (adapted from 24 with permission from Nature Publishing Group).

5. Disaggregate for 20–30 min with shaking periodically.
6. Add 4 mL of DMEM to stop reaction.
7. Centrifuge for 6 min at 800 rpm.
8. Discard the supernatant carefully (save ~0.5 mL).
9. Resuspend in 1 mL of KCM without EGF and plate.

3.2.2. FACS Staining for Purification (Optional)

1. Strain cells into a 50-mL tube to remove clumps of cells.
2. Wash strainer with 5 mL of KCM.
3. Centrifuge for 6 min at 800 rpm and resuspend cell pellet in 2 mL of cold PBS + 3% BSA.
4. Wash cells twice with 1 mL of PBS + 3% BSA.
5. Count cells (all further measurements will be per 10^6 cells).
6. Pellet cells and resuspend in 200 μL of PBS + 3% BSA.
7. Incubate for 15 min on ice.
8. Pellet cells and re-suspend in primary antibody diluted in PBS + 3% BSA (the dilution for each lot of antibody should be determined by the individual laboratory. Typical dilutions range from 1:50 to 1:300).
9. Incubate for 1 h on ice.
10. Wash cells twice with 500 μL of PBS + 3% BSA.
11. Spin down and resuspend with fluorochrome-conjugated secondary antibody diluted in PBS + 3% BSA (dilutions for the secondary antibody should be determined by the individual laboratory. Typical dilutions range from 1:100 to 1:500).
12. Incubate on ice for 30 min in the dark.
13. Wash cells twice with 500 μL of PBS + 3% BSA.
14. Pellet the cells and resuspend in PBS + 1% BSA + 10 mM HEPES (final cell concentration should be $1–2 \times 10^7$ cells/mL).
15. Transfer to a 5-mL FACS tube, cover with aluminum foil, and sort immediately.
16. Sort cells into KCM containing 20% FBS.

3.2.3. Feeder Layer Preparation

1. Seed 2×10^6 J2-3T3 fibroblasts in T75 flask or 2–300,000 cells/well of six-well plate on the day before receiving tissue or prior to subculturing.
2. Next morning, wash cells with PBS and replace the medium with Mitomycin C (15 μg/mL)/DMEM for 2 h at 37°C.
3. Wash cells three times with DMEM.
4. Add 15 mL of KCM without EGF 2 h before adding keratinocytes.

3.2.4. Serial Passage of Keratinocytes (Six-Well Plate)

1. Aspirate off the medium from keratinocytes.
2. To remove the residual medium, wash once with 5 mL of versene.
3. Add 5 mL of versene and incubate at room temperature for 1 min. Shake the plate until all J2-3T3 cells have been dislodged.
4. Aspirate the versene with 3T3-J2 cells.
5. Wash quickly with 5 mL of versene to completely remove 3T3-J2 cells.
6. Add 1.5 mL of 1× trypsin–EDTA and incubate at 37°C for less than 8 min. Check for the plate every 2–3 min for complete detachment of the cells.
7. Neutralize the trypsin by adding 10 mL of DMEM + 10% FBS.
8. Pellet by centrifuging for 6 min at 200 × *g*.
9. Decent off supernatant and resuspend cells in KCM–EGF.
10. Plate cells into new culture dish with 3T3-J2 feed layer.
11. Allow the cells to grow 14 h in a 37°C tissue culture incubator to permit attachment to the plate.
12. The following day, replace the culture media with KCM with EGF.
13. Allow to grow to 75% confluence or 9–14 days.

3.3. Clonogenicity/Colony-Forming Assay

1. Culture 2,000 epidermal stem cells in a six-well plate for 9–12 days. Grow the cells in KCM and coculture with 3T3-J2 feeder cells.
2. At end point, wash the cells twice with PBS. Aspirate off all PBS.
3. Fix the cells for 5 min with 100% methanol.
4. Stain with 0.2% Crystal Violet (Fisher) in 2% methanol/water (v/v) (Fig. 1).
5. Take pictures of the wells using either a digital camera or a desktop scanner.
6. Colony number and size can be determined by counting/measuring manually or by digital analysis.

3.4. Immunohistochemistry and Immunofluorescence

1. For frozen sections or cells cultured on coverslips
   (a) Fix with acetone, cold methanol, or 4% paraformaldehyde for 20 min at 4°C (let air-dry 20 min if cold methanol was used) (see Note 3).
   (b) Rinse in PBS (×3) for 5 min.
   (c) For permeabilization of the cell membranes, add 0.1% Triton X in PBS and incubate at room temperature for 5 min.
   (d) Rinse in PBS (×3) for 5 min. Continue to step 3.

2. For paraffin-embedded sections
   (a) Deparaffinize and rehydrate paraffin-embedded slides: Xylene or xylene substitute for 5 min (×2), 100% EtOH for 3 min, 95% EtOH for 3 min, 70% EtOH for 3 min, 50% EtOH for 3 min.
   (b) Rinse in PBS 3 min.
   (c) Antigen retrieval (AR): antigen unmasking solution (Vector Laboratories, Burlingame, CA), 1 mM EDTA (pH 8) or 10 mM citrate buffer (pH 6). Heat in microwave oven for 10–12 min and cool down for 20 min.
   (d) Rinse in PBS for 5 min. Continue to step 3.
3. Quench: incubate in 0.3% H_2O_2 in water for 30 min at room temperature.
4. Rinse in PBS for 5 min.
5. Blocking: incubate in 1–3% BSA or 5–10% normal goat serum in PBS for 60 min at room temperature.
6. Rinse in PBS (×3) for 5 min (see Note 4).
7. Primary antibody: antibody (or antisera) diluted in 1% BSA or in 5% normal serum/PBS then put 100–300 μL of antibody dilution on the sections.
8. Put glass slides in a humidified chamber and incubate at 4°C for 14 h or 2 h at room temperature.
9. Wash with PBS (×4).
10. Apply 100 μL of appropriately diluted biotinylated secondary antibody (using the antibody dilution buffer) for IHC or fluorescence-tagged secondary for IF (see below) to the sections on the slides and incubate in a humidified chamber at room temperature for 30 min.
11. Wash slides in 300 mL PBS for two changes, 5 min each.

3.4.1. IHC

1. Apply 100 μL of appropriately diluted avidin–HRP conjugates (using the antibody dilution buffer) to the sections on the slides and incubate in a humidified chamber at room temperature for 30 min (keep protected from light).
2. Wash slides in 300 mL of PBS for two changes, 5 min each.
3. Apply 100 μL of DAB substrate solution (freshly made just before use: 0.05% DAB–0.015% H_2O_2 in PBS) to the sections on the slides to reveal the color of antibody staining.
4. Allow the color development for <5 min until the desired color intensity is reached.
5. (Caution: DAB is a suspect carcinogen. Handle with care. Wear gloves, lab coat, and eye protection).
6. Wash slides in 300 mL of PBS for three changes 2 min each.

7. (Optional) counterstain slides by immersing sides in hematoxylin (e.g., Gill's Hematoxylin) for 1–2 min.
8. Rinse the slides in running tap water for >15 min.

3.4.2. IF

1. Secondary antibody (conjugated): incubate in fluorescence conjugated antispecies-specific IgG (H+L) (Molecular Probes) (1:500) in 5% normal serum/PBS for 60 min at room temperature in the dark.
2. Wash with PBS (×3) in dark.
3. Counter-stain with antifade medium with or without DAPI and mount (see Note 5).

3.5. Migration

3.5.1. Preparation

1. Prepare PBSABC solution: PBS supplemented with 1 mM $CaCl_2$ and 1 mM $MgCl_2$. You can make PBSABC by yourself or use DPBS ready solution.
2. Coat flask with 50 μg/mL human collagen IV in DPBS for 14 h at 4°C (see Note 6).
3. Block the collagen with 4–5 mL of 0.5% BSA at 37°C for 30 min.

3.5.2. Wound Assay

1. Seed 8×10^5 epidermal stem cells per well in a six-well plate using SFM (Gibco) with 1% Pen/Strep, 25 μg/mL BPE but without EGF and allow to attach over-night.
2. The next day, wash cells twice with PBS and once with SFM.
3. Treat cells with 1 mL of SFM + 15 μg/mL mitomycin C at 37°C for 90 min.
4. Wash cells twice with PBS and once with SFM.
5. Add 2 mL of SFM + 1% Pen/Strep to each well.
6. Scratch cells with a 10-μL pipette tip as shown in Fig. 2.
7. Wash cells twice with PBS and once with SFM.
8. Add 2 mL of SFM + 1% Pen/Strep to each well and take pictures, marking the view so that the same position can be found for subsequent time-points.
9. Wash cells once with SFM + 1% Pen/Strep.
10. Add 2 mL of SFM + 1% Pen/Strep, 25 μg/mL BPE, and 2 ng/mL EGF.
11. Incubate for 24 h at 37°C.
12. Repeat steps 7–10 as necessary until the desired time-point is reached or wound is closed.

3.5.3. Time-Lapse Video Microscopy

1. Replate cells into a flask pretreated with 50 μg/mL collagen IV.
2. Allow cell to adhere for at least 10 min and then wash with KCM.
3. A final concentration of 10 μM HEPES was added to the media.

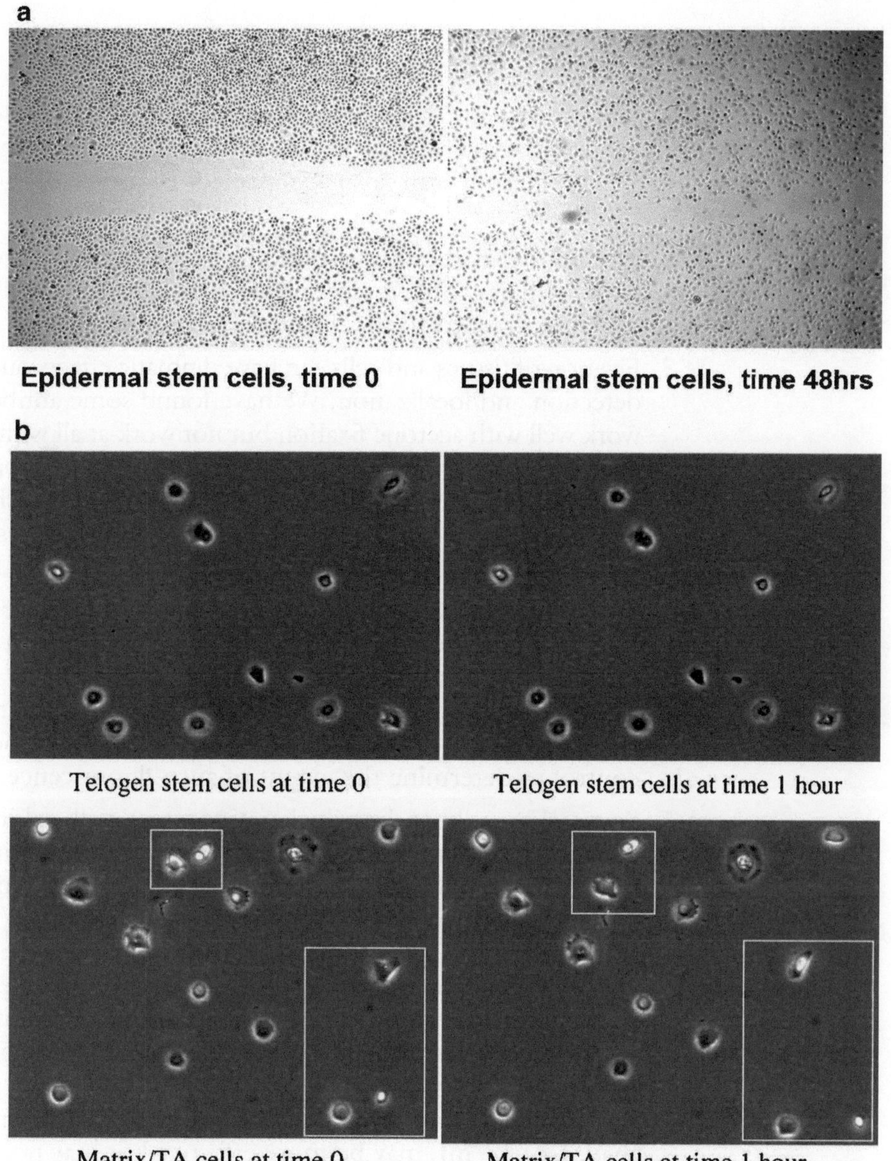

Fig. 2. Panel (**a**): Wound assay performed on human skin stem cells. Confluent culture is "scratched" with pipette tip to activate stem cell migration and imaged at time 0 (*left*) and 48 h later (*right*). Percentage of wound closure can be calculated. Panel (**b**): Migration of stem cells and transit amplifying cells by time-lapse video microscopy. Cells are images every 15 min for 14 h. Primary stem cells show significantly slower rate of migration than transit-amplifying cells consistent with their normal anchorage in the stem cell niche (adapted from 24 with permission from Nature Publishing Group).

4. Infuse the flask with 5% CO_2 and seal.
5. Videotape for 4–24 h with 15 min between frames by time-lapse video microscopy on a heated stage to keep the culture temperature at 37°C.
6. Measure motility using a cell tracking extension (ICRF) written for IPLab (Scanalytics, Inc., Fairfax, VA) (Fig. 2).

4. Notes

1. It is best to cut strips of skin ~1 cm wide, cutting along the angle of hair growth so that the enzyme can access the tissues. The epidermis can also be peeled to isolated epidermal keratinocytes.
2. The epithelial sheath surrounding the hair shaft is removed with the hair. These intact epithelial tissues can be cultured as explants attached to the dish for initial outgrowth of cells.
3. Fixation of tissues and cells can have dramatic effects on antigen detection and localization. We have found some antibodies to work well with acetone fixation but not work at all with methanol fixation. In addition, the cellular localization can appear different (nuclear vs. cytoplasmic) with different fixation techniques. For new antibodies, tissues, or cells a panel of fixatives is recommended to determine optimal conditions for expected results.
4. Skin frequently has high background auto-fluorescence making immunofluorescence difficult (39). The samples can be pretreated with a reducing agent (0.1 M glycine in PBS or 20 mM sodium sulfite) for 5 min to help eliminate background fluorescence. Always perform a no-antibody only control to determine the extent of auto-fluorescence.
5. Some skin cells (primarily keratinocytes) will exhibit high nonspecific staining, especially with rabbit polyclonal antibodies. To reduce such background, incubate sections with unconjugated AffiniPure Fab fragment (H+L) (Jackson ImmunoResearch Labs) for 1 h at room temperature (use an antibody fragments from a species different that those of your primary antibody). Dilute the antibody with PBS and make sure its final concentration is NOT lower than 0.1 mg/mL or 100 μg/mL; the antibody concentration lower than 0.1 mg/mL starts to produce background staining; concentration higher than 0.1 mg/mL may be more effective but may not be necessary. Increasing blocking time to 2 h at room temperature or 14 h at 4°C may be more effective, but in this case, lower concentration may be used.
6. It is very important that the collagen coating of coverslips and culture dishes is even. If it appears that the coating is not consistent, treatment with 1 mM magnesium acetate (Sigma) for 2–3 h and then coating with 0.1 mg/mL poly-D-lysine (Sigma) prior to coating with collagen IV will help generate a more even coat.

References

1. Ohyama, M., and Vogel, J.C. (2003) Gene delivery to the hair follicle *J Investig Dermatol Symp Proc* **8**, 204–6.
2. Sugiyama-Nakagiri, Y., Akiyama, M., and Shimizu, H. (2006) Hair follicle stem cell-targeted gene transfer and reconstitution system *Gene Ther* **13**, 732–7.
3. Stenn, K.S., and Cotsarelis, G. (2005) Bioengineering the hair follicle: fringe benefits of stem cell technology *Curr Opin Biotechnol* **16**, 493–7.
4. Hoeller, D., *et al.* (2001) An improved and rapid method to construct skin equivalents from human hair follicles and fibroblasts *Exp Dermatol* **10**, 264–71.
5. Navsaria, H.A., Ojeh, N.O., Moiemen, N., Griffiths, M.A., and Frame, J.D. (2004) Reepithelialization of a full-thickness burn from stem cells of hair follicles micrografted into a tissue-engineered dermal template (Integra) *Plast Reconstr Surg* **113**, 978–81.
6. Potten, C.S. (1975) Epidermal cell production rates *J Invest Dermatol* **65**, 488–500.
7. Potten, C.S. (1975) Epidermal transit times *Br J Dermatol* **93**, 649–58.
8. Tiede, S., *et al.* (2007) Hair follicle stem cells: walking the maze *Eur J Cell Biol* **86**, 355–76.
9. Kaur, P. (2006) Interfollicular epidermal stem cells: identification, challenges, potential *J Invest Dermatol* **126**, 1450–8.
10. Bieniek, R., Lazar, A.J., Photopoulos, C., and Lyle, S. (2007) Sebaceous tumours contain a subpopulation of cells expressing the keratin 15 stem cell marker *Br J Dermatol* **156**, 378–80.
11. Akiyama, M., Dale, B.A., Sun, T.T., and Holbrook, K.A. (1995) Characterization of hair follicle bulge in human fetal skin: the human fetal bulge is a pool of undifferentiated keratinocytes *J Invest Dermatol* **105**, 844–50.
12. Kobayashi, K., Rochat, A., and Barrandon, Y (1993) Segregation of keratinocyte colony-forming cells in the bulge of the rat vibrissa. *Proc Natl Acad Sci USA* **90**, 7391–7395.
13. Cotsarelis, G., Sun, T.T., and Lavker, R.M. (1990) Label-retaining cells reside in the bulge area of pilosebaceous unit: implications for follicular stem cells, hair cycle, and skin carcinogenesis *Cell* **61**, 1329–37.
14. Wilson, C., *et al.* (1994) Cells within the bulge region of mouse hair follicle transiently proliferate during early anagen: heterogeneity and functional differences of various hair cycles *Differentiation* **55**, 127–36.
15. Bickenbach, J.R., and Mackenzie, I.C (1984) Identification and localization of label-retaining cells in hamster epithelia. *J Invest Dermatol* **82**, 618–22.
16. Braun, K.M., *et al.* (2003) Manipulation of stem cell proliferation and lineage commitment: visualisation of label-retaining cells in wholemounts of mouse epidermis *Development* **130**, 5241–55.
17. Tumbar, T., *et al.* (2004) Defining the epithelial stem cell niche in skin *Science* **303**, 359–63.
18. Lavker, R.M., and Sun, T.T. (1982) Heterogeneity in epidermal basal keratinocytes: morphological and functional correlations *Science* **215**, 1239–41.
19. Cotsarelis, G., Cheng, S.Z., Dong, G., Sun, T.T., and Lavker, R.M. (1989) Existence of slow-cycling limbal epithelial basal cells that can be preferentially stimulated to proliferate: implications on epithelial stem cells *Cell* **57**, 201–9.
20. Moll, I. (1995) Proliferative potential of different keratinocytes of plucked human hair follicles *J Invest Dermatol* **105**, 14–21.
21. Schneider, M.R., Schmidt-Ullrich, R., and Paus, R. (2009) The hair follicle as a dynamic miniorgan *Curr Biol* **19**, R132–42.
22. Allen, T.D., and Potten, C.S. (1974) Fine-structural identification and organization of the epidermal proliferative unit *J Cell Sci* **15**, 291–319.
23. Potten, C.S., and Allen, T.D. (1976) A model implicating the Langerhans cell in keratinocyte proliferation control *Differentiation* **5**, 43–7.
24. Roh, C., Tao, Q., Photopoulos, C., and Lyle, S. (2005) In vitro differences between keratinocyte stem cells and transit-amplifying cells of the human hair follicle *J Invest Dermatol* **125**, 1099–105.
25. Taylor, G., Lehrer, M.S., Jensen, P.J., Sun, T.T., and Lavker, R.M. (2000) Involvement of follicular stem cells in forming not only the follicle but also the epidermis *Cell* **102**, 451–61.
26. Tomic-Canic, M., Komine, M., Freedberg, I.M., and Blumenberg, M. (1998) Epidermal signal transduction and transcription factor activation in activated keratinocytes *J Dermatol Sci* **17**, 167–81.
27. Rizvi, A.Z., and Wong, M.H. (2005) Epithelial stem cells and their niche: there's no place like home *Stem Cells* **23**, 150–65.
28. Watt, F.M., Lo Celso, C., and Silva-Vargas, V. (2006) Epidermal stem cells: an update *Curr Opin Genet Dev* **16**, 518–24.
29. Owens, D.M., and Watt, F.M. (2003) Contribution of stem cells and differentiated cells to epidermal tumours *Nat Rev Cancer* **3**, 444–51.

30. Blanpain, C., and Fuchs, E. (2009) Epidermal homeostasis: a balancing act of stem cells in the skin *Nat Rev Mol Cell Biol* **10**, 207–17.
31. Singer, A.J., and Clark, R.A. (1999) Cutaneous wound healing *N Engl J Med* **341**, 738–46.
32. Cavani, A., *et al.* (1993) Distinctive integrin expression in the newly forming epidermis during wound healing in humans *J Invest Dermatol* **101**, 600–4.
33. Haapasalmi, K., *et al.* (1996) Keratinocytes in human wounds express alpha v beta 6 integrin *J Invest Dermatol* **106**, 42–8.
34. Werner, S., *et al.* (1994) The function of KGF in morphogenesis of epithelium and reepithelialization of wounds *Science* **266**, 819–22.
35. Zambruno, G., *et al.* (1995) Transforming growth factor-beta 1 modulates beta 1 and beta 5 integrin receptors and induces the de novo expression of the alpha v beta 6 heterodimer in normal human keratinocytes: implications for wound healing *J Cell Biol* **129**, 853–65.
36. Nanney, L.B., Sundberg, J.P,. and King, L.E. (1996) Increased epidermal growth factor receptor in fsn/fsn mice *J Invest Dermatol* **106**, 1169–74.
37. Etienne-Manneville, S., and Hall, A. (2002) Rho GTPases in cell biology *Nature* **420**, 629–35.
38. Alberts, A.S., and Treisman, R. (1998) Activation of RhoA and SAPK/JNK signalling pathways by the RhoA-specific exchange factor mNET1 *EMBO J* **17**, 4075–85.
39. Wu, B.P., Tao, Q., and Lyle, S. (2005) Autofluorescence in the Stem Cell Region of the Hair Follicle Bulge *J Investig Dermatol* **124**, 860–2.

Part VI

Pathways Implicated in Stem Cell Migration

Chapter 19

Pathways Implicated in Stem Cell Migration: The SDF-1/CXCR4 Axis

Yaron Vagima, Kfir Lapid, Orit Kollet, Polina Goichberg, Ronen Alon, and Tsvee Lapidot

Abstract

The hallmark of hematopoietic stem and progenitor cells (HSPCs) is their motility, which is essential for their function, such as recruitment upon demand. Stromal Derived Factor-1 (SDF-1, CXCL12) and its major receptor CXCR4 play major roles in stem cell motility and development. In vitro migration assays, implicating either gradients or cell surface-bound forms of SDF-1, are easy to perform and provide vital information regarding directional and random stem cell motility, which correlate with their repopulation potential in clinical and experimental transplantations. In vivo stem cell homing to the bone marrow, their retention, engraftment, and egress to the circulation, all involve SDF-1/CXCR4 interactions. Finally, other stem cell features such as stem cell survival and proliferation, are also dependent on the SDF-1/CXCR4 axis.

Key words: SDF-1, CXCR4, Hematopoietic stem and progenitor cells, HPCSs migration, Bone marrow, Mobilization, CD34+

1. Introduction

SDF-1 (also termed CXCL12) is the only known powerful chemoattractant for both human and murine hematopoietic stem and progenitor cells (HSPCs), which functionally express its major receptor CXCR4 (1, 2). Thus, the SDF-1/CXCR4 axis serves as a principal mediator for directional stem cell motility in vitro and in vivo. SDF-1 is highly expressed in the bone marrow (BM) by various stromal cells, including osteoblasts, endothelial cells, and perivascular reticular cells (3, 4). Endogenous or transplanted circulating HSPCs are therefore recruited to the BM, as part of the multistep "homing" process, in which cells cross the physical endothelial/extracellular matrix (ECM) barrier from the blood to

Marie-Dominique Filippi and Hartmut Geiger (eds.), *Stem Cell Migration: Methods and Protocols*, Methods in Molecular Biology, vol. 750, DOI 10.1007/978-1-61779-145-1_19, © Springer Science+Business Media, LLC 2011

the BM microenvironment (5). The SDF-1/CXCR4 axis is also essential for both steady-state egress and induced "mobilization" of HSPCs from the BM into the circulation, a process required for host defense and repair mechanisms (6). At steady-state conditions, circadian rhythmic changes in murine BM SDF-1 levels determine the levels of circulating progenitors (7), while in stress situations, such as repetitive cytokine treatments, strong reduction in BM SDF-1 expression enables mobilization of cells to the peripheral blood (6). Manipulating SDF-1 levels and CXCR4 expression is one of the key mechanisms controlling HSPCs homing, mobilization, and their consequent localization (1–6). Direct disruption of the interaction between CXCR4 and SDF-1 may lead to mobilization as well, for instance by treatment with the CXCR4 antagonist AMD3100 (8). Notably, SDF-1 is involved in other stem cell functions, including regulation of cell quiescence and survival (4, 9). Motility is the hallmark of HSPCs. Therefore, evaluating HSPC migration capacity toward a gradient of SDF-1 in vitro, which correlates with their in vivo repopulation in preclinical and clinical transplantations (1, 3, 5, 10), is a common and efficient tool to characterize the overall progenitor cell motility features. For example, the lack of CD45 function or inhibition of PKCzeta activity impair SDF-1-induced migration of immature hematopoietic cells and their in vivo repopulation potential (11, 12). These motility features might be translated into improved clinical homing and engraftment efficiencies of transplanted human CD34+ progenitor cells (5). Moreover, aberrant motility of HSPCs is associated with dysfunction and diseases, for example, increased CXCR4 sensitivity in WHIM syndrome (13, 14) and Waldenstrom Macroglobulinemia (15). In summary, examining migration capacities toward a gradient of SDF-1 is an invaluable tool for studying progenitor cell motility and function in both basic and clinical stem cell research.

2. Materials

2.1. SDF-1-Induced Trans-Migration

2.1.1. Cell Culture: Mouse Bone Marrow Mononuclear Cells (mBM-MNC), Human CD34+ Progenitors Cells (hCD34+), Mouse MS-5 (BM Stromal Cell Line), Human Bone Marrow SV-Transformed Endothelial Cell (BMEC)

1. Primary mBM-MNC or enriched hCD34+ are maintained in 25 mM HEPES-containing full RPMI supplemented with 10% fetal calf serum, 2 mM l-glutamine, and 1% Penicillin and Streptomycin antibiotics (full RPMI). MS-5 cells are maintained in full RPMI media (as described above), supplemented with 50 nM β-mercaptoethanol. The hBMEC, a kind gift from Dr. S. Raffi (Weill Medical College of Cornell University, New York) are maintained in full DMEM with low glucose (1 mg/ml).

2.1.2. Transwells and Coating (Fibronectin Coating, Matrigel Coating)

1. Transwells: Costar transwells (6.5 mm/diameter, 5 μm/pore, Corning Costar).
2. Transwell's coating:
   - Fibronectin (FN) 25 μg/ml (Sigma-Aldrich Ltd), store at 4°C.
   - Growth Factor Reduced Matrigel™ Matrix (MG) 60 μg/insert (thick layer) or 10 μg/insert (thin layer) (see Note 1) (BD Biosciences), store at –20°C.

2.1.3. SDF-1 or Conditioned Media (of MS-5) as Chemoattractants

1. SDF-1α 125 ng/ml (see Note 2), keep at –20°C and store at 4°C before usage.
2. Conditioned Media (CM) of any cell type suspected to secrete SDF-1 α due to specific treatment, should be collected and stored in aliquots at –80°C. Exact amount of CM required to attract cells due to its SDF-1 content, should be individually calibrated.

2.1.4. Evaluating the Levels of Migrated Cells by Flow Cytometry

1. Flow cytometry analysis on FACSCalibur (Becton Dickinson) with CellQuest software.

2.1.5. Colony-Forming Units Assay

1. Methylcellulose preparation: 20 g of Methylcelloluse powder (Sigma-Aldrich Ltd), 500 ml of DDW, 500 ml of DMEM (Concentrated ×2). Boil DDW and add Methylcellulose upon stirring. Cool to room temperature, add DMEM with 1% Penicillin and Streptomycin antibiotics. Store in aliquots at –20°C.
2. Supplements: 30% FCS, 50 ng/ml SCF, 5 ng/ml IL-3, 5 ng/ml GM-CSF, and 2 μg/ml Erythropoietin.
3. Culture dish 35 × 10 mm.

2.2. Visualization of Cell Migration to a SDF-1 Gradient

1. Cell Culture: human CD34+ progenitors cells (hCD34+): Primary enriched human CD34+ are maintained in 25 mM HEPES-containing full RPMI supplemented with 10% fetal calf serum, 2 mM L-glutamine, and 1% Penicillin and Streptomycin antibiotics (full RPMI).
2. PBS$^{-/-}$: DPBS$^{-/-}$ ×10 (–)CaCl2 (–)$MgCl_2$ store at 4°C before usage.
3. Coating solution: 50–150 μg/ml hyaluronan (HA) (Sigma-Aldrich Ltd) or 10–50 μg/ml Fibronectin (FN).
4. Microscope cover slips: MatTeck glass-bottom culture plates (MatTek Corporation).
5. Agarose beads: Blue Sepharose 6 Fast Flow, particle size range 45–165 μm.
6. Predesigned imaging chambers: μ-slides, Integrated BioDiognostics, Germany.

3. Methods

3.1. SDF-1-Induced Trans-Migration

SDF-1 induced Trans-Migration is initially designed to evaluate the response and migratory capacity of cells toward a gradient of SDF-1. It can serve for both mature and immature hematopoietic cells. In the course of mobilization from the BM to the blood circulation and vice versa, HSPCs of both human and mouse are fully dependent on a delicate balance between SDF-1 concentration and its major receptor CXCR4. Augmented or impaired chemotactic response of HSPCs to SDF-1 can reflect several biological meanings, such as the surface expression levels of the receptor CXCR4 or its functionality. Further investigation of HSPCs response to SDF-1 can point at CXCR4 signaling transmission capacity. Hence, this assay can also be applied as a pharmacological tool for screening various compounds that can inhibit or enhance the chemotactic response of HSPCs to their major chemoattractant SDF-1.

SDF-1 induced trans-migration assay can also serve as a platform to understand extravasation processes, such as matrigel-barrier-breakage (a model for physiological ECM), by degrading enzymes, which are expressed by HSPCs. Coating the filter with the barrier of interest, such as endothelial cells, stromal cells, or a matrigel, has an implication in evaluating the ability of the cells to penetrate through an artificial barrier, which partially mimics physiological conditions.

Determination of migrated cells is quantitative and can be designed at several levels. Absolute numbers of migrated cells could be evaluated by counting Trypan Blue exclusion of dead cells and live cell counting under a microscope. A more accurate quantification could be achieved by using a flow cytometry machine (FACS) under fixed acquisition time of cell suspension obtained from each well. While using heterogeneous cell population, staining and evaluating the percentage of the subset of interest (e.g., Sca-1+/c-Kit+/Lin−, termed SKLs and defined as fraction enriched for HSPCs) before and after migration, characterizes a specific population. Alternatively, in vitro functional assays of HSPCs such as colony-forming units assay (CFU), can serve as additional readout to evaluate SDF-1-induced migratory capacity of HSPCs, by plating the same number of cells obtained from each transwell after migration and evaluating the number of colonies.

3.1.1. Cell and Transwell Preparation and Cell Loading

1. In order to evaluate trans-migration capacity via mechanical barrier, three different barriers are suggested: (a) For fibronectin coating, transwells are precoated by dropping 200 μl of Fibronectin (25 ng/ml) and kept in 4°C over night. (b) For matrigel coating, 50 μl of cold mixture contains either 60 μg (thick) or 12.5 μg (thin) of matrigel with full RPMI is gently

dropped on each filter and allowed to polymerize for 30 min in 37°C. (c) For BMEC-coating, 24 h before conducting the migration, filters are coated with 100,000–200,000 BMECs by dropping 100–200 μl of the cells suspension. Transwells are soaked in full DMEM for over night to allow cell growth. Upon reaching a confluence of ~80%, the BMEC-coated transwells are ready then for use.

2. Cells of interest (e.g., mBM-MNCs, mSKLs or hCD34+ cells) are collected by a relevant enrichment protocol. At this point any desired treatment of the cells is performed (if necessary, additional step of washing should be applied prior to the loading step). Before loading, the cells are counted and evaluated for viability by trypan blue staining.
3. Using Costar transwells, the lower chamber is filled with 600 μl of full RPMI supplemented with or with out 125 ng/ml SDF-1α (see Notes 2 and 3).
4. After completing the preliminary steps described above, 100–150 μl of medium containing 50,000–100,000 h CD34+/mSKLs or 200,000 mBM-MNCs is gently dropped on the center of the transwells (load the cells carefully in order to avoid perforation of the transwells!) Immediately after loading the cells, both coated or bare transwells are located in the wells previously filled with 600 μl of full RPMI supplemented with the desired concentration of SDF-1, and allowed to migrate at 37°C in a humidified atmosphere containing 5% CO_2 (see Note 4).
5. Migration duration should be individually considered according to the cell type and the performance of the migration assay. For coated filters between 3 and 5 h of migration is recommended. Bellow 3 h could be insufficient for enough cells to migrate, while exceeding 5 h can be resulted in entry to the "Plato phase" of the assay. With regard to "bare filters," between 2 and 3 h is sufficient. In addition, the level of cell's motility should also be taken into consideration. Some of the leukemic cell lines (G2, U937) are hyper motile, hence exhibit higher motility and 2 h of migration are sufficient.
6. In order to stop migration, transwells are gently removed to an empty well. Before collecting the media at the lower chamber, it is recommended to examine appearance of cells in the media, under a microscope, to assure successful migration.
7. Finally, 600 μl of cell suspension from each well are collected to FACS tubes ready to be evaluated for migration capacity.

3.1.2. Evaluating Migrated Cells by Flow Cytometry (FACS)

In order to evaluate the numbers of migrated cells, it is recommended to acquire cell numbers using a flow cytometry machine to achieve a better accuracy.

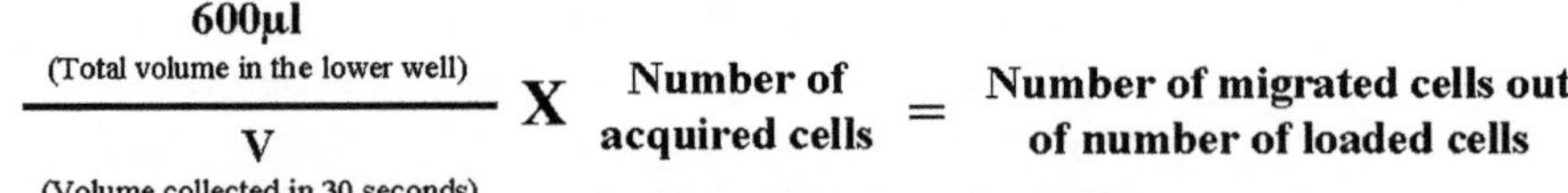

$$\frac{600\mu l \text{ (Total volume in the lower well)}}{V \text{ (Volume collected in 30 seconds)}} \times \text{Number of acquired cells} = \text{Number of migrated cells out of number of loaded cells}$$

Fig. 1. Formula for calculating the absolute number of migrated cells: The total volume filled in the lower well: 600 μl, is divided by the volume collected by the FACS machine in a certain time and rate of acquisition. The ratio is then multiplied by the number of acquired cells to give the total number of migrating cells. The total number of migrating cells can be then represented as "% of migrating cells," by calculating the percent of migrating cells out of the loaded cells.

1. Prior to acquisition, the volume collected by the machine per defined time (e.g., 30 s) and speed (e.g., high speed) should be measured by adding 1 ml of water before measurement followed by 30 s acquisition in high speed rate and re-measurement of the residual remains volume. This factor is called V.
2. Now, the cells are acquired by the FACS machine at a fixed time in a fixed collection rate.

   Numbers of cells are docwumented.
3. The following formula (Fig. 1) is used to calculate the absolute number of migrated cells.

3.1.3. Evaluating Migrated Colony-Forming Cells

It is possible to evaluate the capacity of migrating Colony-Forming Cells (CFCs) by plating equal volume of migrated cell suspension from the lower wells of the different treatments, in a semi-solid culture (16).

1. Equal volume, ranged between 200 and 400 μl of medium containing the cells from each well of different treatment are collected and plated in 1 ml volume of mixture containing: 533 μl of methylcellulose, 167 μl of full RPMI, 300 μl of FCS (30%), 50 ng/ml SCF, 5 ng/ml IL-3, 5 ng/ml GM-CSF, and 2 μg/ml Erythropoietin.
2. Mixture of methylcellulose and cells is gently vortexed for homogeneity and aseptically left in a room temperature to allow clearance of air bubbles.
3. The mixture is collected with a syringe and 1 ml is plated in 35 × 10 mm culture dish.
4. The cultures are incubated in 37°C in a humidified atmosphere containing 5% CO_2 and scored 7 days (murine colonies) or 14 days (human colonies) later by morphology criteria under a microscope.
5. Number of colonies can be compared between different treatments in order to differentially evaluate the chemotactic response of colony-forming cells to SDF-1.

3.2. Visualization of HSPCs Migration to a Gradient of SDF-1

For live imaging of SDF-1-induced migration, HSPCs are plated on glass cover slips (see Note 5) or introduced into chambers adapted for microscopical observations. To establish a time-stable (few hours) gradient of SDF-1, several approaches are employed. For example, application of a small number of SDF-1-soaked agarose beads in a confined area of the imaging chamber. SDF-1 released from the beads attracts cells with a similar efficiency to the recombinant protein (see Note 6). Predesigned imaging chambers with two reservoirs (one of each contains chemoattractant) connected by a narrow observation area are also widely applied to achieve gradient and image directional cell motility. Light microscopy methods (see Note 7) are used to perform real-time imaging of SDF-1-induced changes in cell morphology, extension, and retraction of cell protrusions, membrane ruffling, as well as velocity, directionality, and persistence of cell movement. Specific dynamic information on subcellular compartments is obtained by fluorescent tagging of organelles or molecules using vital dyes, tagged-protein expression or nonblocking Abs for cell-surface receptors. To study function of specific molecules, chemical inhibitors or neutralizing Abs can be applied either before or during the imaging. Highly motile cells are typically imaged for 10–30 min, though up to few hours observation might be performed under appropriate conditions (see Note 8). At the end of the experiment, cells can be fixed and subjected to immunocytochemical analysis.

1. Prepare coating solution, e.g., 50–150 μg/ml hyaluronan (HA) or 10–50 μg/ml fibronectin in phosphate-buffered saline without Ca2+ and Mg2+ (termed herein $PBS^{-/-}$).
2. Microscope cover slips are incubated for 18 h at 4°C with the coating solution and extensively washed in $PBS^{-/-}$. Keep up to 1 day in $PBS^{-/-}$ at 4°C.
3. Prepare SDF-1-soaked agarose beads by 1 h incubation with recombinant SDF-1, wash several times in excess volume of $PBS^{-/-}$.
4. Human $CD34^{+}$ cells in 25 mM HEPES-containing full RPMI are plated on cover slips (1×10^5 cells in 250 μl per 12-mm diameter) and are allowed to adhere for 1 h at 37°C in a tissue culture incubator.
5. Using a micropipettor, a limited number of SDF-1-soaked beads (5–10 particles per 12-mm-diameter cover slip) is added in a confined area.
6. Alternatively, use predesigned imaging chambers (μ-slides, Integrated BioDiognostics) and apply SDF-1 to one of the reservoirs (see Note 9).
7. Image the cells using inverted wide-field microscope by phase contrast or Nomarski interference contrast mode. This could be combined with epi-fluorescent imaging if a fluorescent tag is present.

3.3. Shear Flow Assays on Surface-Bound SDF-1

Bone marrow microvessels express selectins and integrin ligands like VCAM-1 and ICAM-1 (17). Endothelial cell monolayers assembled in flow chambers have been useful in assessing in vitro the ability of endothelial-presented SDF-1 to trigger and modulate adhesive interactions between human CD34+ progenitor cells and these integrin ligands (18). HSPCs can capture and roll on cytokine-activated endothelial cultures (e.g., HUVEC), but fail to spontaneously arrest on VCAM-1 or ICAM-1. However, in the presence of immobilized SDF-1 presented on the endothelial monolayer, HSPCs arrest on activated endothelial cells and develop high resistance to detachment by shear flow conditions that mimic the blood circulation in the BM vasculature. Furthermore, SDF-1 immobilized within substrates reconstituted with physiological densities of either VCAM-1 or ICAM-1, also trigger robust integrin-mediated HSPC adhesion. This approach has been useful in the dissection of rapid SDF-1 triggered integrin activation events on HSPCs without any contribution from additional endothelial signals. Using these systems, the role of Gi-protein signaling and downstream machineries in SDF-1 triggered integrin-mediated HSPC adhesion to endothelial integrin ligands can be readily analyzed (18, 19). SDF-1 has been shown to promote robust motility and transendothelial migration of T an B lymphocytes in response to SDF-1 signals presented either apically or sublumenally by a variety of endothelial monolayers (20–22). However, in our hands, HSPCs arrested on TNF-α stimulated HUVEC overlaid with SDF-1 failed to spread or crawl from their arrest site, and did not cross this endothelial barrier (Peled, Alon, unpublished). HSPC adhesion to and motility on and through freshly isolated BM endothelial cell monolayers must be therefore evaluated as an alternative to HUVEC monolayers. In addition, ectopic expression of additional promigratory molecules on HUVEC or dermal microvascular cells, such as CD44 ligands or SCF, should be considered. These systems can be assembled in flow chambers overlaid with SDF-1 and used to probe the ability of the chemokine to promote adhesive and transendothelial migratory capacities of HSPCs in the context physiological conditions of shear flow.

3.4. In Vivo Assays

3.4.1. Homing into the BM and Specific Recruitment to Other Organs

Homing of HSPC into the BM is a pre-requisite for successful engraftment and as a consequence for durable multilineage hematopoiesis. The process of tissue-specific chemoattraction is realized in vivo as multistep cell homing via the circulation into the organ, involving active directional migration, adhesion interactions, and transendothelial migration (5). Thus, assessing the homing ability of transplanted cells is a valuable tool to study the functional output that follows their migration in the blood (5). The basic principle of homing assays is injecting cells intravenously and rapidly detecting their presence in the organ of interest, several hours afterward. In order to track the cells, one has to either prelabel them (e.g., CFSA-DE) or distinguish donor-type infused cells from the resident recipient cells (e.g., antibodies that bind specifically to the donor cells, such as anti-hCD45+, recognizing human hematopoietic cells in transplanted immune-deficient mice (23)). The common screening tool is flow cytometry, although other means have been utilized (PCR and immunohistochemistry (23), Southern-Blot (1) and in vivo imaging (24)). The time period of assessing homing may start as soon as few minutes (25) but is mostly assessed at 3–24 h (23). Usually, nonspecific homing is diminished by time. However, the quality of assessing homing ability at longer time periods than 24 h may be overrepresented by cell division. The crucial role of SDF-1/CXCR4 in homing of repopulating human stem cells into the murine BM is demonstrated by blocking CXCR4 activity on transplanted cells, either by neutralizing antibodies (1) or by CXCR4 antagonists (8). The SDF-1/CXCR4 axis is involved in both human and mouse settings (5, 26), as well as human leukemic stem cells (27). Moreover, in stress-induced conditions or in the course of tissue repair, stem cells are specifically recruited via the blood circulation to the damaged organ (e.g., liver, ischemic sites undergoing angiogenesis and myocardial infarction (28–30)).These processes have also been shown to require the SDF-1/CXCR4 axis.

3.4.2. Engraftment and Repopulation Assays of Murine Cells

Stem cells are assayed functionally in engraftment and repopulation assays. Beside its role in chemotaxis and homing to the BM, SDF-1 serves as a survival and proliferation factor for stem cells, promoting retention and quiescence of HSPC at their niches (4, 9). Deleting CXCR4 thereby severely impairs hematopoietic reconstitution (4). As a rule of thumb, lethal, total body irradiation (TBI) of recipient mice is required for successful engraftment, evacuating niches for colonizing stem cells (31). Irradiation also increases the levels of SDF-1 within the BM, contributing to engraftment of human HSPC as well (3). Total BM cells, progenitor cells enriched populations or sorted stem cells are derived from the BM of mutant/treated mice/model of interest and then transplanted in recipient mice, either intravenously or by using an intrabone technique (5). Congeneic mice are commonly used to distinguish between donor-type and host-type cells. Assessing engraftment is divided into two categories: noncompetitive repopulation and competitive repopulation. In noncompetitive repopulation assays, the ability of stem cells to reconstitute the entire hematopoietic system is analyzed and therefore eliminating the host milieu is essential. Screening is done by monitoring survival of engrafted mice or donor chimerism in the BM or PB. Competitive repopulation assays allow transplanting low cell numbers, since the host-type cells are mixed with donor-type cells in order to guarantee mice survival. Assessing repopulation levels is done by comparing the donor-type chimerism, using specific antibodies. Sublethal irradiation is another app-roach to limit host-type hematopoiesis, without causing mortality. It should be noted that limiting dilutions are sometimes required in order to observe clear differences between the tested cells and their control counterparts by evaluating CRUs (competitive repopulation units – frequency of a long-term repopulating stem cell in the BM). Assessing short-term engraftment is done usually several weeks following transplantation; however, it does not discriminate activities of the more differentiated short-term stem cells vs. the most primitive long-term stem cells. Long-term engraftment is usually assessed at 4–6 months following transplantation. Nevertheless, the true function of stem cell is assessed in serial transplantation assays, in which BM cells are derived from the primary recipients and transplanted into secondary recipients, and so forth. This assay proves the self-renewal capacity of stem cells.

3.4.3. Engraftment and Repopulation Assays of Human Cells

Functional preclinical models of human hematopoiesis are based on engrafting immunodeficient mice that minimally reject human cells, such as the commonly used NOD/SCID model (32). Sublethally irradiated mice (350 cGy from a cesium source) are engrafted with enriched human CD34+ progenitors or the more primitive CD34+CD38−/low stem cells derived from umbilical CB, BM aspirates, or mobilized PB, and engraftment is assessed several weeks following transplantation. Xenotransplantation of fetal liver and thymus under the kidney capsule of immunodeficient mice is another approach to establish competent humanized models, in order to study engraftment of transplanted human HSPC to a human microenvironment (33). Blocking SDF-1/CXCR4 functions leads to abrogated homing and engraftment, demonstrating the major role of this axis in the regulation of human CD34+ progenitor cells (1). The same model is applicable for testing human leukemic stem cell homing and engraftment (34), in which SDF-1/CXCR4 plays a significant role as well (27). Leukemic cells may be derived from a cell line or patient's primary cells.

3.4.4. Hematopoietic Stem Cell Mobilization

Low levels of HSPCs normally circulate at steady state; however, in stress situations or upon administration of mobilizing agents, cell egress and recruitment from the BM is immensely augmented (6, 35). Despite the opposite direction, in which HSPCs migrate, mobilization is not a mirror-image process to homing. Broad proteolytic activity as well as involvement of various microenvironmental players release HSCs from their adhesion to their niches and direct them out to the circulation (36–38). Disrupting SDF-1/CXCR4 interactions leads to loss of retention and as a result to mobilization of HSPCs (8). In various induced mobilization models, functional SDF-1 is either downregulated, cleaved, or released into the blood, demonstrating its importance (39, 40). Using CXCR4 antagonists, such as AMD3100 (8), cause mobilization, which can be screened by measuring the percentages of primitive populations in the peripheral blood by means of flow cytometry using cell markers specific for HSPCs or by colony and transplantation assays. Mobilization can be induced in chimeric mice pre-engrafted with donor cells, in order to compare mobilization of mutant cells to their wild-type counterparts, or in the case of humanized mice, the specific mobilization of human cells in the murine host (6).

4. Notes

1. Thick or thin layers of Matrigel-coated filters are considered according to the cell type and the assay performed. In case of evaluation activity of membranal proteolytic enzymes, it is recommended to use thick layer as an optimal challenge. Accordingly, the time of migration should be considered and usually takes place between 4 and 5 h.
2. SDF-1 concentration can be modified and reduced in order to evaluate migration of cells highly sensitive to SDF-1 (41).
3. Control of migration toward SDF-1-free-media could reflect spontaneous migration of the cells, and should be considered.
4. Appearance of bubbles in the interface between the media and the bottom of the transwell should be examined. If so, remove the transwell from the well, discard the bubble and relocate the transwell.
5. If required, glass cover slips are precoated with extracellular matrix components or other proteins (e.g., chemokine or adhesion receptor). To improve the coating efficiency, a thin layer of poly-l-lysine can be applied on a glass. However, this might affect cell motility. Finally, various three-dimensional (3D) matrices (e.g., matrigel) are used, as described above.
6. A stable chemokine gradient is established 5–10 min following introduction of the beads. Chemotactic potential of SDF-1 released from the beads could be analyzed in the transwell migration assay in comparison to recombinant SDF-1.
7. Wide-field microscopy in either phase contrast or Nomarski Interference Contrast modes. Images are collected by CCD cameras as single sections or as Z stacks. These time series are then analyzed, and involve image segmentation and reconstruction of 3D surfaces, followed by image analysis to recognize particular motility features and statistical analysis.

8. In the absence of environmental chambers to maintain conditions optimal for cell migration (i.e., temperature of 37°C, 5% CO_2, and humidity), heated objectives and HEPES-containing imaging media are frequently used.
9. Amount of SDF-1 and volume should be calibrated for a specific experiment.

References

1. Peled, A., Petit, I., Kollet, O., et al. (1999) Dependence of human stem cell engraftment and repopulation of NOD/SCID mice on CXCR4 *Science* **283**, 845–8.
2. Wright, D.E., Bowman, E.P., Wagers, A.J., Butcher, E.C., and Weissman, I.L. (2002) Hematopoietic stem cells are uniquely selective in their migratory response to chemokines *J Exp Med* **195**, 1145–54.
3. Ponomaryov, T., Peled, A., Petit, I., et al. (2000) Induction of the chemokine stromal-derived factor-1 following DNA damage improves human stem cell function *J Clin Invest* **106**, 1331–9.
4. Sugiyama, T., Kohara, H., Noda, M., and Nagasawa, T. (2006) Maintenance of the hematopoietic stem cell pool by CXCL12-CXCR4 chemokine signaling in bone marrow stromal cell niche *Immunity* **25**, 977–88.
5. Lapidot, T., Dar, A., and Kollet, O. (2005) How do stem cells find their way home? *Blood* **106**, 1901–10.
6. Lapidot, T., and Petit, I. (2002) Current understanding of stem cell mobilization: the roles of chemokines, proteolytic enzymes, adhesion molecules, cytokines, and stromal cells *Exp Hematol* **30**, 973–81.
7. Mendez-Ferrer, S., Lucas, D., Battista, M., and Frenette, P.S. (2008) Haematopoietic stem cell release is regulated by circadian oscillations *Nature* **452**, 442–7.
8. Broxmeyer, H.E., Orschell ,C.M., Clapp, D.W., et al. (2005) Rapid mobilization of murine and human hematopoietic stem and progenitor cells with AMD3100, a CXCR4 antagonist *J Exp Med* **201**, 1307–18.
9. Nie, Y., Han, Y.C., and Zou, Y.R. (2008) CXCR4 is required for the quiescence of primitive hematopoietic cells *J Exp Med* **205**, 777–83.
10. Voermans, C., Kooi, M.L., Rodenhuis, S., van der Lelie, H., van der Schoot, C.E., and Gerritsen, W.R. (2001) In vitro migratory capacity of CD34+ cells is related to hematopoietic recovery after autologous stem cell transplantation *Blood* **97**, 799–804.
11. Petit, I., Goichberg, P., and Spiegel, A., et al. (2005) Atypical PKC-zeta regulates SDF-1-mediated migration and development of human CD34+ progenitor cells *J Clin Invest* **115**, 168–76.
12. Shivtiel, S., Kollet, O., Lapid, K., et al. (2008) CD45 regulates retention, motility, and numbers of hematopoietic progenitors, and affects osteoclast remodeling of metaphyseal trabecules *J Exp Med* **205**, 2381–95.
13. Kawai, T., Choi, U., Whiting-Theobald, N.L., et al. (2005) Enhanced function with decreased internalization of carboxy-terminus truncated CXCR4 responsible for WHIM syndrome *Exp Hematol* **33**, 460–8.
14. Kawai, T., Choi, U., Cardwell, L., et al. (2007) WHIM syndrome myelokathexis reproduced in the NOD/SCID mouse xenotransplant model engrafted with healthy human stem cells transduced with C-terminus-truncated CXCR4 *Blood* **109**, 78–84.
15. Ngo, H.T., Leleu, X., Lee, J., et al. (2008) SDF-1/CXCR4 and VLA-4 interaction regulates homing in Waldenstrom macroglobulinemia *Blood* **112**, 150–8.
16. Metcalf, D. (1977) Hemopoietic colonies: in vitro cloning of normal and leukemic cells. *Recent Results Cancer Res* (**61**):Title page, 1–227.
17. Wagner, D.D., and Frenette, P.S.. (2008) The vessel wall and its interactions *Blood* **111**, 5271–81.
18. Peled, A., Grabovsky, V., Habler, L., et al. (1999) The chemokine SDF-1 stimulates integrin-mediated arrest of CD34(+) cells on vascular endothelium under shear flow *J Clin Invest* **104**, 1199–211.
19. Hartmann, T.N., Grabovsky, V., Pasvolsky, R., et al. (2008) A crosstalk between intracellular CXCR7 and CXCR4 involved in rapid CXCL12-triggered integrin activation but not in chemokine-triggered motility of human T lymphocytes and CD34+ cells *J Leukoc Biol* **84**, 1130–40.
20. Cinamon, G., Shinder, V., and Alon, R. (2001) Shear forces promote lymphocyte migration

across vascular endothelium bearing apical chemokines *Nat Immunol* **2**, 515–22.

21. Shulman, Z., Pasvolsky, R., Woolf, E., et al. (2006) DOCK2 regulates chemokine-triggered lateral lymphocyte motility but not transendothelial migration *Blood* **108**, 2150–8.
22. Schreiber, T.H., Shinder, V., Cain, D.W., Alon, R., and Sackstein, R. (2007) Shear flow-dependent integration of apical and subendothelial chemokines in T-cell transmigration: implications for locomotion and the multistep paradigm *Blood* **109**, 1381–6.
23. Vagima, Y., Avigdor, A., Goichberg, P., et al. (2009) MT1-MMP and RECK are involved in human CD34+ progenitor cell retention, egress, and mobilization *J Clin Invest* **119**, 492–503
24. Kalchenko, V., Shivtiel, S., Malina, V., et al. (2006) Use of lipophilic near-infrared dye in whole-body optical imaging of hematopoietic cell homing *J Biomed Opt* **11**, 050507.
25. Wright, D.E., Wagers, A.J., Gulati, A.P., Johnson, F.L., and Weissman, I.L. (2001) Physiological migration of hematopoietic stem and progenitor cells *Science* **294**, 1933–6.
26. Christopherson, K.W., 2nd, Hango, C.G., Mantel, C.R., and Broxmeyer, H.E. (2004) Modulation of hematopoietic stem cell homing and engraftment by CD26 *Science* **305**, 1000–3.
27. Tavor, S., Petit, I., Porozov, S., et al. (2004) CXCR4 regulates migration and development of human acute myelogenous leukemia stem cells in transplanted NOD/SCID mice *Cancer Res* **64**, 2817–24.
28. Kollet, O., Shivtiel, S., Chen, Y.Q., et al. (2003) HGF, SDF-1, and MMP-9 are involved in stress-induced human CD34+ stem cell recruitment to the liver *J Clin Invest* **112**, 160–9.
29. Jin, D.K., Shido, K., Kopp, H.G., et al. (2006) Cytokine-mediated deployment of SDF-1 induces revascularization through recruitment of CXCR4+ hemangiocytes *Nat Med* **12**, 557–67.
30. Wojakowski, W., Tendera, M., Michalowska, A., et al. (2004) Mobilization of CD34/CXCR4+, CD34/CD117+, c-met+ stem cells, and mononuclear cells expressing early cardiac, muscle, and endothelial markers into peripheral blood in patients with acute myocardial infarction *Circulation* **110**, 3213–20.
31. Weiss, L., Bullorsky, E., Ashkenazi, and Y.J., Slavin S. (1988) Optimal time interval between myeloablative whole body irradiation and reconstitution with syngeneic bone marrow graft *Bone Marrow Transplant* **3**, 207–10.
32. Meyerrose, T.E., Herrbrich, P., Hess, D.A., and Nolta, J.A.. (2003) Immune-deficient mouse models for analysis of human stem cells *Biotechniques* **35**, 1262–72.
33. Wege, A.K., Melkus, M.W., Denton, P.W., Estes, J.D., and Garcia, J.V. (2008) Functional and phenotypic characterization of the humanized BLT mouse model *Curr Top Microbiol Immunol* **324,** 149–65.
34. Dick, J.E. (2008) Stem cell concepts renew cancer researc. *Blood* **112**, 4793–807.
35. Pelus, L.M. (2008) Peripheral blood stem cell mobilization: new regimens, new cells, where do we stand *Curr Opin Hematol* **15**, 285–92.
36. Velders, G.A., and Fibbe, W.E. (2005) Involvement of proteases in cytokine-induced hematopoietic stem cell mobilization *Ann N Y Acad Sci* **1044,** 60–9.
37. Kollet, O., Dar, A., and Lapidot, T. (2007) The multiple roles of osteoclasts in host defense: bone remodeling and hematopoietic stem cell mobilization *Annu Rev Immunol* **25,** 51–69.
38. Spiegel, A., Kalinkovich, A., Shivtiel, S., Kollet, O., and Lapidot, T. (2008) Stem cell regulation via dynamic interactions of the nervous and immune systems with the microenvironment *Cell Stem Cell* **3**, 484–92.
39. Dar, A., Kalinkovich, A., Netzer, N., et al. (2006) AMD3100 Signals Via the Nervous System, Inducing Release to the Circulation of Bone Marrow SDF-1, Which Is Crucial for Progenitor Cell Mobilization *ASH Annual Meeting Abstracts* **108**, 1315.
40. Petit, I., Szyper-Kravitz, M., Nagler, A., et al. (2002) G-CSF induces stem cell mobilization by decreasing bone marrow SDF-1 and upregulating CXCR4 *Nat Immunol* **3**, 687–94.
41. Spiegel, A., Kollet, O., Peled, A., et al. (2004) Unique SDF-1-induced activation of human precursor-B ALL cells as a result of altered CXCR4 expression and signaling *Blood* **103**, 2900–7.

Chapter 20

The Role of Receptor Tyrosine Kinases in Primordial Germ Cell Migration

Louise Silver-Morse and Willis X. Li

Abstract

During embryonic development in *Drosophila*, rodents, and other organisms, primordial germ cells (PGCs) migrate from their points of origin to the nascent gonads, where they give rise to germ line stem cells. Receptor tyrosine kinase (RTK) activity is required for normal migration of primordial germ cells in both *Drosophila* and rodents. In this chapter, we discuss in vivo as well as in vitro methods which have been used to elucidate the role of the RTK Torso in *Drosophila* germ cell migration. Included are protocols for embryo collection, fixation, and immunostaining; the dominant female sterile technique; in vitro culture and observation of PGCs; pole cell transplantation; and labeling of pole cells for in vivo observation.

Key words: Primordial germ cell migration, Receptor tyrosine kinases, *Drosophila*, Embryo fixation, Immunostaining, Pole cell transplantation

1. Introduction

During embryonic development in *Drosophila*, rodents, and other organisms, primordial germ cells migrate from their points of origin to the nascent gonads, where they give rise to germ line stem cells (1). These stem cells, in turn, move out of their niches in the gonads to differentiate into gametes (2, 3). In both *Drosophila* and rodent embryos the primordial germ cells travel a complex route to the gonadal tissue, and require receptor tyrosine kinase (RTK) activity for normal migration (1, 4, 5).

RTKs comprise a class of cell surface receptors with intracellular kinase domains (6). Upon ligand binding to the receptor, the kinase domain phosphorylates tyrosine residues, activating an intracellular signaling pathway or network which effects changes in gene expression. RTKs include PDGFR, VEGFR, EGFR, Torso, and c-kit; the Torso RTK is required for proper migration

Marie-Dominique Filippi and Hartmut Geiger (eds.), *Stem Cell Migration: Methods and Protocols*, Methods in Molecular Biology, vol. 750, DOI 10.1007/978-1-61779-145-1_20, © Springer Science+Business Media, LLC 2011

of primordial germ cells in *Drosophila* (4), while the c-kit RTK is required for primordial germ cell migration in rodents (5). Both Torso and c-kit have an intracellular kinase domain similar in structure to that of PDGFR; the kinase domain is split by a stretch of approximately 100 hydrophobic amino acids (7). Moreover, Torso, c-kit, and PDGFR are able to activate STAT molecules (4, 8–10) as well as the Ras/MAPK/ERK cascade (11, 12); and reviewed by (7, 13, 14). The similarities in structure and function of *Drosophila* Torso and mouse c-kit suggest conservation of a molecular mechanism in primordial germ cell migration.

Primordial germ cells (PGCs) originate at the posterior pole of the *Drosophila* embryo, where they are referred to as "pole cells." The early embryo is a syncytium of synchronously dividing nuclei. The nuclei which will give rise to PGCs, associate with maternally derived germ plasm at the embryo's posterior pole; reviewed by (15, 16). Cells form as nuclei are surrounded by cell membranes, with the pole cells being the first to form. PGCs divide about two to three times before gastrulation. During gastrulation, the PGCs move into the interior of the embryo in conjunction with the posterior midgut primordium, arriving inside of the posterior midgut epithelium. They then migrate through the epithelium and away from the midgut, toward the adjacent mesoderm, where they associate with somatic gonadal precursor cells.

In this chapter, we will focus on in vivo as well as in vitro approaches that our lab has taken to elucidate the role of Torso in *Drosophila* germ cell migration. We will present protocols for embryo collection, fixation, and immunostaining; the dominant female sterile technique; in vitro culture and observation of PGCs; pole cell transplantation; and labeling of pole cells for in vivo observation. *Drosophila melanogaster* is a model organism which readily lends itself to in vivo experimentation. For more than 100 years, biologists have used genetics to study *Drosophila*. At first, a number of spontaneous mutations were identified. Later, mutagenic chemicals and X-rays were used to induce new mutations. With the advent of molecular biology, genetic systems were devised enabling researchers to upregulate or downregulate a specific gene product in a tissue of choice and at a chosen time in development.

2. Materials

2.1. Fly Strains and Genetics

For an introduction to *Drosophila* genetics and molecular biology, the *Drosophila* life cycle, the care of *Drosophila*, and the basic equipment required for experimenting with *Drosophila*, we recommend *Drosophila* Methods and Protocols, published in 2008 as part of the Methods in Molecular Biology series (17). For extensive experimental protocols we recommend *Drosophila* Protocols (18).

We used flies carrying the following alleles, which are either null alleles (i.e., the mutant gene shows no activity) or strong alleles (i.e., the mutant gene shows a strong mutant phenotype).

1. tor^{XR1} – a null allele of the receptor tyrosine kinase *torso*, in which much of the gene is deleted (19); the allele is a revertant of a recessive gain-of-function mutation, tor^{RL3}, induced by ethylmethane sulfonate (EMS) (20).
2. $Ras^{\Delta C40B}$ – a deletion of the intracellular protein kinase *Ras* gene, which is normally activated by *torso* signaling (21).
3. $Draf^{11\text{-}29}$ – a null allele of the intracellular protein kinase *raf*, due to an insertion of exogenous DNA into the gene (22).
4. $stat92E^{6346}$ (also known as mrl^{6346}) – a null allele of the *stat* gene, due to a p-element insertion (23).
5. hop^{C111} – a strong loss-of-function allele of the *Drosophila* intracellular Janus kinase (JAK), due to an X-ray-induced deletion of about 300 bp of the gene (24).
6. upd^{YM55} – a strong loss-of-function allele due to a nonsense mutation in the *unpaired* gene, which encodes a cytokine-like ligand that activates the JAK/STAT pathway in *Drosophila* (25).

We also used the following gain-of-function alleles (the mutant gene shows more activity than the wild-type gene).

7. tor^{Y9} – an EMS-induced, dominant gain-of-function *tor* allele (20).
8. tor^{RL3} – an EMS-induced, recessive gain-of-function *tor* allele (20).
9. hop^{TumL} – a temperature-sensitive, gain-of-function *hopscotch* allele, the result of a point mutation (26).

To learn whether *torso* was involved in PGC migration, we set up matings which resulted in embryos lacking the *torso* gene product. tor^{XR1}/tor^{XR1} females were mated to wild-type males. We then fixed and immunostained the embryos with anti-Vasa and other antibodies. Anti-Vasa antibodies specifically stain the PGCs (Fig. 1).

2.2. Embryo Collection and Fixation

1. Petri dishes containing apple juice agar. For approximately 100 60-mm apple juice agar plates: 650 ml of distilled water, 30 g of agar, 285 ml of apple juice, 40 ml (~56 g) of molasses, and (optional) 23 ml of 10% Nipagin fungicide (in EtOH). Mix agar in water (tap water is fine) in a 2 L flask and autoclave on a liquid cycle. Let the temperature cool down to ~80°C (so as not to break down Nipagin). Add apple juice, Nipagin (10% in ethanol) and molasses. Mix well slowly and avoid making

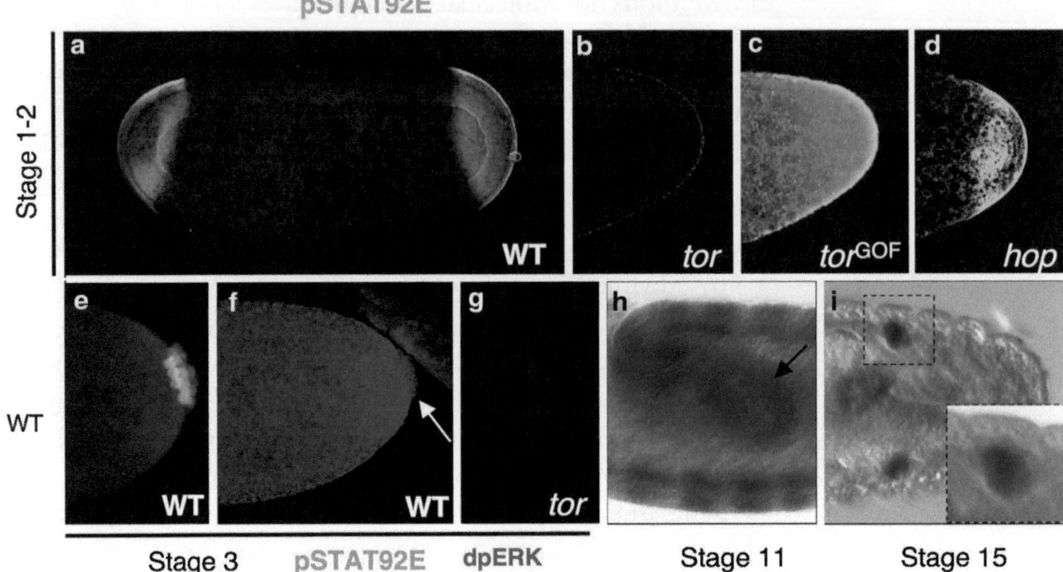

Fig. 1. STAT92E activation by Tor and, in germ cells, detection of STAT92E activation (**a–e**, **h**, and **i**) by anti-pSTAT92E antibody staining ((**a–e**); (**h** and **i**),) in embryos of indicated stages. Activation of the Ras-MAPK pathway was detected by an anti-dpERK antibody (**f** and **g**). (**a**) STAT92E activation is detected in the two polar regions in early stage wild-type embryos. This pattern is reminiscent of Tor activation. (**b**) pSTAT92E staining is absent in embryos from tor homozygous females. (**c**) torGOF embryos exhibit higher levels and broader domains of pSTAT92E staining. (**d**) Wild-type pattern of pSTAT92E staining persists in hop^{mat-} embryos. (**e**) Detection of pSTAT92E in pole cells of early embryos. (**f** and **g**) ERK/MAPK activation was detected in the posterior region and pole cells (*arrow*) in wild-type embryos (**f**). Such staining was absent in tor mutant embryos (**g**). (**h**) STAT92E activation persists in pole cells (*arrow*) once inside the gut pocket, as well as in the gut epithelium and parasegments. (**i**) In late embryos, pSTAT92E is strongly expressed in the gonad. (From (4) with permission).

any bubbles (use stir bar but do not shake). Pour a portion into a 0.5 L or 1 L plastic measuring beaker for easier pouring. Pour an appropriate amount into each plate. Cover the plates with mosquito netting, and allow them to cool at room temperature overnight. The next day, cover the plates with their plastic covers and put them, upside down, into plastic sleeves. Tape, date, and store at 4°C.

2. Yeast paste – dry yeast mixed with distilled H_2O to the consistency of a thick paste.
3. Plastic beaker with air holes. Air holes can be made with a fine needle. They must be small enough to prevent flies from escaping.
4. Rubber bands.
5. PEM – Prepare solution containing 0.1 M PIPES, 2 mM EGTA, 1 mM $MgSO_4$. In order to dissolve the ingredients, titrate the solution to pH 6.95 with NaOH.

6. Paraformaldehyde powder dissolved in PEM: Prepare a 4% formaldehyde stock solution by dissolving 0.4 g of paraformaldehyde in 10 ml of PEM in a 15 ml polypropylene screw-cap tube. Seal the tube tightly with Saran wrap or parafilm, since formaldehyde vapor is harmful. Heat the tube at 80°C for 5 min (a nucleic acid hybridization oven can be used) then vortex; repeat the heating and vortexing until the paraformaldehyde is completely dissolved and the solution is clear. Dissolving the paraformaldehyde in this fashion should prevent its inactivation.
7. 10× PBS (phosphate-buffered saline). Dissolve 80 g of NaCl (1.37 M), 2 g of KCl (27 mM), 14.4 g of Na_2HPO_4 (100 mM), 2.4 g of KH_2PO_4 (20 mM) in 1 L of distilled H_2O. Do not adjust the pH.
8. Saran wrap or parafilm.
9. Triton X-100 detergent.
10. Squirt bottle.
11. Small paintbrush.
12. Chlorine bleach (laundry bleach).
13. Embryo strainer made of strong nylon mesh with 100 μM pores (BD Falco).
14. Heptane.
15. Methanol.
16. Pasteur pipettes.

2.3. Immunostaining of Fixed Embryos

1. Methanol.
2. PBS-T – Prepare PBS containing 0.1% Tween 20 or Triton X-100.
3. PBS-BT – Prepare PBS containing 0.1% Tween 20 or Triton X-100 and 0.5% bovine serum albumin (BSA).
4. Normal goat serum.
5. Mowiol (27): To prepare mowiol, add 2.4 g of MOWIOL 4–88 to 6 g (~5 mL) of glycerol in a 50-ml centrifuge tube while stirring with a stir bar. To avoid lumps, mix the MOWIOL and glycerol well before adding water. Add 6 ml of distilled H_2O and continue stirring. Then, add 12 ml of 0.2 M Tris, pH 8.5 to the mixture, and stir briefly. Heat in a hybridization oven set at 53°C until the MOWIOL is completely dissolved (this takes hours and can be done overnight). Centrifuge for 20 min at 4,000–5,000 rpm. Add 0.7 g of Dabco to the supernatant to obtain a 2.5% solution. Aliquot 500 μL of the supernatant into 1.5 ml Eppendorf tubes and store at –20°C or –80°C, where the aliquots will be stable for 12 months. Once defrosted, the aliquots are stable at room temperature for 1 month.

6. Glass slides, coverslips, nail polish.
7. Fine forceps.

2.4. In Vitro Culture and Observation of PGCs

1. Fixed embryos (refer to Embryo Collection and Fixation)
2. *Drosophila* Schneider's medium (BRL/Gibco) supplemented with L-glutamine, 10% bovine serum, and 30 μg/ml penicillin/streptomycin
3. Laminin-coated culture dishes

2.5. Pole Cell Transplantation

1. Fixed embryos (refer to Embryo Collection and Fixation)
2. Bibulous paper
3. Silica gel desiccation chamber
4. Double-sided sticky tape (see Note 1)
5. 24 × 60 mM glass coverslips
6. Glass microscope slides
7. Fine forceps
8. Halocarbon oil 400
9. Glass capillaries (Frederick Haer and Co)
10. Siliconized glass micropipettes, with a 45° opening at the tip (see Notes 2 and 3)
11. Syringe for filling micropipettes and dispensing their contents (see Note 3)
12. Micromanipulator (see Note 3)
13. Moist chamber
14. Food vials
15. Fluorescein-dextran–Mr 70,000 (Molecular Probes)

3. Methods

3.1. Embryo Collection and Fixation

The *Drosophila* embryo is encased in a shell which consists of an outer chorion and an inner vitelline membrane (Fig. 2a). The fixation procedure involves removal of this shell.

1. Spread a dime-size glob of yeast paste on each apple juice agar dish (one per fly stock). Anesthetize the flies using CO_2 and dump them into a plastic beaker with air holes. Quickly cover them with the apple plate. Make sure that there is a tight fit between the beaker and plate. Use a rubber band to secure the beaker to the plate, and wait until the flies have awakened before turning the contraption right side up (Fig. 3a).
2. Put the contraption at the desired temperature for egg collection.

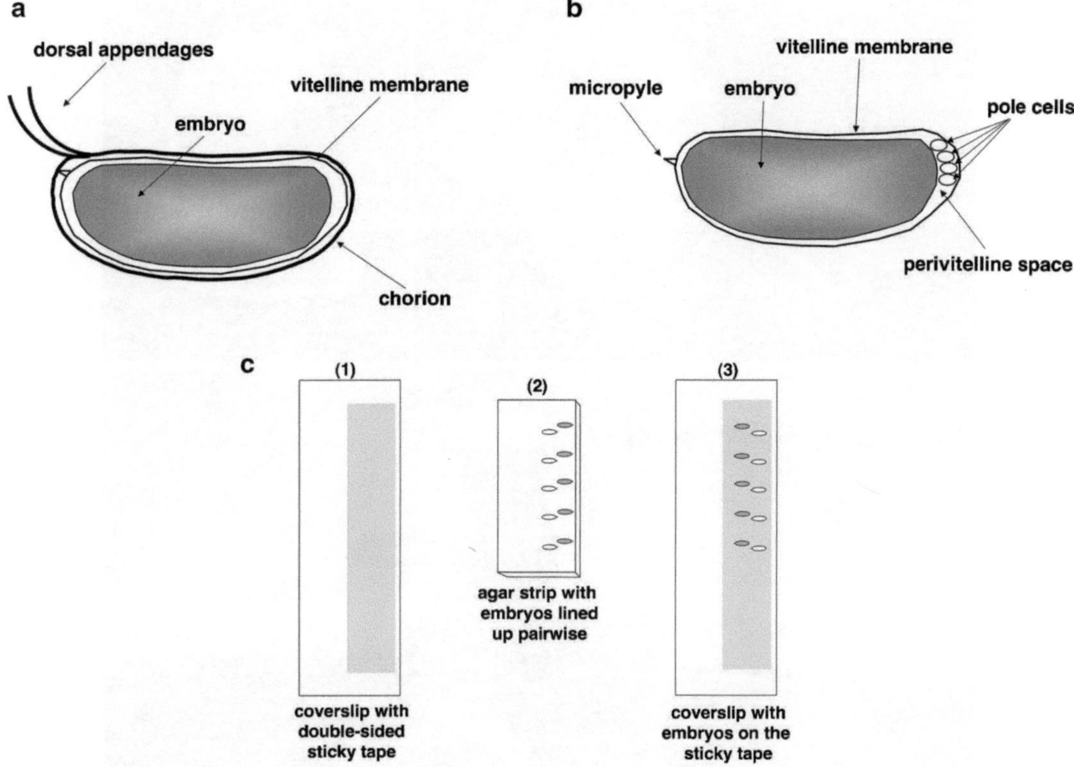

Fig. 2. The Drosophila embryo. (**a**) The shell of the *Drosophila* embryo consists of the outermost chorion and the inner vitelline membrane. The dorsal appendages are found near the anterior end (to the left). The dorsal aspect is upward. (**b**) A dechorionated stage 4 embryo showing pole cells at the posterior end and the micropyle at the anterior end. The dorsal aspect is upward. (**c**) Lining up of embryos for pole cell transplantation. (1) A strip of double-sided sticky tape was put onto a 24 × 60 mM coverslip. (2) Donor and recipient embryos were lined up pairwise on an agar strip. Recipient embryos are to the left and donor embryos are to the right. The anterior ends of all of the embryos are to the right. (3) The coverslip with adherent embryos is shown embryo side up. The coverslip had been gently put on top of the embryos on the agar strip, sticky tape side down, to allow the embryos to adhere. The posterior ends of the embryos are now to the right.

3. When the collection is complete, anesthetize the flies and replace the plate that you want to study with a new plate.
4. Prepare the 4% formaldehyde stock solution half an hour prior to embryo dechorionation.
5. Loosen the embryos from the agar on which they have been deposited, using a small volume of 0.1% triton X-100 detergent/water in a squirt bottle, and with a small paintbrush, transfer the embryos to an embryo strainer. Rinse the embryos in 0.1% triton X-100 detergent/water. Then rinse them under a gentle stream of water for 1 min.
6. Dechorionate the embryos in a 1:1 mixture of chlorine bleach and water, gently moving the embryo strainer back and forth for approximately 2 min, until the dorsal appendages (Fig. 2a) disappear and the embryos have a shiny appearance. Rinse the embryos in the strainer with plenty of water for at least 3 min.

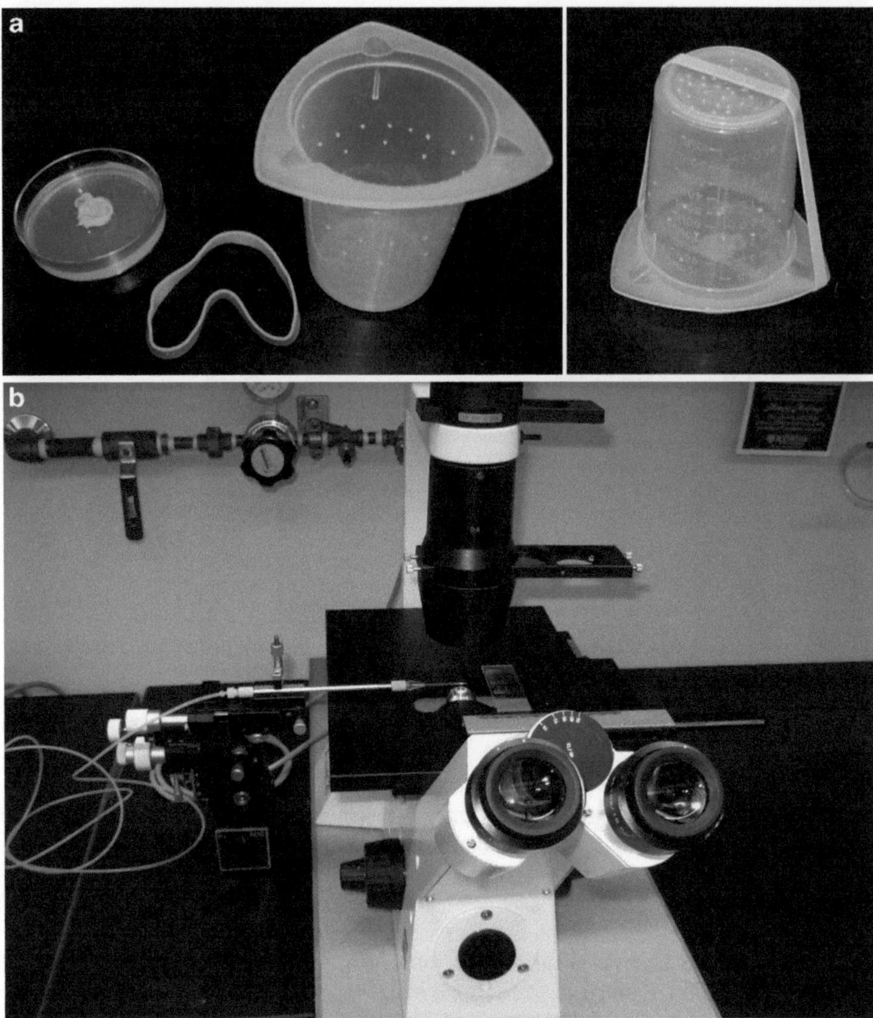

Fig. 3. Equipment used to collect embryos and to microinject them. (**a**) For embryo collection, male and female flies are anesthetized and put into a plastic beaker with tiny air holes. The beaker is covered with a petri dish (containing apple juice agar and a bit of yeast paste), which fits snugly onto the top of the beaker. A rubber band is put around the beaker and dish to secure them to each other. When the flies have become active again, the contraption is turned beaker side up. (**b**) A microinjection setup. A micromanipulator is holding the needle; the coverslip with sticky tape is on a slide on the microscope stage.

7. Transfer the embryos to a 7 ml glass scintillation vial containing a two-phase mixture of 3 ml of heptane and 3 ml of 4% paraformaldehyde (freshly made). Shake at 200 rpm for 20 min on a platform shaker. The fixation time may have to be varied depending upon the target tissue and the particular antibody being used.
8. Use a Pasteur pipette to completely remove the lower aqueous phase. Add more heptane until the vial is nearly half full. Add an equal volume of COLD methanol (kept at −20°C). Shake vigorously until the majority of the embryos sink to the

bottom (about 30 s). Carefully remove about 75% of the heptane and methanol and replace with 5 ml of methanol. Shake once more. All or most of the embryos should now have sunk to the bottom of the vial. The vitelline membrane is removed at this step. Remove all the liquid and then rinse three times with 5 ml methanol. Embryos can be stored in methanol at −20°C for several months.

3.2. Immunostaining of Fixed Embryos

1. Gently transfer fixed embryos to a 500 μL eppendorf tube, using a Pasteur pipette. Approximately 50 μL of embryos works well.
2. To rehydrate the embryos, wash them for 5 min with methanol:PBS-T (1:1), and for 10 min, three times, with PBS-T, on a nutating shaker.
3. To prevent nonspecific sticking of antibodies to the embryos, wash them for 10 min in PBS-BT+1% NGS (blocking solution).
4. Incubate the embryos in primary antibody (diluted in PBS-BT) at 4°C overnight, or for 2 h at room temperature, on a nutating shaker. Various dilutions of antibody may have to be tried to determine which gives the best results.
5. Wash for 5 min, five times, with PBS-T. If a secondary antibody is not required for visualization, go to step 9.
6. If a secondary antibody is required, wash once for 5 min in PBS-BT.
7. Incubate with the secondary antibody at 4°C overnight, or for 2 h at room temperature, on a nutating shaker.
8. Wash for 5 min, five times, with PBS-T.
9. To mount embryos, gently transfer them to a glass slide and remove excess liquid with a Pasteur pipette; use fine forceps to arrange the embryos for analysis; blotting paper can be used with care to remove liquid outside of the field of embryos. To avoid dehydration of the embryos, quickly add Mowiol (15 μL for a 60×60 mm coverslip) with a micropipetter, avoiding air bubbles. Carefully lower a glass coverslip onto the embryos with fine forceps, again avoiding air bubbles, and seal the edges of the coverslip with nail polish.

3.3. The Dominant Female Sterile Technique

To determine whether Torso acts through the Ras kinase pathway or through STAT92E, or both, in promoting PGC migration, another genetic approach was employed, the dominant female sterile technique (28). A different approach was required because homozygous mutants of ras and STAT92E are not viable, so it was not possible to use homozygous mutant females to generate embryos with no maternal contribution of these gene products. The dominant female sterile technique employs the mutation ovo^{D1}.

Females carrying even one copy of this mutation do not lay eggs. If males carrying this mutation are mated to normal females, the female progeny that receive the *ovo*^{D1} mutation will be sterile (Fig. 4a), unless a crossing-over event occurs in the DNA that eliminates *ovo*^{D1} from some of the female germ cells. If these females carry a copy of a recessive mutation on a chromosome homologous to the one carrying *ovo*^{D1}, a crossing-over event may result not only in the loss of *ovo*^{D1} but also in the generation of progeny that lack the maternal contribution of the mutant gene product. Normally, this crossing-over occurs at levels too low to be experimentally useful. X-irradiation will increase the levels of crossing-over somewhat, but they are still very low. However, when the ovo^{D1} mutation is combined with the FLP-FRT system, a useful number of maternally mutant progeny are generated.

The yeast FLP-FRT site-specific recombination system consists of the FLP yeast recombinase and FLP recombination target sequences (FRTs), with FLP effecting recombination between chromosomes that carry FRT sequences. Using plasmids, the DNA sequence encoding FLP, as well as FRT sequences, have been transferred into the Drosophila genome. The FLP recombinase has been put under the control of a heat-shock promoter and can be activated by elevating the ambient temperature to 37°C at a desired time in development.

We generated *stat92E*$^{mat-}$ embryos (embryos lacking the maternally contributed *stat92E* gene product) as follows. Since the stat92E gene is found on the third chromosome, we used females in which one of the third chromosomes carried the *stat92E*^{6346} allele as well as an FRT sequence on the same chromosome arm, at position 82B; *stat92E*^{6346} and FRT^{82B} were gotten onto the same chromosome via recombination. The other third chromosome was a "balancer" (TM3), in which the order of the genes is scrambled, thus preventing spontaneous crossing-over of homologous chromosomes. We mated these females to males carrying an FRT sequence at position 82B, and the *ovo*^{D1} allele on the same arm, of one of their third chromosomes; the other third chromosome was the same balancer employed in the females; the males also carried heat-shock Flp on their X-chromosome. The third chromosome balancer contains a mutation which shows a recognizable phenotype (short bristles) in adult flies (Fig. 4b). Third instar larvae resulting from this mating were subjected to heat shock in a 37°C water bath for 2 h daily (see Note 4). When the progeny reached adulthood, females carrying both the *ovo*^{D1} and *stat92E*^{6346} mutations were identified by the normal phenotype of their bristles, indicating that they lacked a third chromosome balancer. These females were collected and mated to males carrying the *stat92E*^{6346} allele on one of their third chromosomes, balanced by TM3, for

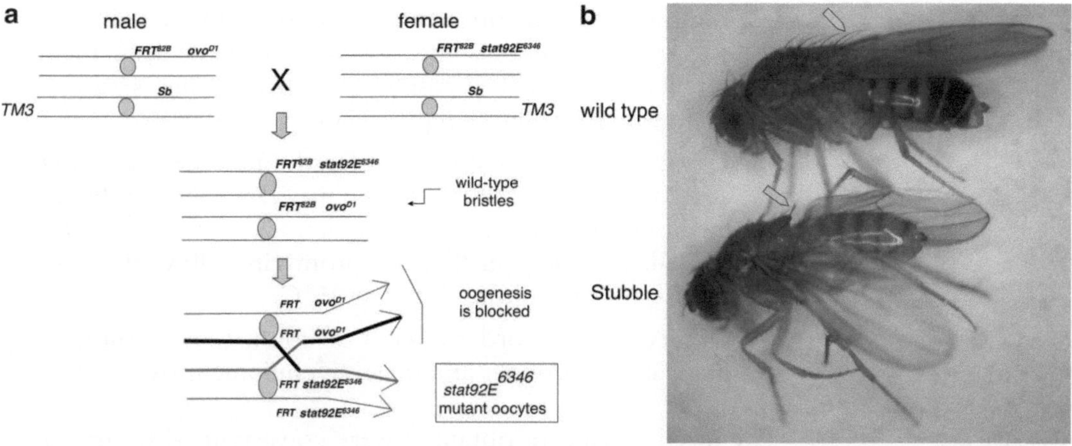

Fig. 4. (**a**) In this example, *STAT* mutant females are mated to males carrying the ovo^{D1} allele. Males carry *FRT* and ovo^{D1} on the same arm of one of their third chromosomes; their first chromosome carries *hsFLP* (not shown). Females carry an homologous *FRT* as well as $stat92E^{6346}$ on the same arm of one of their third chromosomes. The other third chromosome of males and females is a balancer (TM3) carrying wild-type *stat92E* (not shown) as well as the dominant phenotypic marker *Stubble* (*Sb*; short bristles). Adult female progeny carrying both the $stat92E^{6346}$ and ovo^{D1} alleles can be identified by their wild-type bristles (of normal length); these females will be sterile, unless an appropriate crossing over event occurs in their germline cells, in which case, the females will produce oocytes that are mutant for *STAT* ($stat92E^{6346}/stat92E^{6346}$), and, upon mating, will give rise to maternal $stat92E^{6346}$ mutant embryos. Other progeny are not shown here, and only third chromosomes are shown. The FLP/FRT system insures that appropriate crossing over events occur with sufficient frequency. To activate the FLP recombinase, which is under control of a heat shock promoter (*hsFLP*), larvae resulting from the mating of *STAT* mutant females to ovo^{D1} and *hsFLP* carrying males, are subjected to heat shock. (**b**) A wild-type fly has normal bristles (*upper arrow*). A *Stubble* mutant has shorter bristles (*lower arrow*).

production of $stat92E^{mat-}$ embryos (see Note 5). Embryos lacking the maternal product of the *Ras1* gene were produced in a similar fashion using the $Ras1^{\Delta C40B}$ allele; embryos lacking both the *stat92E* and *Ras1* maternal gene products were produced using a recombinant chromosome carrying both the $stat92E^{6346}$ and $Ras1^{\Delta C40B}$ alleles. These embryos were immunostained, as above.

3.4. In Vitro Culture and Observation of PGCs

Data obtained from our immunostaining experiments indicated that the *torso*, *ras1*, and *STAT92E* gene products are required for normal PGC migration (4). To determine whether PGCs require these gene products in a cell autonomous manner, we removed pole cells from cellularization stage wild-type, *hop* gain-of-function (which have elevated STAT92E levels), and *tor* embryos, and put them into culture. Embryos were staged according to ref. (29).

1. Dechorionate cellularization stage embryos (refer to the Embryo Collection and Fixation procedure Subheading 3.1, steps 1–6) (see Note 6).

2. Rupture the dechorionated embryos with fine forceps to release the pole cells into *Drosophila* Schneider's medium supplemented with l-glutamine, 10% bovine serum, and 30 μg/ml penicillin/streptomycin.
3. Using a Pasteur pipette, transfer the pole cells to laminin-coated culture dishes containing *Drosophila* Schneider's medium supplemented with l-glutamine, 10% bovine serum, and 30 μg/ml penicillin/streptomycin. Allow the cells to settle onto the culture dishes at 25°C.
4. Observe and record the cells at room temperature, using phase contrast optics and imaging equipment (see Note 7).

3.5. Pole Cell Transplantation

The in vitro data we obtained were consistent with our observations of pole cells in fixed embryos: Pole cells isolated from *hop* gain-of-function embryos dispersed more actively than those from wild-type embryos, while pole cells from *tor* embryos dispersed hardly at all. We went on to investigate the issue of cell autonomy in vivo, by doing pole cell transplantations (4), using the procedure essentially as described by (30). We took pole cells from stage 4, cellularization stage (29) embryos lacking the maternal *stat92E*, *Ras 1*, or *tor* gene products, as well as from wild-type embryos (see Note 8). We put the pole cells into stage 4–6 recipient embryos that carry the ovo^{D1} mutation (see Note 9), and thus only able to give rise to progeny of the transplanted pole cells. We also transplanted wild-type pole cells into *tor* homozygous mutant host embryos. When the transplant recipients developed into adult flies, test crosses were done in order to examine the phenotypes of the resulting progeny and ascertain whether they were descended from donor germ cells (see step 12 below). In other experiments, done to observe the actual migration of transplanted germ cells, we labeled donor cells with fluorescein-dextran before transplanting them into host embryos.

3.5.1. Transplantation Procedure

1. Select pole cell donor and recipient embryos that are at a stage just before their pole cells have become well separated from the blastoderm. After several minutes, these pole cells will have aggregated into a polar cap and the perivitelline space can be seen easily around the cap (Fig. 2b).
2. Dechorionate donor and host embryos (refer to the Embryo collection and fixation procedure Subheading 2.2, steps 1–3 and 5–6) (see Note 10).
3. Desiccate the embryos by blotting them for 30 s on bibulous paper, and putting them into a silica gel desiccation chamber for 3–8 min, depending on temperature and humidity, just prior to transplantation (see Note 11).

4. Cut out a rectangular strip of apple juice agar. Transfer embryos to the agar with a small paintbrush. Using fine forceps, line up host and donor embryos pairwise on the agar, with the long axis of each embryo perpendicular to the longer sides of the agar strip, and with the anterior ends of the embryos oriented in one direction (Fig. 2c). The anterior end of the embryo can be identified by the micropyle (Fig. 2b).
5. Press a strip of double-sided sticky tape onto a 24×60 mM coverslip, leaving a margin of about 1 mM between the edge of the tape and the edge of the coverslip. Transfer the lined-up embryos to the sticky tape by very gently placing the coverslip onto the embryos, tape side down. Turn the coverslip embryo side up, and cover the embryos with halocarbon oil (Fig. 2c).
6. To support the coverslip, put a drop of halocarbon oil onto a microscope slide and place the slide under the coverslip, with the embryos facing upward. Mount the slide on the stage of an inverted, phase contrast microscope, so that the long axes of the embryos are parallel to the injection micropipette.
7. Attach a siliconized glass micropipette to a water-filled syringe, mount the micropipette on a micromanipulator able to move in three dimensions (Fig. 3b), and fill the tip of the micropipette with halocarbon oil.
8. Draw up the donor pole cells slowly into the micropipette, taking care to avoid drawing up blastoderm cells. Upon withdrawing the micropipette from the donor embryo draw up a small amount of oil into the micropipette tip before raising the micropipette out of the oil droplet. This will protect the pole cells from dehydration.
9. Reposition the coverslip so that the recipient embryo is directly below and parallel to the loaded micropipette. Lower the micropipette into the oil again and expel the oil plug just before penetrating the posterior pole of the recipient, and injecting the donor cells into the perivitelline space at the posterior pole.
10. After the pole cell transplantations have been completed, keep the recipient embryos in a moist chamber at 18°C, for better recovery.
11. Transfer the hatched larvae to a food vial for further development.
12. Individually test-cross surviving adults. An example of a test-cross is as follows: Surviving adult recipients of *stat92E*$^{mat-}$ donor pole cells, both male and female, were individually crossed to flies carrying the *white* and *yellow* mutations on the X-chromosome(s). Because the females that were test-crossed developed from host embryos carrying an *ovo*^{D1} allele, they were identified as germline mosaic females. Since mosaic males that were

test-crossed developed from host embryos that were wild type for the *white* and *yellow* genes, mosaics were identified as those males that produced female offspring with mutant white eyes and yellow body color.

3.5.2. Labeling of Pole Cells

To observe the actual migration of transplanted pole cells in host embryos, we labeled donor pole cells with fluorescein-dextran prior to transplantation, essentially as described in (31).

1. Using the same setup as for transplantations (see previous section), inject stage 2–3 embryos (29) at the posterior pole with fluorescein-dextran.
2. Following cellularization, the pole cells will contain fluorescein-dextran in their cytoplasm. Take the pole cells from these embryos and transplant them into stage 4–6 host embryos as described above. Transplant up to five donor pole cells at a time. Follow the migration of the transplanted germ cells in live host embryos using a microscope with fluorescent optics and photographic capability, taking care that the embryos are sufficiently covered with Halocarbon oil 400.

The data obtained from in vitro observation of pole cells and from pole cell transplantations indicate that PGCs require RTK activity in a cell autonomous fashion for their normal migration in the *Drosophila* embryo (4). Since rodent PCGs also require RTK activity for their normal migration (5), the results suggest the evolutionary conservation of a role for RTKs in PCG migration.

4. Notes

1. Some double-sided sticky tape can be toxic to embryos. Take the tape you plan to use and test it ahead of time to see if larvae hatch from embryos which have developed on it.
2. Micropipettes can be obtained commercially or can be pulled from glass capillaries using a commercial puller. A 45° opening at the tip can be obtained either by carefully breaking the tip (the method used in our lab) or by grinding on a rotating grinder. The tip need not be precisely at a 45° angle to the shaft of the pipette, but has to be sharp enough to penetrate the embryos and at the same time have a wide enough internal diameter to accommodate a pole cell.
3. A helpful discussion of procedures for microinjection of *Drosophila* embryos may be found in *Drosophila* Protocols (18).
4. Heat shock of larvae: It is important that during heat shock the vials containing the larvae are submerged up to the level

of the plug at the top of the vial, to prevent the crawling larvae from reaching regions of cooler temperatures. This can be accomplished by covering the vials with a weight with interstices that allow for air flow to the vials.

5. Half of these *stat92E*$^{mat-}$ embryos received the paternal *stat92E*^{6346} allele and thus were zygotically as well as maternally mutant for *stat92E*. The other half received the paternal TM3 chromosome, carrying a wild-type *stat92E* allele.
6. For greater viability when live embryos are to be used for experiments, leave the embryos in bleach for only 1.5 min, rather than the usual 2 min.
7. We recorded images every 5 s for 1 h.
8. All "wild-type," *stat92E* mutant, and *Ras* mutant donor pole cells carried the following two recessive mutations on their X-chromosomes: the *white (w)* mutation, which gives a white eye color rather than the wild-type red, and the *yellow (y)* mutation, resulting in a yellow body color which is lighter than the wild-type brown. These mutations allowed for the identification of male progeny derived from donor cells (refer to the Transplantation Procedure Subheading 3.5.1, step 12).
9. The recipient embryos did not carry the *w* or *y* mutations.
10. For greater viability when live embryos are to be used for experiments, leave the embryos in bleach for only 1.5 min, rather than the usual 2 min.
11. It is important to dehydrate the embryos sufficiently to prevent leakage of cytoplasm during and after injections. However, too much drying can be lethal.

References

1. Santos, A., and Lehmann, R. (2004) Germ cell specification and migration in *Drosophila* and beyond *Current Biology* **14**, R578–R89.
2. Dadoune, J. (2007) New insights into male gametogenesis: what about the spermatogonial stem cell niche? *Folia Histochemica et Cytobiologica* **45a**, 141–7.
3. Fuller, M., and Spradling, A. (2007) Male and female *Drosophila* germline stem cells: two versions of immortality *Science* **316**, 402–4.
4. Li, J., Xia, F., and Li, W. (2003) Coactivation of STAT and Ras is required for germ cell proliferation and invasive migration in Drosophila *Developmental Cell* **5**, 787–98.
5. Orth, J., Qui, J., Jester, W.J., and Pilder, S. (1997) Expression of the c-kit gene is critical for migration of neonatal rat gonocytes in vitro *Biology of Reproduction* **57**, 676–83.
6. Schlessinger, J. (2000) Cell signaling by receptor typrosine kinases *Cell* **103**, 211–25.
7. Li, W. (2005) Functions and mechanisms of receptor tyrosine kinase Torso signaling: Lessons from *Drosophila* embryonic terminal development *Developmental Dynamics* **232,** 656–72.
8. Brizzi, M., Dentelli, P,. Rosso, A., Yarden, Y., and Pegoraro, L. (1999) STAT protein recruitment and activation in c-Kit deletion mutants *Journal of Biological Chemistry* **274,** 16965–72.
9. Deberry, C., Mou, S., and Linnekin, D. (1997) Stat1 associates with c-kit and is activated in response to stem cell factor *Biochemical Journal* **327**, 73–80.
10. Ning, Z., Li, J., McGuinness, M., and Arceci, R. (2001) STAT3 activation is required for

Asp(816) mutant c-Kit induced tumorigenicity *Oncogene* **20**, 4528–36.

11. Lu, X., Chou, T., Williams, N., Roberts, T., and Perrimon, N. (1993) Control of cell fate determination by p21ras/Ras1, an essential component of torso signaling in Drosophila *Genes and Development* 7, 621–32.
12. De Miguel, M., Cheng, L., Holland, E., Federspiel, M., and Donovan, P. (2002) Dissection of the c-Kit signaling pathway in mouse primordial germ cells by retroviral-mediated gene transfer *Proceedings of the National Academy of Sciences (USA)* **99**, 10458–63.
13. Duffy, J., and Perrimon, N. (1994) The *torso* pathway in *Drosophila*: lessons on receptor tyrosine kinase signaling and pattern formation *Developmental Biology* **166**, 380–95.
14. Perrimon, N., Lu, X., Hou, X., et al. (1995) Dissection of the Torso signal transduction pathway in Drosophila. *Molecular Reproduction and Development* **42**, 515–22.
15. Starz-Gaiano, M., and Lehmann, R. (2001) Moving towards the next generation *Mechanisms of Development* **105**, 5–18.
16. Wylie, C. (2000) Germ cells *Current Opinion in Genetics and Development* **10**, 410–3.
17. Dahmann, C, ed. (2008) Drosophila Methods and Protocols. *Totowa, NJ, Humana Press.*
18. Sullivan, W., Ashburner, M., and Hawley, R. (2008) Drosophila Protocols. *Cold Spring Harbor, NY, Cold Spring Harbor Laboratory Press.*
19. Sprenger, F., Stevens, L., and Nusslein-Volhard, C. (1989) The Drosophila gene *torso* encodes a putative receptor tyrorsine kinase *Nature* **338**, 478–83.
20. Klingler, M., Erdelyi, M., Szabad, J., and Nusslein-Volhard, C. (1988) Function of *torso* in determining the terminal anlagen of the Drosophila embryo *Nature* **335**, 275–7.
21. Hou, X., Chou, T., Melnick, M., and Perrimon, N. (1995) The *torso* receptor tyrosine kinase can activate Raf in a Ras-independent pathway *Cell* **81**, 63–71.
22. Ambrosio, L., Mahowald, A., and Perrimon, N. (1989) Requirement of the *Drosophila raf* homologue for *torso* function *Nature* **342**, 288–91.
23. Hou, X., Melnick, M., and Perrimon, N. (1996) Marelle acts downstream of the *Drosophila* HOP/JAK kinase and encodes a protein similar to the mammalian STATs *Cell* **84**, 411–9.
24. Binari, R., and Perrimon, N. (1994) Stripe-specific regulation of pair-rule genes by *hopscotch*, a putative Jak family tyyrosine kinase in *Drosophila Genes and Development* **8**, 300–12.
25. Harrison, D., McCoon, P., Binari, R., Gilman, M., and Perrimon, N. (1998) *Drosophila unpaired* enclodes a secreted protein that activates the JAK signaling pathway *Genes and Development* **12**, 3252–63.
26. Harrison, D., Binari, R., Nahreini, T., Gilman, M., and Perrimon, N. (1995) Activation of a *Drosophila* Janus kinase (JAK) causes hematopoietic neoplasia and developmental defects *EMBO Journal* **14**, 2857–65.
27. Heimer, G., and Taylor, C. (1974) Improved mountant for immunofluorescence preparations *Journal of Clinical Pathology* **27**, 254–6.
28. Chou, T., and Perrimon, N. (1992) Use of a yeast site-specific recombinase to produce female germline chimeras in Drosophila *Genetics* **151**, 643–53.
29. Campos-Ortega, J., and Hartenstein, V. (1997) The Embryonic Development of Drosophila melanogaster. *Berlin, Springer-Verlag.*
30. Van Deusen, E. (1977) Sex determination in germ line chimeras of Drosophila melanogaster *Journal of Embryology and Experimental Morphology* **37**, 173–85.
31. Jaglarz, M., and Howard, K. (1994) Primordial germ cell migration in *Drosophila melanogaster* is controlled by somatic tissue *Development* **120**, 83–9.

Chapter 21

Rho GTPases in Hematopoietic Stem/Progenitor Cell Migration

Wei Liu, Yuxin Feng, Xun Shang, and Yi Zheng

Abstract

Rho GTPases including RhoA, Rac1, and Cdc42 are a class of intracellular signaling proteins critical for the regulation of cytoskeleton organization, adhesion, and migration. Molecular mechanisms of mammalian cell migration were first revealed in fibroblasts where RhoA, Rac1, and Cdc42 facilitate in the multistep process including establishment and maintenance of polarity, formation of actin-rich protrusions, remodeling of adhesive contacts, and generation of force. In hematopoietic stem/progenitor cells, Rho GTPases relay signals from chemokines and cytokines such as SDF-1α and SCF to the actin and microtubule cytoskeleton through effector kinases and/or adaptor molecules that affect adhesion or transcription. Comprehensive use of murine conditional gene knockout technology combined with biochemical approaches in recent studies allows for physiologically relevant investigations of the involvement of Rho GTPases in hematopoietic stem/progenitor cell migration, providing important mechanisms for the stem/progenitor maintenance.

Key words: Rho GTPases, SDF-1α/CXCR4, SCF/c-Kit, Hematopoietic stem and progenitor cells, Migration, F-actin cytoskeleton

1. Introduction

Cell migration is a multistep process involving the establishment and maintenance of polarity, formation of actin-rich protrusions, remodeling of adhesive contacts and generation of force. This is orchestrated by a large number of receptors and intracellular signaling proteins, and Rho GTPases are major intracellular components in this process.

The Rho guanosine triphosphatases (GTPases) family includes RhoA, Rac1, and Cdc42 (1). They are molecular switches that cycle between the guanosine diphosphate (GDP)-bound inactive state and the GTP-bound active in response to diverse receptor signals.

Marie-Dominique Filippi and Hartmut Geiger (eds.), *Stem Cell Migration: Methods and Protocols*, Methods in Molecular Biology, vol. 750, DOI 10.1007/978-1-61779-145-1_21,

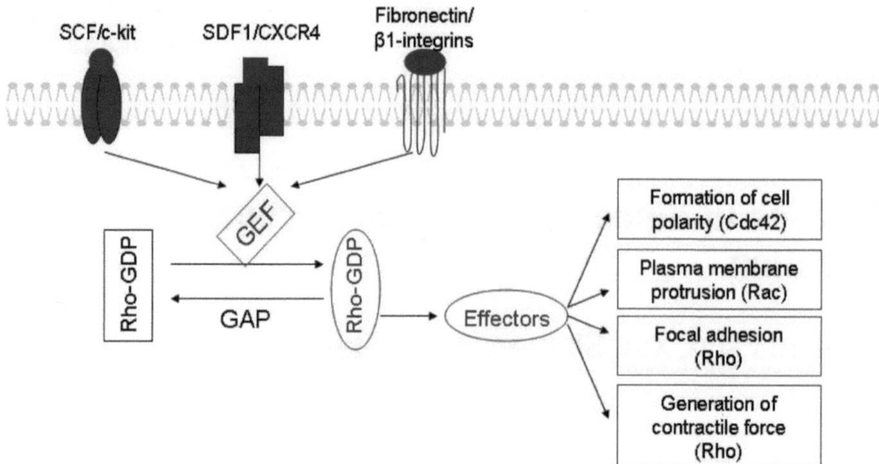

Fig. 1. Signal Transduction through Rho GTPases. Rho GTPases transduce signals from multiple surface receptors in HSPCs to regulate the multistep process of cell migration including establishment of cell polarity, plasma membrane protrusion and focal adhesion, and generation of contractile force.

In the active state, Rho GTPases recognize effector targets and generate cellular responses including assembly of actin and microtubule cytoskeleton structures (2), motility (3), superoxide production (4), cell cycle progression (5), and gene expression (6).

The mechanistic contributions of Rho GTPases to mammalian cell motility are best understood in fibroblasts. RhoA, Rac1, and Cdc42 have been shown to affect different aspects of cell migration (Fig. 1). Cdc42 is associated with cell polarity establishment and maintenance, and asymmetrical distribution of active Cdc42 and related signaling at filopodia leading edge is critical for the directionality of cell movement. Rac1, on the other hand, regulates lamellipodia formation where extensive actin polymerization occurs at the moving edge of the cell to provide the driving force for membrane protrusion (7–9). RhoA is known to regulate focal adhesion assembly and promotes the formation of stress fibers, thereby affecting cell contractility (10). However, RhoA can also counter-react with Rac1 activity at the site of cell movement to produce dynamic cytoskeleton flows and regulate adhesion complex assembly and disassembly that propels cell directional migration (11, 12).

Directional migration is important in the homing, lodging, and retention of hematopoietic stem/progenitor cells (HSPCs) in the bone marrow microenvironment, where self-renewal and differentiation of these cells occur (13). Rho GTPases transduce signals from chemokine and cytokine receptors at the plasma membrane to the actin and microtubule cytoskeleton, through effector kinases or adaptor proteins that affect adhesion or transcription. Among the chemokines upstream of Rho GTPase signaling, stromal-derived factor-1α (SDF-1α) is expressed by both human and murine BM endothelium and stroma and is essential for HSPC trafficking (14). SDF-1α induces β1 integrin-mediated arrest of hematopoietic progenitor cells and facilitates their transendothelial migration that is intimately associated with homing, lodging, and engraftment of HSPCs in the bone marrow (15). The essential cytokine stem cell factor, SCF, is also implicated in HSPC migration. Expressed by BM stromal cells, it is the ligand for the receptor tyrosine kinase, c-kit (13).

In general, the overall migration potential is lower in primitive HSPC when compared with mature hematopoietic cells. Additionally, it was shown that this migration can be boosted by lysophospholipid in combination with SDF-1α (16). While RhoA, Rac1, and Cdc42 are ubiquitously expressed, Rac2 expression is confined to hematopoietic lineages. The role of Rho GTPases in this process were first demonstrated by using Clostridium difficile toxin B which inactivates all major Rho GTPases (Rho, Rac, Cdc42) and by using Y27632, an inhibitor of one of the downstream targets of RhoA, Rho kinase. Rac1, Rac2, and Cdc42 are activated by SDF-1α and SCF in HSPCs, and Rac1/Rac2- or Cdc42- deficient HSPCs do not respond to SDF-1α in directional migration or transendothelial migration (17–19). Rac1/Rac2 or Cdc42-deficient HSPCs are defective in F-actin polymerization and β1 integrin-mediated adhesion to fibronectin, which are thought to be critical events in HSPC migration. Indeed, Rac1/Rac2 or Cdc42 gene targeting results in massive mobilization of HSPCs from the BM to the periphery, and renders the cells deficient for homing and engraftment (18, 19).

Given the accepted role of the Rho GTPases in F-actin cytoskeleton reorganization, adhesion, and directional migration of HSPCs, that are important cellular mechanisms involved in homing, lodging, and engraftment, assays to accurately measure and compare the endogenous Rho GTPases activity under chemokine and/or cytokine stimulation, to define the migration potential of HSPCs and the rearrangement of the cytoskeleton have been developed. We describe here the protocols that have been successfully adopted in studying Rac1, Rac2, and Cdc42 functions related to HSPC migration.

2. Materials

2.1. Cells

1. Freshly isolated wild-type or genetically modified (e.g., gene-deleted) murine low-density bone marrow cells are maintained in Iscove's Modified Dulbecco's Media (IMDM) containing 10% fetal bovine serum (FBS), 2 mM glutamine, 100 U penicillin, and 2 μg/ml streptomycin.
2. Endothelial cells (see Note 1).

2.2. Effector Domain Pulldown Assay for Rho GTPase Activity

1. Glutathione-agarose beads (Sigma-Aldrich).
2. Anti-Rac1, anti-Cdc42 antibodies (BD Transduction), RhoA antibody (Santa Cruz Biotechnology).
3. GST-PAK1 PBD is the PBD domain of GST-PAK1 expressed in *Escherichia coli*. GST-Rhoteckin RBD is the RBD domain of GST-Rhoteckin expressed in *E. coli*.
4. Cell lysis buffer: 20 mM Tris–HCl pH 7.6, 100 mM NaCl, 10 mM $MgCl_2$, 0.5% Triton-X100, 2 mM PMSF, 10 μg/ml leupeptin, 10 μg/ml aprotinin, 0.5 mM DTT, 0.2% Sodium Deoxcholate.
5. Washing buffer: 20 mM Tris–HCl pH 7.6, 100 mM NaCl, 10 mM $MgCl_2$.
6. Various stimuli can be used: 100 ng/ml rrSCF, 20 ng/ml IL3, or 100 ng/ml SDF1-α.

2.3. Transendothelial Migration Assays Using Transwell Chambers

1. Transwells (6.5 mm diameter, 5.0 μm pore size, Costar 3421, Corning, NY).
2. Chemotaxis buffer: 0.5% bovine serum albumin (BSA) in IMDM.
3. Methylcellulose medium (MethoCult® GF M3434, Stemcell Technologies).
4. Retronectin (Sigma-Aldrich).

2.4. Analysis of Migration Using Fluorescence-Marked Murine Bone Marrow Cells

1. Chemotaxis buffer with or without chemokines (i.e., SDF1) as in Subheading 2.3 but without phenol red.
2. Calcein/AM; Molecular Probes.
3. Transwells (NeuroProbe: 5.7 mm diameter, 5 μm pore size, polycarbonate membrance).
4. Microplate spectrofluorometer with filter cube with excitation, 485 nm and emission, 530 nm.

2.5. Isolation of HSC Based on Lin, c-Kit, and Sca Surface Markers

1. PBS containing 3% FBS (PBS buffer).
2. Antibodies: anti-Lineage antibody cocktail. The cocktail includes anti-B220, anti-CD3e, anti-Gr-1, anti-Mac-1, and anti-TER119 (e.g., all biotin labeled). Anti-Sca-1 and anti-c-Kit (CD117) antibodies. We use a Anti-Sca-1-PeCy7-conjugated antibody, c-Kit PE-conjugated, streptavidine-FITC-conjugated (All from BD Pharmingen).

2.6. Analysis of Cytoskeleton Reorganization

1. HBSS containing 1 mM Ca^{2+} and 1 mM Mg^{2+} and supplemented with 0.1% bovine serum albumin (BSA) called HBSS+ (see Note 2).
2. Retronectin (Sigma-Aldrich).

3. Fixation with 2% paraformaldehyde (PF). Make a stock solution of 4% of PF, pH 7.4. Weight out enough powder to make a 4% solution. Add 50–75% of total volume to be used of ddH_2O. Heat carefully to 55°C and start adding 1, 5, or 10 M NaOH (the larger the volume the higher the concentration of NaOH you may want to use) until solution clears. Make sure the solution does not rise above 60–65°C. If it does, start over! Filter the solution with filter paper and vacuum system to remove large impurities. Add appropriate amount of 10× PBS and bring up to final volume with ddH_2O. Adjust the Ph to 7.4 or slightly higher (PF fixes better at higher pH's). Aliquot and store at −20°C. Thaw the solution at room temperature or 37°C, but never higher than 50°C. You can refreeze and thaw up to three times.
4. Permeabilization solution: Triton X-100. Make a 0.1% triton solution in PBS.
5. Blocking and washing solution: PBS containing 2% BSA.
6. Staining for F-actin structures: phalloidin-conjugated with Rhodamine or AlexaFluor^{R} dyes (Invitrogen). Phalloidin binds specifically to F-actin (i.e., polymerized) but not to G-actin (i.e., nonpolymerized).
7. Staining for microtubule structures: alpha-tubulin and beta-tubulin from mouse (Sigma) and secondary anti-mouse antibody labeled with AlexaFluor^{R} 488 or any other dyes depending on the staining combination. Or use antibodies to identify any protein of interest.
8. Mounting solution: Slowfade Gold antifade reagent (Invitrogen).
9. Glass chamber slides (Nunc).
10. Immunofluorescence microscope.

3. Methods

3.1. Effector Domain Pulldown Assay for Endogenous RhoA, Rac1, and Cdc42 Activities in Low-Density Bone Marrow Cells

The endogenous activities of RhoA-GTP, Rac1-GTP, and Cdc42-GTP in low-density bone marrow cells in responses to chemokine or cytokine stimulation can be examined by a protocol using effector domain (PBD or RBD) of GST-PAK1 or GST-Rhoteckin to pulldown the active form of the respective GTPases from cell lysates, adapted from previous applications to fibroblasts or neutrophils (20). Rac1-GTP and Cdc42-GTP proteins bind to GST-PAK1, and RhoA-GTP binds to GST-Rhoteckin, immobilized on glutathione-agarose beads with high affinity, and can be collected by centrifugation. The relative amounts of the Rho GTPase activities can be detected by Western blotting of the pulldowns with appropriate antibodies.

1. The isolated low-density bone marrow cells were cultured at 2×10^6/ml in serum/cytokine-free IMDM overnight, and challenged with 100 ng/ml rrSCF, 20 ng/ml IL3, or 100 ng/ml SDF1-α for 0, 2, 5, or 15 min (see Note 3).
2. Collect the cells by centrifugation at 500 g for 5 min and wash the cell pellets twice with PBS buffer. Add lysis buffer to the cell pellets, vigorously vortex for 2 min, and then put on ice for 20–30 min.
3. Centrifuge the sample at maximum speed for 1 min at 4°C. Transfer the supernatant containing the protein to a clean microcentrifuge tube and put on ice.
4. Measure the concentration of the proteins in the cell lysates. Dilute the protein concentration to 1 μg/μl and use 500–1,000 μg of total protein for each pulldown sample. Leave aside 30 μl of protein lysates at 4°C in clean microcentrifuge tubes; these samples will be analyzed by SDS-PAGE in step 6 as the total input protein.
5. Add 10 μg of GST-PAK1-PBD or GST-Rhoteckin-RBD-conjugated glutathione-agarose beads to each sample and incubate the sample at 4°C for 1 h (for Rac1 or Cdc42 detection) or 2 h (for RhoA detection) under constant agitation (see Note 4).
6. Collect the beads by centrifugation at 2,000 rpm (400 g) for 2 min. Discard the supernatants. Wash the beads with 1 ml ice-cold washing buffer three times.
7. Analyze the beads and the input proteins (from step 4) by SDS-PAGE in a 12% gradient gel. The relative amounts of active Rac1, RhoA, and Cdc42 are detected by probing the Western blot with monoclonal anti-Rac1, anti-Cdc42, and RhoA antibodies (see Note 5).

3.2. Analysis of Migration of Murine Low-Density Bone Marrow Cells

3.2.1. Transendothelial Migration of Murine Low-Density Bone Marrow Cells Using Transwell

To perform the transendothelial transwell-chamber migration assay, a monolayer of endothelial cells is grown onto the filter of the transwell. The transwells are then utilized for the migration assay. The cells are placed in the top chamber of the transwell containing endothelial cells, while a chemoattractant is placed in the bottom chamber. Cells migrated to the bottom chamber are quantified after a period of incubation in tissue culture. The results can be expressed as percentage migration or converted to a relative migration index for comparative studies.

1. Coat transwells with 25 μg/ml fibronectin before seeding endothelial cells to prevent endothelial cells from detaching (14).
2. Seed the endothelial cells (5×10^5) on transwells and incubate for 3 days before migration assay (see Note 1).

3. The monolayers achieve full confluency after 3 days and are suitable for transmigration studies.
4. The formation of confluent monolayers on transwell filters can be verified by visual inspection using an inverted microscope (see Note 6).
5. Wash the freshly isolated low-density bone marrow cells once with chemotaxis buffer (see Note 7) and resuspend at 10^6 cells/ml in the chemotaxis buffer. The cell concentration may vary and depends on the scale of the experiments.
6. Add 0.5 ml of chemotaxis buffer in the bottom chamber (24-well plates) with SDF-1 at a concentration of 100 ng/ml.
7. Load 1×10^5 cells resuspended into 100 μl of chemotaxis buffer onto the top chambers of the transwells, in triplicate. Place the same amount of cells (100 μl) in tubes containing 0.4 ml of chemotaxis buffer and keep them in a humidified incubator at 37°C, 5% CO_2. This will be used later to enumerate input cells.
8. Carefully place the transwells into the 24-well plate (see Notes 8–12).
9. Incubate the plates in a humidified tissue culture incubator at 37°C, 5% CO_2 for 4 h (see Note 9).
10. After 4 h, carefully remove the top chambers.
11. Collect the migrated cells in the bottom chamber and resuspend the cells in 50 μl of PBS.
12. Enumerate the cells with hemocytometer (see Note 11).
13. The input cells are counted similarly.
14. The percentage of migration can be calculated by dividing the number of cells migrated to the bottom chamber by the total input events multiplied by 100. The degree of migration can also be expressed as migration index: fold change over background migration in the absence of any chemoattractants. This will be obtained by dividing the migrated cell events by the background migration (see Note 12).
15. Alternatively, migration of hematopoietic progenitor cells can be evaluated by quantifying colony-forming units (CFU) in the migrated bone marrow cells using a semisolid culture.
16. Wash migrated and input cells, and resuspend them in 4 ml of MethoCult® GF M3434 containing hematopoietic growth factors that support CFU proliferation.
17. Enumerate colonies 1 week later. The percent of migrated CFU is calculated by dividing the number of CFU in the lower chamber by the number of input CFU multiplied by 100.

3.2.2. Analysis of Migration Using Fluorescence-Marked Murine Bone Marrow Cells

Chemotaxis assays can be performed using 96-well chemotaxis chambers and fluorescence-tagged cells (21). This method makes it easy to quickly quantify migrated cells based on fluorescence signal and minimizes the use of chemokines and cells. This is especially beneficial for analyzing rare populations, such as Lin^-c-Kit^+ cells. However, this protocol is not practical for conducting CFU-assay or analyzing intracellular or cell-surface markers of migrated cells.

1. Add 300 μl of chemotaxis buffer with or without chemokines (i.e., SDF-1α) but without phenol red in the lower chamber of 96-well flat bottom tissue culture plates.
2. Incubate bone marrow cells with 4 μg/ml of Calcein/AM and place 20,000 fluorescence-tagged cells in 50 μl of medium to the upper side of the membrane.
3. Incubate the chamber at 37°C for 4 h.
4. Measure fluorescence (excitation, 485 nm; emission, 530 nm) of both migrated cells and input cells on a microplate spectrofluorometer.
5. Calculate the number of cells using samples with known cell concentration and fluorescence and determine percentage of migration by dividing the number of cells in the lower well by the total cell input multiplied by 100.

3.2.3. Migration of Phenotypically Defined Hematopoietic Progenitor Cells

Hematopoietic progenitor cells can be defined by cell-surface markers such as Lin^-c-Kit^+. To identify migration of subpopulation of phenotypically defined hematopoietic progenitor cells, cells can be sorted by fluorescence-activated cell sorting (FACS) based on these parameters followed by their use in migration assays. However, there is always substantial cell loss during isolation. Therefore, as an alternative strategy, murine bone marrow cells can be stained with specific markers after migration.

1. Perform migration assay and count migrated and input cells as described in Subheading 3.2.1.
2. Transfer the cells to 1.5-ml microtubes and centrifuge for 2 min.
3. Wash the cells once with chilled 0.5% BSA in PBS.
4. Incubate input and migrated cells with Fc blocking reagent on ice for 10 min.
5. Split the cells into two replicates and stain them with cell-surface antibodies or an appropriate isotype control in 0.5% BSA in PBS on ice for 30 min.
6. Wash the cells once with 0.5% BSA in PBS.
7. If secondary staining is necessary, stain the cells with secondary Ab for 20–30 min on ice.

8. Wash cells once with 0.5% BSA in PBS and resuspend in PBS or fix with 1% PF in PBS.
9. Calculate the number of positive or negative cells for the marker in input and migrated cells by multiplying the number of total cells (migrated and input) by the percentage of positive or negative cells.
10. Calculate percentage migration.

3.3. Analysis of Cytoskeleton Regulation of Purified HSC

Hematopoietic stem cell migration is essentially regulated by Rho GTPase proteins and necessitates the rearrangement of cytoskeleton structures, i.e., F-actin and microtubules. Therefore, assessing the rearrangement of cytoskeleton structures following various chemokine stimulations is a good functional assay of Rho GTPase activity. In vivo, the cells migrate either across endothelial barriers or within tissue. The migration process is thus dependent on both chemokine stimulation and cell–cell or cell-matrix interaction (22). The rearrangement of the cytoskeleton and the associated migratory responses will depend on both the type of chemokine and the type of interaction, for example, SDF-1 and interaction with fibronectin, i.e., beta-1 integrin-dependent, respectively (18). Cytoskeleton regulation may be specific of the cell type that is analyzed. It is thus important to analyze a purified population of hematopoietic stem cells.

3.3.1. Isolation of HSC, Based on Lin, c-Kit, and Sca-1 Surface Marker

1. Incubate bone marrow cells with anti-lineage antibodies (1 μg/4×10^6 cells in 100 μl of PBS buffer), on ice for 30 min. Wash out the unbound antibody with at least 2 or 3 ml of cold PBS buffer. Spin cells down (400 g) at 4°C.
2. Incubate the cells with streptavidin, anti-c-Kit and anti-Sca antibodies (1 μg/4×10^6 cells in 100 μl of PBS buffer) on ice for 30 min. Wash out the unbound antibody with at least 2 or 3 ml of cold PBS buffer. Spin cells down (400 g) at 4°C.
3. Resuspend at 2×10^7 cells/ml in PBS buffer. The sample is now ready for sorting by flow cytometry (other isolation procedures of HSC are also well described in Chapters 3, 13, and 15 of this book).

3.3.2. Staining Procedures of the Cytoskeleton

1. Coat glass chamber slides with 25 ng/ml fibronectin for 2 h at 37°C. Wash the slides with PBS and block with 2% BSA for 20 min at room temperature. Keep the slides until use at room temperature.
2. Wash freshly isolated HSC cells with HBSS+ and deposit 1×10^4 cells (fewer cells can be used) in 100 μl of HBSS+ containing the chemokine stimuli into the chamber. For SDF-1α, the optimum concentration is 100 ng/ml. Incubate the cells for 10–30 min at 37°C. The cells will start migrating on the slide.

3. Fix the cells by carefully removing 150 μl of medium and quickly adding 200 μl of 2% PF. Incubate for 20 min (see Note 13).
4. Remove PF with a 200 μl pipette (DO NOT USE VACUUM) and wash two or three times with PBS using a pipette. This is critical to avoid detaching the cells from the slides, as HSC adhesion can be weak.
5. Permeabilize with 100 μl of 0.1% triton for 1 min at room temperature. Remove triton with pipette (DO NOT USE VACUUM) and wash two or three times with PBS.
6. Block the nonspecific binding sites with 2% BSA for 20 min at room temperature.
7. Stain the cells to identify cytoskeleton structure or any intracellular protein of interest. The staining procedure will depend on the type of antibody. Classically, the cells are stained with rhodamine-labeled phalloidin or anti-tubulin antibodies for 1 h at room temperature.
8. Wash with PBS, and stain for the secondary antibody is necessary.
9. Wash the slides vigorously with PBS to remove excess of fluorescently labeled antibodies (see Note 14). The chambers are then removed and the slides are counter-stained with antifade medium with or without DAPI. We use Slowfade Gold antifade reagent.
10. Analyze the cells with confocal or epifluorescence microscope.

4. Notes

1. Besides BMEC-1 cells (23), other endothelial-derived cells such as HUVEC (24), STR-10, STR-12, and LE1SVO cells (25) have been used. The number of cells to be plated onto the transwells needs to be determined experimentally and can vary depending on the cell types and culture conditions.
2. Calcium and magnesium are required for efficient cytoskeleton rearrangement. It is thus critical to use a medium containing 1 mM $CaCl_2$ and 1 mM $MgCl_2$. Check whether the medium you use already contain $CaCl_2$ and $MgCl_2$, if not you will have to add it.
3. Either a single stimulus or a mixture of chemokines and cytokines can be used to stimulate the cells.
4. 20 μl of a 50% slurry of glutathione-agarose bead is the maximum beads that should be added into each lysate sample. GST protein bound to the beads can be used as a negative control.
5. The relative amounts of active RhoA, Rac1, or Cdc42 can be normalized by dividing the signals from the Western blots for

RhoA, Rac1, or Cdc42 proteins by the total input proteins in the cell lysates.

6. In addition, the ability of these monolayers to act as a barrier may be tested by measuring permeability to ^{125}I-labeled human serum albumin (^{125}I-HSA). Briefly, wash confluent monolayers on transwell filters and place aliquots of ^{125}I-HSA in assay medium in the upper chamber. Incubate at 37°C for the expected time frame to be used for the migration assay. Collect the contents of the lower chamber at the end of the incubation period and measure radioactivity. If the equilibration of ^{125}I-HSA is less than 10% across endothelium-covered transwell filters over the migration period, the integrity of the layer can be considered to be intact (23, 24).
7. It is very important that no gradient exists in the chemotaxis buffer except those produced by the chemokine or cytokine to be evaluated. Cells need to be washed thoroughly with chemotaxis buffer to remove potential contamination from serum used to culture the cells. Use the same batch of chemotaxis buffer throughout the experiment.
8. When transwells are placed into the wells in the plates, hold the plates at an angle. This will help prevent bubble formation beneath the wells which can impede cell migration. Do not shake the plates.
9. The best-characterized chemokine for murine low-density bone marrow cells is SDF-1α. The optimal concentration of SDF-1α for chemotaxis is usually between 50 and 200 ng/ml. If multiple wells are to be prepared under the same conditions, the chemokine should be prepared in the medium as a pre-mixture for allocation to different wells to minimize differences between wells. Incubation for 4 h works best in our experience; however, incubation time may vary and researchers should determine the appropriate incubation time for a specific experiment, particularly when a different chemokine/cytokine is used. The chemokine gradient normally reaches plateau after 24 h owing to the diffusion effect from bottom to top wells or desensitization of the chemokine receptor. However, if chemokinetic activity is to be determined, cells can be incubated for 24 h or longer in the transwell, as diffusion of the chemoattractant does not affect chemokinesis.
10. Because cell migration varies from well to well, it is highly recommended to prepare multiple wells with the same treatment in order to obtain the appropriate number of data points for statistical analysis.
11. If the migrated cell concentration is too low for reliable counting by hemocytometer, flow cytometry should be used for enumeration. Counting migrated cells and input cells using flow cytometry gives a more accurate cell number.

Flow cytometry permits measurement of cell events/time for both migrated and input cells. In addition, dead cells can be excluded based on forward and side scatter.

12. Generally, 100 ng/ml of SDF-1α stimulates maximal migration of murine low-density bone marrow cells. At lower or higher SDF-1α concentration, migration number or index can be lower. Some investigators may use serum in the chemotaxis buffer; however, serum contains factors that normally increase cell migration. If the effect of the chemokine or cytokine is small, the presence of serum may mask the effect. If this is the case, avoid using serum in the chemotaxis buffer.
13. The fixation conditions may vary depending on the cytoskeleton structure that is analyzed. To ensure integrity of the microtubule network, it is very important to perform the fixation procedure at 37°C.
14. At this stage, a vigorous wash is critical to avoid high level of fluorescence in the slide background.

References

1. Hall, A. (1998) Rho GTPases and the actin cytoskeleton *Science* **279**, 509–14.
2. Ridley, A.J., and Hall, A. (1998) The small GTP-binding protein rho regulates the assembly of focal adhesions and actin stress fibers in response to growth factors *Cell* **70**, 389–99.
3. Ridley, A.J. (2001) Rho GTPases and cell migration *J Cell Sci* **114**, 2713–22.
4. Abo, A., Pick, E., Hall, A., Totty, N., Teahan, C.G., and Segal, A.W. (1991) Activation of the NADPH oxidase involves the small GTP-binding protein p21rac1 *Nature* **353**, 668–70.
5. Welsh, C.F., Roovers, K., Villanueva, J., Liu, Y., Schwartz, M.A., and Assoian, R.K. (2001) Timing of cyclin D1 expression within G1 phase is controlled by Rho *Nat Cell Biol* **3**, 950–7.
6. Cantrell, D. (1998) Lymphocyte signalling: a coordinating role for Vav? *Curr Biol* **8**, R535–8.
7. Sells, M.A., Pfaff, A., and Chernoff, J. (2000) Temporal and spatial distribution of activated Pak1 in fibroblasts *J Cell Biol* **151**, 1449–58.
8. Etienne-Manneville, S., and Hall, A. (2001) Integrin-mediated activation of Cdc42 controls cell polarity in migrating astrocytes through PKCzeta *Cell* **106**, 489–98.
9. Nobes, C.D., and Hall, A. (1995) Rho, rac, and cdc42 GTPases regulate the assembly of multimolecular focal complexes associated with actin stress fibers, lamellipodia, and filopodia *Cell* **81**, 53–62.
10. Burridge, K., and Chrzanowska-Wodnicka, M. (1996) Focal adhesions, contractility, and signaling *Annu Rev Cell Dev Biol* **12**, 463–518.
11. Rottner, K., Hall, A., and Small, J.V. (1999) Interplay between Rac and Rho in the control of substrate contact dynamics *Curr Biol* **9**, 640–8.
12. Ren X.D., Kiosses W.B., and Schwartz M.A. (1999) Regulation of the small GTP-binding protein Rho by cell adhesion and the cytoskeleton. EMBO J **18**, 578–85.
13. Lapidot, T., Dar, A., and Kollet, O. (2005) How do stem cells find their way home? *Blood* **106**, 1901–10.
14. Peled, A., Petit, I., Kollet, O., et al. (1999) Dependence of human stem cell engraftment and repopulation of NOD/SCID mice on CXCR4 *Science* **283** , 845–8.
15. Peled, A., Kollet, O., Ponomaryov, T., et al. (2000) The chemokine SDF-1 activates the integrins LFA-1, VLA-4, and VLA-5 on immature human CD34(+) cells: role in transendothelial/stromal migration and engraftment of NOD/SCID mice *Blood* **95**, 3289–96.
16. Whetton, A.D., Lu, Y., Pierce, A., Carney, L., and Spooncer, E. (2003) Lysophospholipids synergistically promote primitive hematopoietic cell chemotaxis via a mechanism involving Vav 1 *Blood* **102**, 2798–802.

17. Yang, F.C., Atkinson, S.J., Gu, Y., et al. (2001) Rac and Cdc42 GTPases control hematopoietic stem cell shape, adhesion, migration, and mobilization *Proc Natl Acad Sci USA* **98**, 5614–8.

18. Gu, Y., Filippi, M.D., Cancelas, J.A., et al. (2003) Hematopoietic cell regulation by Rac1 and Rac2 guanosine triphosphatases *Science* **302**, 445–9.

19. Yang, L., Wang, L., Geiger, H., Cancelas, J.A., Mo, J., and Zheng, Y. (2007) Rho GTPase Cdc42 coordinates hematopoietic stem cell quiescence and niche interaction in the bone marrow *Proc Natl Acad Sci USA* **104**, 5091–6.

20. Benard, V., Bohl, B.P., and Bokoch, G.M. (1999) Characterization of rac and cdc42 activation in chemoattractant-stimulated human neutrophils using a novel assay for active GTPases *J Biol Chem* **274**, 13198–204.

21. Levesque, J.P., Leavesley, D.I., Niutta, S., Vadas, M., and Simmons, P.J. (1995) Cytokines increase human hemopoietic cell adhesiveness by activation of very late antigen (VLA)-4 and VLA-5 integrins *J Exp Med* **181**, 1805–15.

22. Ridley, A.J., Schwartz, M.A., Burridge, K., et al. (2003) Cell migration: integrating signals from front to back *Science* **302**, 1704–09.

23. Mohle, R., Moore, M.A., Nachman, R.L., and Rafii, S. (1997) Transendothelial migration of CD34+ and mature hematopoietic cells: an in vitro study using a human bone marrow endothelial cell line *Blood* **89**, 72–80.

24. Yong, K.L., Watts, M., Shaun Thomas, N., Sullivan, A., Ings, S., and Linch, D.C. (1998) Transmigration of CD34+ cells across specialized and nonspecialized endothelium requires prior activation by growth factors and is mediated by PECAM-1 (CD31) *Blood* **91**, 1196–205.

25. Imai, K., Kobayashi, M., Wang, J., et al. (1999) Selective transendothelial migration of hematopoietic progenitor cells: a role in homing of progenitor cells *Blood* **93**, 149–56.

Index

A

B

C

Marie-Dominique Filippi and Hartmut Geiger (eds.), *Stem Cell Migration: Methods and Protocols*, Methods in Molecular Biology, vol. 750, DOI 10.1007/978-1-61779-145-1, © Springer Science+Business Media, LLC 2011

M

N

O

P

Q

R

S

T

U

V

W

Y

Z

MIX
Papier aus verantwortungsvollen Quellen
Paper from responsible sources
FSC® C105338

If you have any concerns about our products,
you can contact us on
ProductSafety@springernature.com

In case Publisher is established outside the EU,
the EU authorized representative is:
Springer Nature Customer Service Center GmbH
Europaplatz 3, 69115 Heidelberg, Germany

Printed by Libri Plureos GmbH
in Hamburg, Germany